Packaging Materials and Processing for Food, Pharmaceuticals and Cosmetics

SCIENCES

Agronomy and Food Science, Field Directors – Jack Legrand and Gilles Trystram

Packaging and Recycling, Subject Head – Frédéric Debeaufort

Packaging Materials and Processing for Food, Pharmaceuticals and Cosmetics

Coordinated by
Frédéric Debeaufort
Kata Galić
Mia Kurek
Nasreddine Benbettaieb
Mario Ščetar

iSTE

WILEY

First published 2021 in Great Britain and the United States by ISTE Ltd and John Wiley & Sons, Inc.

ISTE Ltd
27-37 St George's Road
London SW19 4EU
UK

www.iste.co.uk

John Wiley & Sons, Inc.
111 River Street
Hoboken, NJ 07030
USA

www.wiley.com

Library of Congress Control Number: 2020951536

British Library Cataloguing-in-Publication Data
A CIP record for this book is available from the British Library
ISBN 978-1-78945-039-2

ERC code:
LS9 Applied Life Sciences, Biotechnology, and Molecular and Biosystems Engineering
LS9_5 Food sciences (including food technology, food safety, nutrition)

FSC www.fsc.org MIX Paper from responsible sources FSC® C013604

Contents

Preface

Frédéric Debeaufort[1], Kata Galić[2], Mia Kurek[2], Nasreddine Benbettaieb[1] and Mario Ščetar[2]

[1] *Institute of Technology, University of Burgundy, Dijon, France*

[2] *Faculty of Food Technology and Biotechnology, University of Zagreb, Croatia*

Packaging today has a significant impact on our society, as well as on our economic sector.

A huge variety of packaging materials, commonly used in food, pharmaceuticals, cosmetics, etc., all with different characteristics, are available. These include so-called traditional (wood, glass, paper-based, metal) materials and "novel" (polymers, bio-based materials, edible, active, etc.) materials. Polymers are put in a special group when they are considered as a source of many different single (monofilm) and complex (laminate) materials. From this, it is obvious that the selection of the optimal packaging material for a specific food product is a task that is harder than ever. Furthermore, the proper selection of packaging is of great importance to food manufacturers when considering the economy, marketing, logistics, distribution, consumer demands and environmental impact of the packaging.

This book provides the most comprehensive and current information in the field of food packaging. It can serve as a valuable source for not only food science and technology students and professionals, but also packaging technicians and engineers who need to know the characteristics of different packaging materials and their applications, as well as what can happen to food when it is in contact with these materials.

Packaging Materials and Processing for Food, Pharmaceuticals and Cosmetics, coordinated by Frédéric DEBEAUFORT, Kata GALIĆ, Mia KUREK, Nasreddine BENBETTAIEB and Mario ŠČETAR.

The book is divided into 13 chapters. The first four chapters cover traditional packaging materials starting with wood (Chapter 1), followed by paper and cardboard (Chapter 2), and then glass (Chapter 3) and metal (Chapter 4). Among the so-called "novel" packaging materials, plastics are covered in two chapters (Chapters 5 and 6), while Chapter 6 is devoted to complex materials (laminates). The following chapters deal with bio-based materials (Chapter 7) and active and smart packaging (Chapter 8), while important parts of packaging, such as caps and closures and auxiliary materials, are covered in Chapters 9 and 10, respectively. At the end of each chapter a possible food–packaging interaction is given. All chapters are presented with a comprehensive list of references. Information on different food packaging methods is presented in Chapter 11, while the marking and labeling of packaging can be found in Chapter 12. The book ends with Chapter 13, where information on the dependence of packaging material selection on the characteristics of food products (fresh, frozen, chilled, dried, etc.) is presented.

November 2020

Acknowledgements

Frédéric Debeaufort[1], Kata Galić[2], Mia Kurek[2], Nasreddine Benbettaieb[1] and Mario Ščetar[2]

[1] *Institute of Technology, University of Burgundy, Dijon, France*

[2] *Faculty of Food Technology and Biotechnology, University of Zagreb, Croatia*

The authors of this book sincerely thank Professor Audrey Bentz, a colleague from the IUT of Dijon-University, Burgundy, for her careful and efficient proofreading and for improving the English with a lot of kindness.

Introduction to Food Packaging

Frédéric Debeaufort[1] and Kata Galić[2]

[1] *Institute of Technology, University of Burgundy, Dijon, France*

[2] *Faculty of Food Technology and Biotechnology, University of Zagreb, Croatia*

I.1. Introduction

Packaging is one of the aspects that is part of the daily life of modern companies. It provides many services in support of the product and the various users, whether the packaging company, logisticians, users or consumers. Often disparaged when it is emptied of its contents, the packaging, apparently banal to quote some, is the fruit of human intelligence at the service of all.

Today, packaging is the result of the use of various modern technologies over long development processes (computer-aided design – CAD, 2D/3D digital printing, connected packaging, sustainable packaging and industry 4.0). The packaging world generates sophisticated jobs that require training and learning and schools need to meet this challenge. The packaging industry is "ahead" of many other industries; we are talking about industry 4.0 with high technicality in packaging, its mechanization and its level of robotization, without forgetting the numerous patents filed. Indeed, since 2012, patents filed by the packaging industry (all sectors combined) have represented 2.7% of total patents, that is, twice as much as the economic activity of the sector (1.3%) (CNE 2020). The baby boom and easier access to products,

For a color version of all figures in this book, see www.iste.co.uk/debeaufort/packaging.zip.

Packaging Materials and Processing for Food, Pharmaceuticals and Cosmetics, coordinated by Frédéric DEBEAUFORT, Kata GALIĆ, Mia KUREK, Nasreddine BENBETTAIEB and Mario ŠČETAR.

especially food, thanks to the development of large stores (retailing, supermarkets), have been a source of innovation in terms of packaging materials and machines, in order to mass supply products at the right time and at the lowest cost. The arrival of plastic in the 1960s allowed innovation in the processes for implementing packaging associated with functionalities. The arrival of mass distribution in the 1970s accelerated innovations serving the consumer. The integration of use by the user or the consumer is a source of creativity for the benefit of the population, in particular, for the elderly, if it is easy to open, for example. Finally, the regulations, the quest for traceability and the fight against counterfeiting have made it possible to generate packaging and processes for branding and identification purposes.

The global packaging market consumption in 2020 covers five main materials: paper and boards (31.06%), plastic (flexible 24.85% and rigid 22.28%), metal (12.64%), glass (6.81%) and others (2.35%), with approximately 70% of all packaging used in the food industry (WPO 2008; ALL4PACK 2016). In 2015, the global packaging industry value was US$839 billion and is predicted to reach US$998 billion in 2020 (ALL4PACK 2016). The global packaging machine market should grow at an average annual rate of 4.9% in the coming years, reaching a value of US$42 billion in 2018, US$48 billion in 2020 and an estimated 55 billion in 2025 according to Technavio (2020).

I.2. Definition

The definition of "packaging" in EU Directive 94/62/EC (European Commission 1994) is presented as "all products made of any materials of any nature to be used for the containment, protection, handling, delivery and presentation of goods, from raw materials to processed goods, from the producer to the user or the consumer".

The International Packaging Institute (in the *Glossary of Packaging Terms*, 1988) defined packaging as the enclosure of products, items or packages in a wrapped pouch, bag, box, cup, tray, can, tube, bottle or other container to perform one or more of the following functions: containment, protection, preservation, communication, utility and performance (Robertson 2013). Other definitions of packaging include a system that coordinates the preparation of goods for transport, distribution, storage, retailing and end use, a way to ensure its delivery to the consumer in a safe and sound condition. This also includes a techno-commercial function in order to optimize the costs of delivery while maximizing the profits (Coles 2011).

The Glossary of the International Trade Centre (ITC 2020), in the packaging sector, gives the following definitions:

> (i) Pack (noun): Bundle of items wrapped up, tied together or otherwise contained for carrying; (ii) Pack (verb): To put items into a box, bundle, bag, bale, wrap, etc. for storage or transportation; (iii) Package (noun): A sealed wrapping or box containing either a retail-sale quantity of a product (consumer package) or a product or a number of items or smaller packages in transport quantities, for transportation and storage (transport package), and (iv) Packaging, the general term for the function, materials and overall concept of a coordinating system of the preparation of goods for handling, shipment, storage and marketing. Distribution and use at optimum cost, and compatible with the requirements of the product.

Thus, packaging serves as a material-handling tool (containing the desired amount of food within a single container or gathering several identical units into aggregates), a processing aid (e.g. sterilization of food products in metal cans) and protection for items from damage and waste, which is an important marketing tool.

I.3. Levels of packaging

Packaging can be distinguished in regard to its "levels" (Figure I.1). A **primary package** (e.g. metal can, glass bottle and plastic wrap or pouch) is the most important as it is in direct contact with the product, providing the major protective barrier. The primary package is the one that the consumer usually purchases in supermarkets. A **secondary package** contains different numbers of primary packages, for example, a plastic pouch containing unidose packed sweets or biscuits. A **tertiary package**, also known as the transport package, is made up of a number of secondary packages that facilitate national and international trade. In other words, it represents the exact number of secondary packages put on a pallet to fill the space most economically. A **quaternary package** facilitates the handling of tertiary packages and is usually a large metal container (up to 40 m in length) that can accommodate many pallets during transport by ships or trains. When required, the conditions inside the container (temperature, humidity, gas composition and light) can be regulated. Traceability is at the forefront of food safety and is particularly important for perishables such as fresh fruits and vegetables, chilled meats and frozen foods. This also presents the source of innovation of real-time loggers aimed at delivering real-time distribution chain insights from any location (monitoring in-transit shipments, temperature, security and location details).

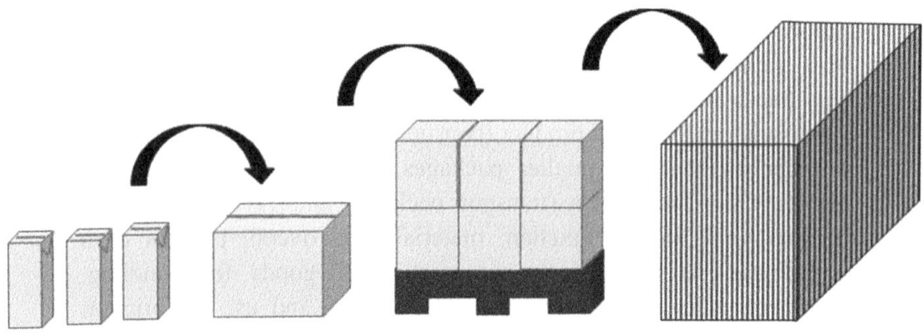

Figure I.1. *Levels of packaging*

I.4. Functions of packaging

Packaging, as an essential element of the product–packaging pair, fulfills various functions, such as the provision of a product to users and consumers, its conservation, its protection and its transport, whether this product is consumed by households, artisans or manufacturers.

Throughout history, packaging has enabled humans to free themselves from both time and space:

– Time, because, thanks to the conservation of a packaged product, humans are no longer obliged to immediately consume what they have just produced.

– Space, because, with the transportability and therefore the availability of the packaged product anywhere, modern humans consume where they want. With packaging, the place of production is separated from the place of consumption.

Four main functions of packaging are thus emphasized: containment, protection, convenience and communication (Figure I.2), which are interrelated, and must all be considered in the package development process (Robertson 2013).

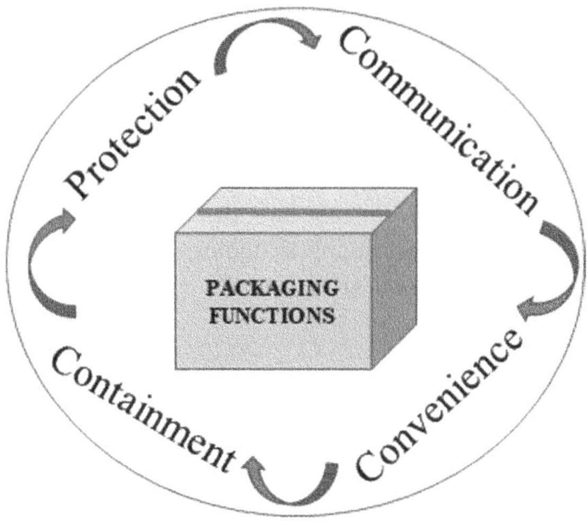

Figure I.2. *Packaging functions*

Containment

In order to perform its basic function successfully, the package must contain the product before it is transferred from one place to another. In case this task is not fulfilled (e.g. due to damage of the package), this can result in content spillages, economic losses and, in some cases, serious damage to the environment.

Protection

Product protection is the most important function of packaging. The package must protect the product from any external condition leading to damage (mechanical, poor environmental conditions, contamination and infestation) during handling, distribution and storage.

Thus, packaging is designed to ensure that the product reaches the consumer in good condition, through its entire journey from the manufacturer to the ultimate consumer.

Where required, packaging can also provide additional product protection. This is achieved through cushioning using different materials (such as sheets of corrugated paperboard, shredded paper, foam polystyrene or wrappings).

Food safety and quality is ensured through tamper-proof packaging. Special child-resistant closures, on items such as household chemicals (cleaning liquids, detergents, etc.) and pharmaceuticals, are developed to protect this vulnerable population. It is evident that packaging technology has made a huge contribution to improving food science and food safety and the reduction of food spoilage, as well as food waste.

Communication

The expression “a package must protect what it sells and sell what it protects” is applied to all packaging levels, from the primary to quaternary package, helping all involved actors to perform their tasks. Package communication helps consumers to select a product among a number of similar ones, and get all relevant information. It serves as an important marketing and selling tool that often influences the consumer when making their buying choice. It also ensures that warehouses and distribution centers are efficient in carrying and storing secondary and tertiary packages, based on the details on the attached labels. When international trade is involved and different languages are spoken, the use of adequate and clear symbols on the distribution packaging is essential.

Convenience

Convenience characteristics, which are much appreciated by consumers, include those which enable easy access to products, simplify usage or consumption and make it easy to hold, open or reclose. Appropriate packaging levels (secondary, tertiary and quaternary) facilitate the transport of packaged goods in interstate and international trade (Scholderer and Grunert 2005; Robertson 2013).

Packaging materials are often taken for granted as not-so-important actors in food protection. Consumers often do not even think about all of the above-mentioned principal packaging roles, not to mention all of the newly designed special functionalities of packaging materials that the broad population is not familiar with. Packaging is more than just a plastic bag leftover after its use.

I.5. Introduction to packaging materials

Apart from not being toxic, important requirements for food packaging materials also include: a) sanitary protection; b) barrier (moisture, gas, odor, light, fat) protection; c) resistance to impact; d) transparency; e) tamper-proofness; f) ease of opening and reclosing; g) ease of disposal; h) size, shape, weight limitations; i) appearance, printability; j) low cost; and k) special features.

The most common food packaging materials are: plastic (monofilms, laminates), glass, paper and board, metal and wood. Each of these materials offers specific advantages and disadvantages that have to be considered in order to select an adequate material for the specific food product.

Considered as a protective barrier or for carriage purposes, packaging has been known as such and used throughout history, even in Egyptian times. Some of the best examples of packaging can be found in nature, such as chestnuts, egg shells and orange skins (Figure I.3).

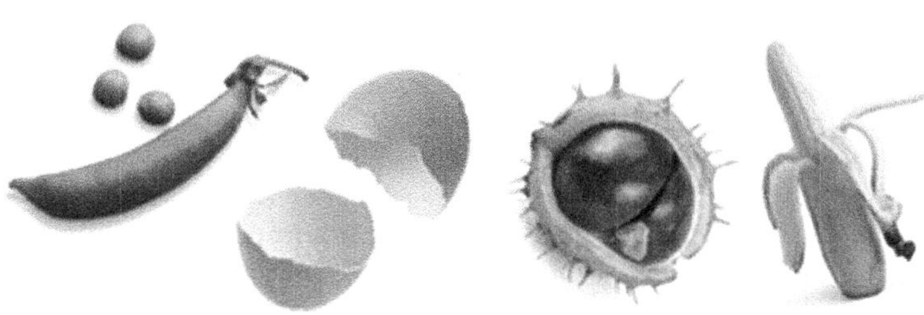

Figure I.3. *Packaging in nature*

It has been found that the first forms of packaging used by humans were flax and banana leaves, and animal products such as leather and stomachs, which are still used today. Some of the packaging-related developments throughout history are presented in Table I.1.

Year	Package development
1800–1850s	– 1809, Nicolas Appert (France) produced hermetically sealed glass jars to thermally preserve food – 1813, In England, handmade cans of "patent preserved meats" were produced – 1824, Canned foods were used by the British Navy
1870s	– 1875, Sardines were first packed in cans – 1879, Robert Gair (USA) produced the first machine-made folding carton
1880s	– 1884, The first cereal was packaged in a folding box (Quaker® Oats)
1890s	– 1892, William Painter (USA), patented the Crown cap for glass bottles
1900s	– 1906, Paraffin wax-coated paper milk containers were sold by G.W. Maxwell
1910s	– 1915, Pure-Pak® filled with milk was commercialized
1920s	– 1921, Zinc compounds in enamel cans were used – 1923, Frozen foods in cartons with wax paper wrappers were commercialized
1930s	– 1935, American brewers began to sell canned beer – 1939, Ethylene was polymerized commercially
1940s	– 1940, Carbonated soft drink canning began – 1946, Saran (PVDC) was used as a moisture barrier
1950s	– 1950, Cellophane was commercialized and used for packaging – 1950, Aluminum foil containers were developed – Polypropylene (PP) was invented – The retort pouch for heat-processed foods was developed – 1956, Tetra Pak® launched its tetrahedral shape multilayer form for milk
1960s	– 1960, Easy-open cans were introduced – 1965, Beverage cans made from aluminum were introduced – 1965, Tin-free steel (TFS chromium) cans were developed – 1967, Ring-pull openers were developed for canned drinks – Tetra Pak launched Tetra Brik® Aseptic (TBA) system for UHT milk
1970s	– The bar code system for retail packaging was introduced (USA) – Boil-in-the-bag frozen meals and bag-in-box systems were developed – MAP retail packs were introduced (USA, Scandinavia and Europe) – 1973, PET bottles were used for colas and other carbonated drinks – 1973, Antimicrobial wrappers were used to extend food shelf life – 1976, Iron-based O_2 scavengers were commercialized
1980s	– Packaging was produced for microwave use and modified atmosphere packaging (MAP) was developed for ready-to-eat fresh fruits and vegetables – 1986, The first use of the terms "Smart packaging" and "Interactive packaging"
1990s	– 1997, Ethanol-generating films or sachets were patented – Shrink-sleeve plastic labels for glass bottles were used
2000–2010	– 2006, Nanotechnology was used to modify the internal surface properties of squeezable plastic bottles – 2007, The world's first 100% recycled PET bottle was used for fruit drinks – Polylactic acid (PLA) bottles were used for water

Table I.1. *Some developments in packaging during the past 200 years (Coles 2011; Robertson 2013)*

I.6. Sociological, psychological and economical aspects of packaging

Packaging strongly influences the consumer's perception of the packed products, consciously or unconsciously, and thus their consumption behavior and habits. People have different lifestyles, which lead to differences in their perceptions of the products they need. Some people who are trying to have a healthy lifestyle may prefer certain products that are perceived as low fat and calorie restricting, whereas most consumers are mainly influenced by the price, which is related to color and shape attractiveness (Deng and Kahn 2009). Food products and their marketing are everywhere we look, on the way to school or work, on the sports courts, on TV and other media. Looking at shelves filled with various food items, color and product identification originating from the package design have significantly more impact on the choice of the product than its positioning on the shelves. The packaging of fast-moving consumer goods has changed frequently and dramatically for years, especially since the 1960s, as producers have fought for positioning, brand recognition and customer loyalty. In this context, it has become crucial to understand more about how consumers perceive and respond to changes in packaging. Package design and impressions play an important part in consumption patterns, and when consumers' impressions do not match expectations, disappointment and dissatisfaction may not only have an immediate impact on sales and profitability, but also lead to long-term damage to brand credibility.

I.6.1. *Packaging characteristics and consumer behavior*

"The clothes do not make the man". It is by the external appearance that we recognize an object or a person. This expression reflects the significant influence of packaging on stimulating the purchase of a product. Does improving the packaging of a product have an impact on its image? How do you measure the psychological impact of this approach? How strong is the impact of health and nutrition claims? Is there any effect on sales because of it?

The goods have been able to dominate the thoughts of the consumer to influence the act of purchase. Ranging from playful to functional, without forgetting the humorous or tendentious (misguided), the packaging of the product takes on all its finery to seduce the buyer. A determining factor in marketing action, the criteria for choosing product packaging, is of paramount importance for brands. The visual aspect associated with the functional parameters of a product determines the sale. But that is not enough, because it is necessary to play on the psychology of the client.

To succeed in a marketing sector, it is important to study the habits of the consumer, since the goal is to encourage them to buy products. It is not an automatic mechanism beforehand, but it is possible to maneuver so that it becomes one. To achieve this, we must bring together a set of ingredients, each subtler than the next, which interact in the cerebral cortex of the individual, potential consumer. Science has demonstrated this through Pavlov's conditioning experience (Wells 2014). It is therefore necessary to do some work to arouse the intention to buy. As previously presented, the marketing function of packaging as a communication tool is the key to selling the product.

However, packaging is the first thing consumers see when they are ready to buy a product. It is the last opportunity for the company to communicate about the product it markets. This moment is therefore decisive and deserves special attention. Successful packaging knows how to attract the attention of the consumer in the hubbub that surrounds it and responds, quickly and easily, to questions that arise. In any case, even if the packaging of competing products is adequate, in this context, the consumer will abandon them, for lack of having really seen them.

To attract the attention of the consumer, to retain them and to make them take the product in their hands, the packaging uses several techniques, among which are the color, the images, the typography, the brand, the design and the finish.

The **color** black is associated with luxury products, white with household products, green with organic or natural products and almost transparent sky blue with water bottles. It can be interesting to break these codes to surprise the consumer. Carbonated water is thus found in red bottles, and do not go unnoticed in a department usually filled with blue bottles.

Images easily convey a message. They do not need to be translated. Sparkling, bright white teeth on a tube of toothpaste immediately speaks to the consumer. It is also the most frequently used technique for products intended for children. This is the reason why brands invest so much in mascots, so that young people can spontaneously associate with a product.

Typography also plays an essential role. For example, elegant typography helps visually reinforce the luxurious character of the product it describes. Typography that resembles handwriting instead gives it an authentic appearance. The labels of some jars of jam are based on this relationship.

The **brand and its logo** are signs that are immediately recognized by the consumer when they have been the subject of intensive publicity. This is why they appear prominently on packaging. They speak to consumers as much as images.

The **design**, in other words, the shape of the product offered to consumers, is fundamental to the perception they may have of it, in particular, with regard to its practicality or its playfulness.

The **finish** adds to the impression given by the product. Varnished, shiny packaging is interpreted as going hand in hand with a quality product. The use of cheap-looking packaging must therefore be clearly explained to the consumer if the product it contains is exceptional.

For instance, the consumer perceptions of packed foods are influenced by the shape and size of the packaging. For example, elongated containers are often seen as larger than equivalent wide and short containers. In addition, people generally underestimate the changes in package volume, especially when packaging changes along two or three spatial dimensions as opposed to just one dimension (Ordabayeva and Chandon 2013). Over the past several decades, people have become accustomed to the supersized packaging in many product categories that reflect a sense of affluence and abundance. The supersizing trend has been especially pronounced in the food industry, where supersized fast food and snack portions have become the norm in many places. However, unforeseen negative side effects are beginning to take their toll. In addition to increased waste disposal issues, supersizing is considered to have contributed to over-consumption, weight gain and a rise in obesity to epidemic proportions. Public health authorities in Western countries have therefore become concerned about the influence of supersizing on consumer health.

I.6.2. *Packaging trends*

According to McTigue Pierce (2020) from Packaging Digest, future trends for packaging will satisfy both the consumer's wishes and food industry requirements. That is:

– Sustainable packaging, or even reaching a zero-waste packaging concept, is in preparation, because companies must promote packaging that is both safe for the consumer and respectful to the environment, e.g. recyclable packaging (paper, glass and metal materials, see Chapters 2, 3 and 4, respectively), bio-sourced and biodegradable packaging (bio-based packaging, see Chapter 7).

– Transparency towards consumers, because they demand honesty regarding the composition of food products as well as their containers (additives, endocrine disruptors) and the way they are made. Traditional packaging is reinvented to adopt a clear and precise formulation and, where appropriate, transparent information that reveals what is inside.

– Sophistication, and therefore a visible (marketing) message, refined with bold and simple, but sophisticated, colors and large print to communicate trust and respect (packaging marking and labeling, see Chapter 12).

– Consistency with the brand image of the product.

– State-of-the-art packaging, because consumers love their smart devices, and the increased use of technology, in turn, puts pressure on companies to offer "smart packaging" (active and intelligent packaging, see Chapter 8).

Sustainability, branding and "smart" packaging technologies resonate with global packaging professionals, based on the leading stories from early 2020. For example, using paper-based bags made from recycled paper with a transparent window made from polylactic acid biopolymer film, with an easy handling design and simple print information could satisfy almost all of the previous trends.

Green packaging has become attractive both for consumers and retailers in the past decade and is also in line with increasing consumer awareness of environmental sustainability. Packaging does not only serve to protect the main product, but is also expected to be environmentally friendly to reduce environmental problems due to packaging waste (Auliandri *et al.* 2018). The purchase intention of young consumers towards green packaging was positively affected by attitude, personal norms and the willingness to pay. The environmental concern positively influenced the purchase intention through the mediation of attitude. According to Kaufmann *et al.* (2012), the consumers' green purchasing behavior directly depends on demographic variables (age, gender, income level, education level, ethnicity, occupation), and could also be influenced by sociological/psychological variables such as altruism, environmental awareness, environmental concern and attitude, the belief about product safety for use and availability of product information and product availability, the perceived consumer effectiveness, the collectivism and the transparency/fairness in trade practices (customer care, product adulteration, unfair pricing, black marketing, misleading advertising, deceptive packaging).

The business sector needs to consider green packaging as one of the company's competitive strategies, as well as a substitute for recycling and waste, but also to the circular economy.

I.6.3. *Packaging and the circular economy*

Unlike the current linear economy, the circular economy forms a cycle. It is based on a model of reasoned production, of a change in consumption influenced by the population, and seeks to revive products by various means (repairing, recycling and transformation). Its simple objective is to produce goods and services while subscribing to a logic of sustainable development. Three areas apply to the circular economy, one of the key players of which is the packaging industry. The first area concerns supply and economic players (sustainable supply, eco-design, etc.). The second area is based on consumer demand and behavior (responsible consumption, duration of use). Finally, the last area concerns waste management, with the concept of recycling. There are seven pillars that make up the circular economy, which are divided into three areas (supply of economic players, consumer demand and behavior and waste management): supply, eco-design, industrial and territorial ecology, economy of functionality, responsible consumption, increased usage time and waste prevention. Packaging industries can and must contribute to all of the seven pillars of the circular economy. The circular economy has become a goal for many governments, especially in Western countries. For example, France aims to double the rate of the incorporation of recycled plastic by 2025 (i.e. more than 400,000 tons for the packaging sector), to make 100% of plastic packaging reusable, recyclable or compostable by 2025, to use 100% recycled glass bottles by 2030 and 90% recycled cans by 2025 in the brewery sector, which means that 100% of French people will be affected by sorting instructions in 2022 (ANIA 2020).

The packaging industries have been practicing the circular economy for a very long time, long before European regulations forced them to do so. The results of material recycling are there to bear witness to this. Indeed, the packaging world has worked significantly in this area in the past, without it being called the circular economy, and is still working on the subject: it is a great opportunity for each link in the value chain of the packaging industry to highlight its good practices (whether for primary, secondary or tertiary packaging). They often feature a lot of bottlenecks, especially passing from laboratory to pilot and industrial scale that needed to be resolved. The industries are engaged in this approach not because of ideology, but often because it makes economic sense and, more recently, ecological sense. It also answers consumer wishes.

The annual production of household waste in the world today exceeds 2 billion tons per year (ANIA 2020), 44% of which is from food or plant sources, and almost 70% is related to packaging. This is the reason why many countries defined regulations for dealing with packaging and packaging waste. The European parliament published the first directive 94/62/EC (European Commission 1994) specifying the first rules for packaging material recycling and waste management early. On December 2, 2015, the European Commission put forward a plan to support the EU's transition to a circular economy. On March 4, 2019, the Commission reported on the complete execution of the action plan. All 54 actions included in the 2015 plan have now been delivered or are being implemented. This will help to boost Europe's competitiveness, modernize its economy and industry to create jobs, protect the environment and generate sustainable growth (European Commission 2020). Many items in this plan concern the packaging sector.

The circular economy, for the packaging sector, is not limited to recycling; it covers all stages of the life of the packaged product, namely: design, production, distribution and use, without forgetting recovery of the packaging. It includes the notions of anchoring in the territories and proximity. It includes resource savings (material, water, energy), in particular, by the eco-design of the product and packaging, the optimization of the use of all resources, the reuse of packaging (especially in business to business), the prevention of packaging waste, the reduction of product losses (also by reducing food waste), the improvement of recyclability and the closure of material flows by its reuse. It also includes any initiative allowing changes in behavior and/or market codes for the best in packaging.

The packaging activities must be mainly local and not easily outsourced. The packaging sector is an emblematic example of a circular economy, where production and recycling generate economic activities synonymous with territorial roots. The packaging industry is generally an economic activity that meets the needs of principals who are in proximity. The reasons for this proximity are historical as well as economic.

There are many details linking the world of packaging and the circular economy. This is a complicated step, especially in a society where this type of operation is not yet the standard. However, the priority of large organizations is no longer based on renewable energies or on the rational use of resources. We must now decide to consider the origin of the problem at its source. If "ecological" packaging does not exist as such, it is up to companies to make strategic choices according to the objectives that they wish to achieve (avoid waste, reduce the carbon footprint, etc.). There are no good or bad practices. A company that does not use plastic will not necessarily have a lower ecological impact than another company that uses it, especially if the latter integrates a circular approach to its activity and continues to optimize its services or products. The development of new materials and technologies by the packaging industry may contribute to a better circular and ecological economy for the food industry.

I.7. References

ALL4PACK (2016). Packaging: Market and challenges in 2016. The global marketplace for packaging, processing, printing & handling. Trade Show, Paris.

ANIA (Association Nationale des Industries Alimentaires) (2020). Economie circulaire et gestion des emballages : les entreprises alimentaires en action [Online]. Available: https://www.ania.net/developpement-durable/economie-circulaire-et-gestion-des-emballages-les-entreprises-alimentaires-en-action and https://ania.net/Livret-dengagements-emballages-ANIA-oct2019.pdf [Accessed 13 April 2020].

Auliandri, T.A., Thoyib, A., Rohman, F., and Rofiq, A. (2018). Does green packaging matter as a business strategy? Exploring young consumers' consumption in an emerging market. *Problems and Perspectives in Management*, 16(2), 376–384.

CNE (Conseil National de l'Emballage) (2020). L'emballage en France [Online]. Available: https://conseil-emballage.org/lemballage-en-france/ [Accessed 13 April 2020]

Coles, R. (2011). Introduction. In *Food and Beverage Packaging Technology*, Coles, R. and Kirwan, M. (eds). Blackwell Publishing Ltd., Chichester.

Deng, X. and Kahn, B.E. (2009). Is your product on the right side? The "Location Effect" on perceived product heaviness and package evaluation. *Journal of Marketing Research*, 46(6), 725–738.

European Commission (1994). Packaging and packaging waste. European Parliament and Council Directive 94/62/EC [Online]. Available: https://eur-lex.europa.eu/legal-content/en/TXT/?uri=CELEX:31994L0062.

European Commission (2020). Internal market, industry, entrepreneurship and SMEs: Sustainability and circular economy [Online]. Available: https://ec.europa.eu/growth/industry/sustainability/circular-economy_en [Accessed 18 April 2020].

ITC (International Trade Centre) (2020). Packaging. The Glossary of the International Trade Centre [Online]. Available: http://www.intracen.org/packaging/glossary/P/ [Accessed 8 April 2020].

Kaufmann, H.R., Panni, M.F., Khan, A., and Orphanidou, Y. (2012). Factors affecting consumers' green purchasing behavior: An integrated conceptual framework. *Amfiteatru Economic Journal*, 14(31), 50–69.

Lee, S.M. (1989). *Dictionary of Composite Materials Technology*. Technomic Publishing Company, Lancaster.

McTigue Pierce, L. (2020). How packaging trends will shape 2020. *Packaging Digest* [Online]. Available: https://www.packagingdigest.com/packaging-design/how-packaging-trends-will-shape-2020-2020-02-27 [Accessed 20 April 2020].

Ordabayeva, N. and Chandon, P. (2013). Predicting and managing consumers' package size impressions. *Journal of Marketing*, 77, 123–137.

Robertson, G.L. (2013). *Food Packaging: Principles and Practice*, 3rd edition. CRC Press, Boca Raton.

Scholderer, J. and Grunert, K.G. (2005). Consumers, food and convenience: The long way from resource constraints to actual consumption patterns. *Journal of Economic Psychology*, 26(1), 105–128.

Wells, V.K. (2014). Behavioural psychology, marketing and consumer behaviour: A literature review and future research agenda. *Journal of Marketing Management*, 30(11–12), 1119–1158.

WPO (2008). Market statistics and future trends in global packaging. World Packaging Organisation [Online]. Available: www.worldpackaging.org [Accessed 6 April 2020].

1

Wood-based Packaging

Frédéric Debeaufort

Institute of Technology, University of Burgundy, Dijon, France

1.1. Introduction

Wood packaging is used to pack, transport, handle, preserve, present and add value to many food products and sectors (fruits and vegetables, fish and seafood, wines and spirits, oils, cheese and dairy, raw meat, cured meat and delicatessen, bread and bakery, dried fruits, etc.). There is a special focus on wood barrels and casks, which are considered as the primary packaging for both the storage and the aging of many beverages, such as wine, whiskey, sherry and cognac.

Wood packaging represents only between 9 and 12% of the tonnage of the overall production of packaging materials and only 2–3% of the value. Nevertheless, wood is substantially less expensive than other raw packaging materials, and, unlike alternative materials, such as plastics, wood is a sustainable resource (FEFPEB 2020). Wooden packaging is reusable, repairable and recyclable. At the end of its life, it becomes a renewable energy source. Finally, wood is the only raw material that is endlessly renewable. It is an essential part of the circular economy.

Wood packaging material, sometimes referred to as non-manufactured wood packing (NMWP) or solid wood packing material (SWPM), is both hardwood (deciduous trees like oak, maple or hornbeam) and softwood (evergreen trees,

For a color version of all figures in this book, see www.iste.co.uk/debeaufort/packaging.zip.

Packaging Materials and Processing for Food, Pharmaceuticals and Cosmetics, coordinated by Frédéric DEBEAUFORT, Kata GALIĆ, Mia KUREK, Nasreddine BENBETTAIEB and Mario ŠČETAR.

mainly conifers and poplars). Wood packaging materials are different from wooden products, for example, plywood, particle board, oriented strand board, veneer and wood wool. These are created by using glue, heat and pressure, or a combination of these methods. Wood packaging materials are used to support, protect and carry goods. Different examples of wood packaging materials include pallets, crates, boxes, cases, bins, reels, drums, load boards, skids, pallet collars and containers.

The wood packaging manufacturing sector includes four major separate areas: pallets, light packaging, cooperage and industrial packaging. Wood packaging is often not considered as packaging by the consumer because it is almost invisible in shops, as it is mainly used in tertiary packaging, dunnage and transport packaging (pallets, boxes). On the other hand, crates or trays are often used for the transportation and storage of fruits, vegetables and cheese, and wood barrels are used for storing and aging wines and distilled beverages such as whiskey or cognac. Wood boxes, which are also considered as secondary packaging, are often used for luxury packaging, for instance, for wine bottles.

Packaging is one of the most important parts of the wood industry, and there is an increasing demand for wooden packaging these days, as awareness is rising about the negative environmental impact of the use of plastic. The upcoming stringent government regulation for plastic packaging has had a positive impact on the wood packaging market. The global wood packaging market is predicted to grow by US$5.42 billion during the 2019–2023 period, with a compound annual growth rate (CAGR) estimated from 4 to 6.6%, and it is estimated that Europe will contribute to about 35% of this growth. In the wood packaging market, pallets represent more than 65% of the value, reaching about US$60 billion in 2018, and more than 90% of these pallets are made from wood (Parobek *et al.* 2019; Technavio 2020). One of the key trends for this market is the growing innovation in wooden pallets. Indeed, distribution centers, production units and manufacturing units are focusing on reducing the cost of shipping during the export and import of goods, which also includes the costs of trade packaging, sanitization, floor spacing, utilization and labor during storage and transportation. Wood packaging, such as pallets and wooden boxes, is a cost-effective option when compared to other products, such as plastic pallets.

1.2. Wood as a raw material for food packaging: characteristics, requirements, limitations and regulations

The term "wood" refers to the solid fabric of trees and shrubs. Wood is thus a renewable natural resource. There are two types of wood, coming from two different types of trees: hardwoods and softwoods (mainly conifers), which are distinguished by the shape of their leaves. The multitude of properties wood has, some common to both hardwood and softwood and some specific to the different types (depending on their structure and chemical composition), allows for great diversity in its use.

Wood is mainly composed of cellulose (40–50%), which constitutes the basic structure and absorbs the tensile forces that are exerted on the tree. Cellulose is a polysaccharide, made up of a long chain of linked glucose molecules. The remaining parts of the wood consist of hemicellulose and lignin. These two substances serve as filling and support material and absorb the pressure forces. Hemicellulose is also a polysaccharide, but its branched chains are made up of different sugars, rather than just glucose. Its chains are also shorter and less solid than the long chains of cellulose. Lignin is the component that gives wood its strength. Chemically, lignin is made up of many different constituent parts. Most are phenols, that is, chemical structures with aromatic rings. Lignin is incorporated into the wall of plant cells and is responsible for the transformation of cells into wood, also known as lignification. In the cell wall, lignin fills the space and binds together other components, such as cellulose and hemicellulose. Without lignin, plants would not be able to rise several meters above the ground. Wood contains high-molecular-weight compounds, which are considered as biopolymers (cellulose, lignin), as well as extractable and volatile compounds, most of which are soluble in aqueous solutions, hydro-alcoholic solutions, or beverages. These influence the properties of the wood, such as the acidity, hygroscopicity, color, odor, mechanical properties and durability.

The interaction of these three materials creates the mechanical properties of wood and determines its flexibility and resistance. As wood has different compositions depending on the species of tree, the types of wood are distinguished by their structure. Because of these differences, each wood has specific properties that make it suitable for one application rather than another. When it comes to food contact and packaging applications, only a limited number of species are allowed to be used (ISPM15 2013). These are birch, fir, Douglas fir, acacia, poplar, alder, aspen, hornbeam, chestnut, ash, olive, maritime pine, scots pine, sycamore and oak. However, solid foods are restricted to poplar, beech, walnut and elm. There is a lack of references on the authorization of resinous and tropical woods for contact with food, the well-established use of properly cured softwood without health problems notwithstanding. Fruit and vegetable pallets and boxes, or tables used to cure cheese are good examples of this.

The main limitations of wood for contact with food and food packaging are related to the risk of splinters, which create risks in handling and ingestion, and hygroscopicity and porosity, which makes wood susceptible to the absorption of chemical compounds, and thus increases microbial contamination and chemical risks. Moreover, the heterogeneous structure of wood (e.g. cellular structure and fiber orientation) depends on genetics, species and climatic conditions, which make it variable as a raw material, meaning the selection and processes must be adapted.

The main argument for not using wood is that it is difficult to clean. Wooden cutting boards, for example, tolerate less cleaning than plastic ones and can be quickly ruined in the dishwasher. Unlike plastic and steel, wood is also a porous material. This means that bacteria can enter the wood and survive until they are released, contaminating food the next time the board is used (Kim *et al.* 2017). Several studies have confirmed that pathogenic bacteria can survive in wood. It has also been demonstrated that bacteria that have been absorbed into wood can be released again when the wood becomes damp (Aviat *et al.* 2016). Other studies have also shown that the effect of certain disinfectants (including quaternary ammonium compounds) is lower on wood than on plastic and that the bacteria count remains high, even after washing. The study by Deza *et al.* (2007) showed that the bacteria count on wood is not reduced by rinsing with water, while the same procedure can reduce the bacteria count on plastic by 99%.

Wood is likely to interact with foodstuffs both bacteriologically and chemically. However, very few problems have been encountered up until now in actual use. Wood is also a material that can have a desired effect in agri-food technology (aging spirits in wooden barrels, maturing cheese with wooden churns, etc.). The assessment of the suitability of wooden equipment or articles that will be in contact with food must therefore take into account their use, and the interactions desired.

Wood has a number of obvious advantages compared with plastic or steel. Wood absorbs impacts better than many other materials, ensuring that objects are less likely to skate across its surface, meaning, in the case of cutting boards, that knives slip less, which increases safety and reduces wear on the knives. Wood has good resistance to compressive and tensile forces, high flexural strength, the ability to split, hardness and low thermal and electrical conductivity. Wooden shelving has the property of being able to absorb and release moisture (from 7 to 20% water content), which is of great significance when maturing cheese, for example (Gaglio *et al.* 2018). There are also arguments in favor of wood from a hygienic point of view. The main argument is that bacteria die more quickly on wood than on other surfaces. The first scientific studies to show that wood is not necessarily less hygienic than plastic were released in 1994. The studies showed that fewer bacteria were demonstrated on wooden boards than on plastic ones. There has since been speculation that the reduced survival of bacteria in wooden materials may be due to them drying out. Several studies have shown that bacteria die quickly in wood, while surviving well on other materials like plastic or glass. However, other studies found no difference in bacteria survival rates on wood and other materials. In contrast to Schonwalder *et al.* (2002), who wrote that the effect of disinfectants is lower on wood than on plastic, another group later found that bacteria on wood can be more easily killed with disinfectants than bacteria on glass or plastic. The hygienic properties of a veneer were also investigated and found to be similar to those of plastic and stainless steel. The transfer of bacteria from wood to food has also been tested, and the studies showed less transfer from wood than from plastic.

Whether wood has an antimicrobial effect itself has long been a subject of discussion. Some studies have compared different types of wood, and some indications have been found that bacteria survive less on some types of wood, especially pine, than on others, and that this effect is regardless of the age of the wood (Filip *et al.* 2012; Fink *et al.* 2013). Extracts of some hardwoods (especially white ash and also, to some extent, cherry and oak, but not many others) have also shown an antimicrobial effect in model trials. Other studies have showed that bacteria have a better survival rate on hardwood than on softwood types, but it is not known whether this is due to the wood itself or to the pre-treatment that the softwood has been exposed to Aviat *et al.* (2016) and Kim *et al.* (2017).

All restrictions and obligations for the use of wood for food and non-food packaging are shown in regulations and directives by the European Commission:

– 995/2010, dated October 20, 2010, dealing with illegally harvested timber and timber products that are prohibited in the EU market, that is, the obligations of operators who place timber and timber products on the market;

– 2005/15/EC, dated February 28, 2005, concerning the European requirements for the import of wood packaging material and dunnage, regarding, particularly, the phytosanitary measures (plant health and biosecurity) and hygiene detailed in the ISPM-15 (2013) procedures (heat-treated or fumigated woods, stamp marking, debarking, etc.);

– and, of course, the general regulation for food packaging, the regulation on materials and articles intended to come into contact with food 1935/2004, dated October 27, 2004, and 10/2011, dated January 14, 2011.

Sanitation of wood materials for food packaging and for coming into contact with food is the most important operation involved. Indeed, the wood surface could carry and transfer microbial and chemical contamination to foods. The treatments applied to wood and wood materials for packaging uses are heat treatments, washing, steam heating, super steam pasteurization, radiation and microwave technology. The cleaning and sanitation efficacy of the wood surface and bulk (due to the porosity) depends on the origin of the wood, the thickness and the exposure to potentially harmful environments.

1.3. Pallets

All pallets are made from the cores of trees. However, because the lumber used to make pallets is a mixture of hardwood and softwood, the pallet producers usually cannot specify the type of wood. The production of pallets accounts for 43% of the hardwood and 15% of the softwood usage in the USA and is almost the same in Europe. However, there are two species of wood used to manufacture pallets that are found in high proportions. These are oak and pine.

The cheapest pallets are made of softwood and are often considered to be expendable, to be discarded as trash along with other wrapping elements at the end of the trip. These are simple stringer pallets and are liftable from two sides. Wooden pallets can pose serious bio-hazard risks as they are susceptible to bacterial and chemical contamination, such as *E. coli* in food and produce transportation, and even insect infestation, and thus, there is a need for ISPM-15 (2013), which is also adopted by the US government (Early 2012). Nevertheless, the risk is very well controlled.

Wooden pallet construction specifications can depend on many different aspects: the pallet's intended use: general, FDA (Food and Drug Administration of USA), storage, chemical or export; the expected load weight; the type of wood desired: recycled, hard, soft, kiln-dried or combo (new and recycled); and even the type of fasteners desired to hold the pallet together: staples or nails (NWPCA 2014). Pallet design and dimensions depend on the countries and transportation requirements (Figure 1.1). The most produced and used pallet in the EU is the Europallet, of which the dimensions 800 x 1200 are adapted to the dimensions of truck trailers for optimized filling. The entry corresponds to the direction where the forks of the handling equipment can be inserted. Each piece of the pallet has a specific function, for instance, the mechanical resistance to handling, carriage, stacking, etc. (Figure 1.2). For special carriage or storage in bulk, or large amounts of some food products, such as apples, pallet boxes or pallet crates are also used (Figure 1.3).

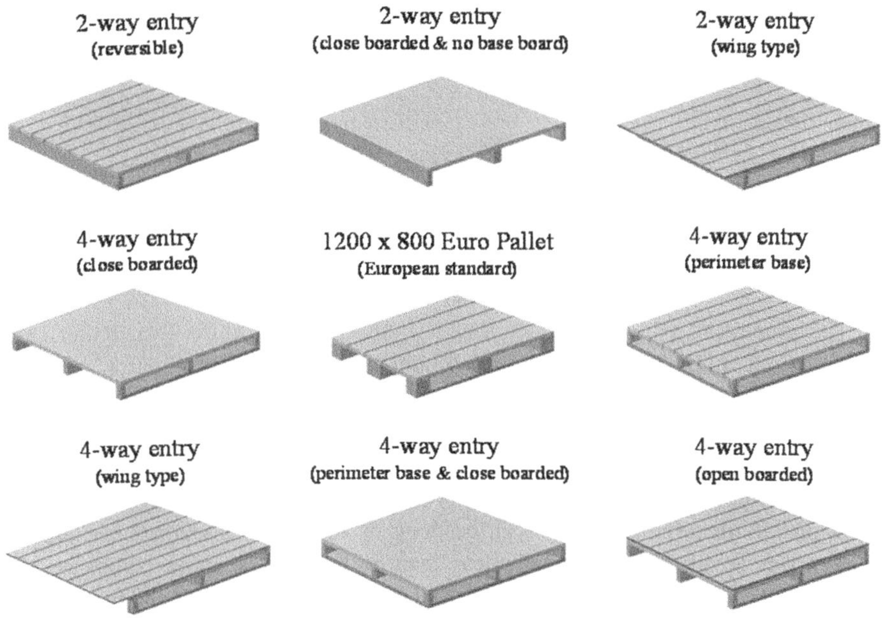

Figure 1.1. *Pallet types (adapted from https://www.homestratosphere.com/types-of-pallets/)*

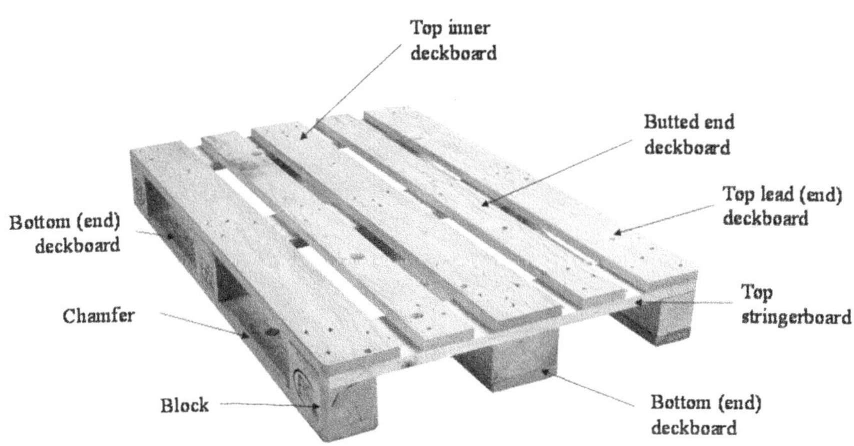

Figure 1.2. *Pallet components and structure*

Figure 1.3. *Pallet box and pallet crates*

The process of pallet manufacturing is quite simple (Figure 1.4). Pallet cants from local sawmills are trimmed to the proper length, ripped into boards and sawn into the proper-sized deck boards and stringers for the type of pallet needed. Next, notches can be cut into stringers to allow four-way entry with forklifts or two-way entry with pallet jacks. If the pallet is going to be used for international shipping or in sanitary applications, the wood may be treated prior to assembly. Finally, the wood pieces are assembled. This can be done by hand with pneumatic nailers and staplers (for small orders, typically less than 200 units) or, more commonly, by automated machines that can nail together and produce pallets at high volumes.

Pallets are submitted to tests of strength (bending, forklifting, compression, stacking, shearing, corner loads, impact, load, stiffness, etc.), durability (moisture content, fungi contamination, etc.) and compatibility with production lines, according to the standards ISO8611 (2011a, 2011b, 2011c) and EC 2005/15/EC (European Commission 2005).

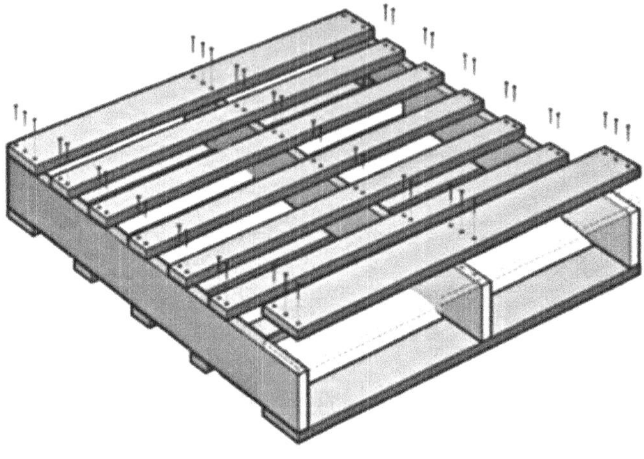

Figure 1.4. *Pallet assembly (adapted from http://jusworx.co.za/)*

1.4. Boxes, crates and trays

Crates and trays are open boxes, baskets or packing cases, used especially to transport fragile goods such as fruits and vegetables or glass containers. Crates are usually large, whereas trays are of medium or small size. Wooden boxes are closed crates. Fruits and vegetables are still mainly packed, carried and displayed on shelves in wooden trays or light crates (Figure 1.5). Their function is similar to a gasket, not to increase preservation or shelf life, but to carry and display fragile food products.

Figure 1.5. *Wooden trays/crates for fruits and vegetables*

Wooden crates have often been used to carry and deliver bottles of milk, wine, water and other beverages (Figure 1.6). They are still widely used for added value beverages (wine), or for heavy fruits and vegetables, such as melons or apples.

Figure 1.6. *Boxes and crates for wine or milk bottles*

1.5. Barrels and casks

Wooden barrels were, for a long time, the general packaging for the transportation and storage of water, wine, vinegar, rum, salted meats or salted fish, fruits, etc., for example, on big sailboats. They have been used for centuries to store and transport wine, and to this day are used for the fermentation and aging of the majority of red wines – and a good number of white wines too. Oak barrels are now vital in wine production.

In the past, cherry, pine, walnut, chestnut and other wood have been made into barrels or casks and have been used to age wine, but none compare to oak. Put simply, wine and oak have an affinity for one another. Oak takes a wine and "calms" it down, matures it, makes it soft on the palate and beautiful to drink. But by no means assume that this means that oak "tames" a wine into being boring: it also makes the wine more interesting and gives it complexity and depth that cannot evolve on its own. A French barrel costs roughly $900 to $1,000 and an American oak barrel costs roughly $400. According to a recent research by Future Market Insights (2020), the global wine barrel market is expected to register a CAGR of 4.7% during the forecast period 2017–2027. The global wine barrel market was estimated to be valued at about US$3.2 billion in 2016 and was expected to reach US$5.3 billion by the end of 2027. The value of barrels and casks remains quite interesting after they have been used for wine-making. The "aged" barrels are usually sold to smaller production wineries or to whiskey or cognac producers to age distilled alcohols.

The manual manufacturing of barrels is today only operated in the craft industry, which always dominates the market in France and Italy. Nowadays, the profession of cooper is still exercised, but with significant machine assistance, in particular, for the hard and tedious work (stave assembly, tightening, etc.). Typically, a good cooper can make one barrel per day, and a quality barrel is used for about five years in higher-end wineries, where they are used to age the winery's top wines for the first two years.

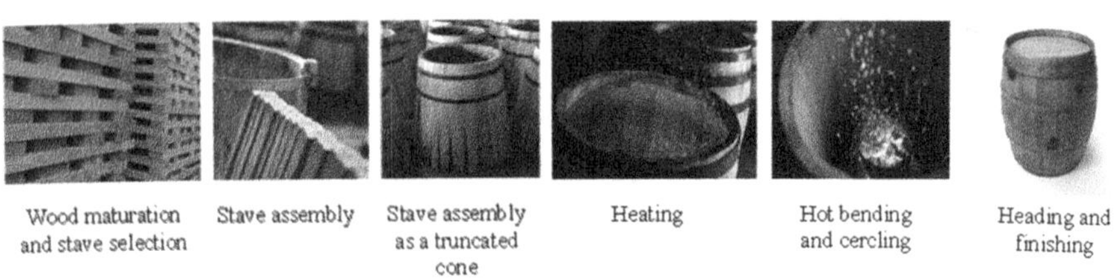

Figure 1.7. *Main stages of wood barrels and casks manufacturing*

The steps of barrel and cask manufacturing are schematized in Figure 1.7. First, tree trunks with zero to few defects (wood knots) are required; usually oak is preferred (Vivas 2002). Wood maturation is a very important step prior to its use. Wood is slowly dried in the open air for one to two years, to decrease the water content to 14–15%, as well as to eliminate the most astringent tannins and refine the aromatic profile. The water content of the wood allows us to ensure that the future barrel is watertight. Trunks are cut in the form of logs, which are split into quarters and, finally, shaped into staves. The next stage, the pinking step, really deals with barrel-making. It consists of mounting the mature staves manually, joined in a truncated cone in a metal ring. Provisional rings, adjusted with a flush and a hammer, hold the staves together. The barrel is then heated on an oak wood brazier for bending. Once its final shape has been obtained, there comes the very important stage of heating once more, in order to cook the wood in-depth and give it all the aromatic qualities that will allow it to match the wine or other beverage. A groove is carved at either end of the barrel, in which the end pieces, called the counter and the chants, are embedded. Rings of wood or strips of metal maintain the strength of the barrel enough that it can be moved by rolling or rocking. The tightness of each barrel is checked after manufacture by introducing pressurized hot water. It will comply if it does not reveal any leaks. Each different piece constituting a barrel or cask has a specific function related to watertightness, carriage, mechanical resistance, opening or cleaning (Figure 1.8).

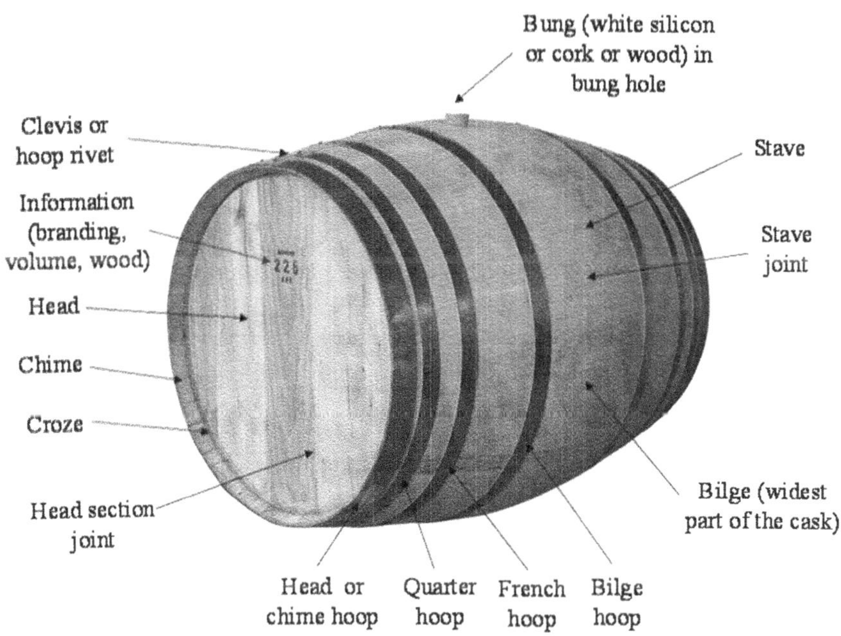

Figure 1.8. *Wood barrel and cask structure*

The barrel's wood composition, size and volume, and its method of manufacture, in particular, heating, have a great influence on the taste of the wine during its aging. A finished standard barrel has a weight close to 45 kg, for a size of about 90 centimeters long and a diameter of 60 to 70 centimeters, corresponding to a volume of 225 liters, which is the traditional cask volume used in French wine production (Boakes 2001). According to the countries and uses targeted, the volume of wood barrels and casks is different, and adapted to both product and function (Figure 1.9). For wine production, barrels with volumes ranging from 200 to 250 liters are usually preferred, whereas cognac production uses more 300-liter casks, 450- to 650-liter casks are used to age sherry and port, and the biggest (700 to 982 liters) are used for beer or fruit fermentation. In North America, the 700-liter cask called a "gorda" is preferred for the marrying or vatting of whiskies. For instance, some producers of Modena-type balsamic vinegars use different-size casks made of different woods (oak, cherry, chestnut or pear) changing from one to another according to the volume of vinegar reduction (evaporation) and the taste the wood imparts onto the final vinegar (Figure 1.10).

Figure 1.9. *Size and functions of casks and barrels*

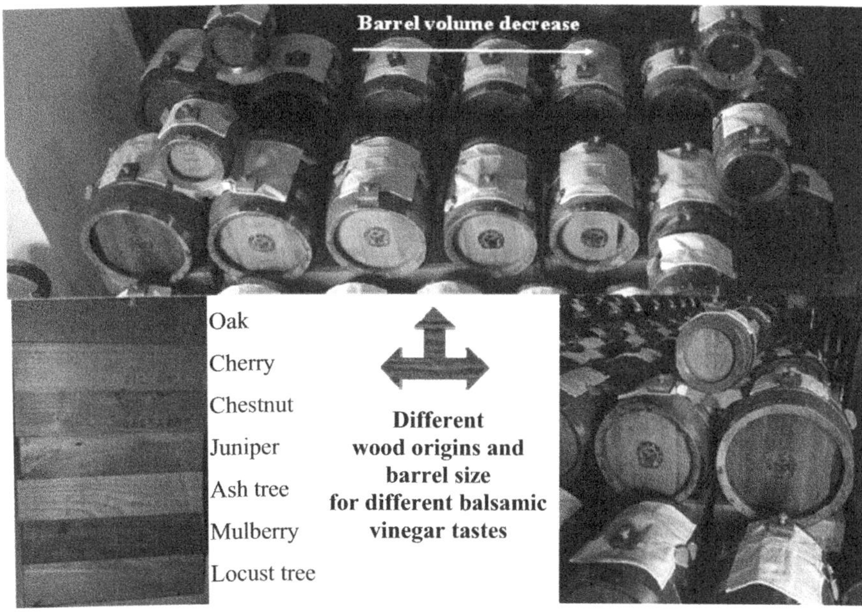

Figure 1.10. *Casks and barrels for balsamic vinegar aging at Midolini Balsameria (© F Debeaufort)*

1.6. Wood package–food interactions

If wood packaging has limited uses due to the risks of splinters and transferring toxins, as described in section 1.2, wood packaging could also interact with food in a beneficial way. Wood has been safely used for centuries in direct contact with food. Fruit and vegetables, as well as fresh and smoked fish have been stored in wooden crates. In cheese- and wine-making, wooden boards and barrels have been indispensable in traditional production (Aviat *et al.* 2016).

In some countries, the use of wooden vats or boxes, or the practice of ripening cheese blocks or wheels on wooden shelves, is considered critical in moisture exchange and in defining the final texture and flavor of the cheese. The use of wooden shelves is also required in the specifications of many PDO (Protected Designation of Origin) cheeses. This example of cheese displays the fact that wood packaging can contribute not only to the packaging but also to the food-making process. The limitations and regulations on the use of wood in contact with food are discussed in section 1.2. Cheeses with bloomy or washed rinds (complex surface microflora) promote the development of a biofilm on the surface of the wood. The composition of biofilms is determined by the type of cheese and its microflora, and by the cleaning or disinfection of the wood (Imhoff *et al.* 2017). The biofilms formed can have a major technological role in the manufacture of cheese, such as acidification, in the case of wooden boxes, or colonization with a surface flora, in the case of wooden shelves. In addition to these symbiotic properties, the material of the wood itself may have interesting properties in terms of food safety. Its hygroscopic properties allow rapid drying, which inhibits microbial growth. Some species of wood have antimicrobial properties, probably linked to the presence of polyphenols. The most effective antimicrobial species are oak and pine. For example, spruce shelving, which is widely used for ripening cheeses, is less effective than oak, but much better than plastic in preventing unwanted or contaminating bacterial growth (Milling *et al.* 2005). Colonization with an adequate biofilm inhibits the establishment and growth of Listeria monocytogenes, probably due to nutritional competition (Guillier *et al.* 2008).

For centuries, wood has been considered natural packaging for the ripening of various food products, especially cheese. New food safety regulations are contributing to the substitution of wood by other materials, such as polypropylene, high-density polyethylene or stainless steel. However, the replacement of wooden utensils with utensils of other materials changes the characteristics of cheeses, affecting the traditional flavor and texture, often not positively (Galinari *et al.* 2014). This can also be observed in other food products.

In the factories making dried egg pasta such as screw-shaped pasta (Fuslilli) and spirals (spirali), wooden trays are used to dry the fresh pasta. Only a part of these trays are in direct contact with the pasta. Even if wood is now gradually being replaced by plastic, Filip *et al.* (2012) showed that the total number of microorganisms (CFU/ 20 cm^2) was significantly lower on the wooden frames, compared to the plastic frames, and that 30% of swabs sampled from the plastic frames exceeded 200 CFU/20 cm^2, whereas for the wooden frames it was only 3% of the swabs.

Wood contributes to the aroma and taste of cheese, but it is one of the key parameters in the quality of the taste of some fermented beverages. Swaffield *et al.* (1997) studied the effect of the biofilms on wood in cider production. This study identified bacteria (lactic acid and acetic acid bacteria) and yeasts that they isolated from wooden cider fermentation vats. They concluded that these kinds of stable biofilms could positively influence the organoleptic profiles of cider when well controlled.

Of course, wood is a fundamental contact material for wine production, and particularly, the fermentation and aging steps. Wood is used as a tool to develop the characteristics of fruity grapes (or hide imperfections such as dilution or vegetable notes), while seeking to avoid possible defects such as notes of sap or dryness, for example (Delabre *et al.* 2011). In general, wooden barrels allow the natural widening of the aromatic palette of wines (the wood offering aromas of vanilla, spices, toast, etc.), the strengthening of the wines' structure (enrichment in tannins) and sweetness on the palate. Unlike the stainless steel tank, the micropores in wood allow a little air to pass through. Thanks to this micro-oxygenation, the wine can continue to breathe to "round" its tannins and reduce the primary aromas, allowing new aromas to be expressed. The barrel, which is in contact with the wine, also releases its own tannins, which make it possible to intensify and stabilize the color of a wine. Its aromas will contribute to the complexity of the wine by bringing notes of wood, toasted almonds, toasted bread, cloves or vanilla (oak, for example, contains vanillin). The use of wood barrels or casks must be reasonable, in order to supplement the initial quality of the harvest and to not mask the intrinsic quality of the pressed grape. Indeed, more and more winegrowers are opposing the fashion of "all new oak barrels", which, by insisting excessively on artificial wood, will denature and standardize wines. This excessive insistence sometimes gives us the impression of drinking plank juice; the barrel has soaked the wine with so many aromatic woody notes. The quality and manufacturing process of the wood barrel creates a high variation of flavor and aromatic compounds that are transferred into the wine in order to make it high or low quality. However, the specifics of any particular barrel company's method or stave selection process are less significant than the fact that, by these various methods, coopers are trying to produce more consistent barrels that have a more predictable sensory impact on the wines aged in them (Phillips 2016). Wine-making and aging in wooden barrels are unique skills mastered by generations of winemakers. It is a century-old collaboration between coopers and winegrowers, which is accompanied by a scientific approach in understanding the interactions between the wood and the wine (Barrera-Garcia *et al.* 2006). A stopper made from cork, which is made from the bark of the trunk of the cork oak, could also affect wine quality, particularly due to its permeability to oxygen, alcohol, sulfur dioxide and aroma compounds (Karbowiak *et al.* 2010). Natural stoppers are presented and discussed in Chapter 9.

1.7. References

Aviat, F., Gerhards, C., Rodriguez-Jerez, J.J., Michel, V., Le Bayon, I., Ismail, R., and Federighi, M. (2016). Microbial safety of wood in contact with food: A review. *Comp. Rev. Food Sci. Food Safety*, 15(3), 491–505.

Barrera-García, D., Gougeon, R.D., Debeaufort, F., Voilley, A., and Chassagne, D. (2006). Transfer of volatile phenols at oak wood/wine interface in a model system. *Dev. Food Sci.*, 43, 449–452.

Boakes, J. (2001). *Making a Barrel Paperback.* Dalesman Publishing Company Ltd, Skipton, North Yorkshire, UK.

Delabre, S., Pic, L., Bonnefond, C., and Granes, D. (2011). Le bois en vinification : quels bénéfices dans le sud de la France ? *Bois en vinification*, 12, 1–2.

Deza, M.A., Araujo, M., and Garrido M.J. (2007). Efficacy of neutral electrolyzed water to inactivate *Escherichia coli*, *Listeria monocytogenes*, *Pseudomonas aeruginosa*, and *Staphylococcus aureus* on plastic and wooden kitchen cutting boards. *J. Food. Prot.*, 70(1), 102–108.

Early, R. (2012). Raw materials. In *Food Hygiene Auditing*, Chesworth N. (ed.). Springer, Tyne and Wear, UK.

European Commission (2004). Materials and articles intended to come into contact with food and repealing Directives 80/590/EEC and 89/109/EEC. Regulation 1935/2004 [Online]. Available: https://eur-lex.europa.eu/legal-content/EN/ALL/?uri=CELEX%3A32004R1935.

European Commission (2005). Amending Annex IV to Directive 2000/29/EC on protective measures against the introduction into the community of organisms harmful to plants or plant products and against their spread within the community. Council Directive 2005/15/EC [Online]. Available: https://eur-lex.europa.eu/legal-content/EN/ALL/?uri=CELEX%3A32005L0015.

European Commission (2010). The obligations of operators who place timber and timber products on the market. Regulation 995/2010 [Online]. Available: https://eur-lex.europa.eu/legal-content/EN/TXT/?uri=CELEX%3A32010R0995.

European Commission (2011). Plastic materials and articles intended to come into contact with food. Commission Regulation 10/2011 [Online]. Available: https://eur-lex.europa.eu/legal-content/EN/ALL/?uri=CELEX%3A32011R0010.

FEFPEB (European Federation of Wooden Pallet and Packaging Manufacturers) (2020). Properties of Wood [Online]. Available: https://www.fefpeb.eu/wooden-packaging/properties-of-wood [Accessed 23 March 2020].

Filip, S., Fink, R., Oder, M., and Jevsnik, M. (2012). Hygienic acceptance of wood in food industry. *Wood Sci. Technol.*, 46(4), 657–665.

Fink, R., Filip, S., Oder, M., and Jevsnik, M. (2013). Wood in food industry – Potential applications and its limitations. *Microb. Pathogens Strat. Combating Them: Sci., Technol. Educ.*, 188–194.

Food and Agriculture Organization (2013). FAO ISPM-15: Wood packaging material in international trade, Regulation. Food and Agriculture Organization of the United Nations and International Plant Protection Convention, Rome.

Future Market Insights (2020). Wooden Plywood Packaging Market – Wooden & Plywood Packaging Market: Global Industry Analysis 2014–2018 and Opportunity Assessment 2019–2029 [Online]. Available: https://www.futuremarketinsights.com/reports/wooden-plywood-packaging-market [Accessed 23 March 2020].

Gaglio, R., Cruciata, M., Scatassa, M.L., Tolone, M., Mancuso, I., Cardamone, C., Corona, O., Todaro, M., and Settanni, L. (2018). Influence of the early bacterial biofilms developed on vats made with seven wood types on PDO Vastedda della valle del Belice cheese characteristics. *Int. J. Food. Microbiol.*, 291, 91–103.

Galinari, E., Escarião da Nóbrega, J., de Andrade, N.J., and de Luces Fortes Ferreira, C.L. (2014). Microbiological aspects of the biofilm on wooden utensils used to make a Brazilian artisanal cheese. *Braz. J. Microbiol.*, 45(2), 713–720.

Guillier, L., Stahl, V., Hezard, B., Notz, E., and Briandet, R. (2008). Modelling the competitive growth between Listeria monocytogenes and biofilm microflora of smear cheese wooden shelves. *Int. J. Food Microb.*, 128(1), 51–57.

Imhof, R., Schwendimann, L., and Riva Scettrini, P. (2017). Sanitising wooden boards used for cheese maturation by means of a steam-mediated heating process. *J. Consumer Prot. Food Safety*, 12, 255–263.

International Standard Organization (2011a). ISO 8611/2011-1, Pallets for materials handling – Flat pallets – Part 1: Test methods [Online]. Available: https://www.iso.org/standard/.

International Standard Organization (2011b). ISO 8611/2011-2, Part 2: Performance requirements and selection of tests [Online]. Available: https://www.iso.org/standard/.

International Standard Organization (2011c). ISO 8611/2011-3, Pallets for materials handling – Flat pallets – Part 3: Maximum working loads [Online]. Available: https://www.iso.org/standard/.

Karbowiak, T., Brachais, L., Alinc, J.-B., Gougeon, R.D., Debeaufort, F., Voilley, A., and Chassagne, D. (2010). White wines oxidation and the role of cork. *Crit. Rev. Food Sci. Nut.*, 50, 20–52.

Kim, K.J., Ryu, J.A., and Eom, T.J. (2017). Characteristics of the heat treated wood packaging materials according to international standards for phytosanitary measures and verifiability of heat treatment. *Wood Res.*, 64(4), 647–658.

Milling, A., Kehr, R., Wulf, A., and Smalla, K. (2005). Survival of bacteria on wood and plastic particles: Dependence on wood species and environmental conditions. *Holzforschung*, 59(1), 72–81.

National Wooden Pallet and Container Association (2014). NWPCA Uniform Standard for Wood Pallets [Online]. Available: www.palletcentral.com [Accessed 23 March 2020].

Parobek, J., Loucanova, E., Nosalova, M., and Dovcikova, A. (2019). Perception of wood-based packaging innovation applying smart solutions. In *Proceedings of the 12th WoodEMA Annual International Scientific Conference on Digitalisation and Circular Economy: Forestry and Forestry-based Industry Implications*, Chobanova, R. (ed.). WoodEma, Zagreb, Croatia.

Phillips, C. (2016). New barrel technology uses science to produce more consistent product. *Wine Business Monthly*, 5, 18–23.

Schonwalder, A., Kehr, R., Wulf, A., and Small K. (2002). Wooden boards affecting the survival of bacteria? *Holz als Roh- und Werkstoff*, 60(4), 249–257.

Swaffield, C.H., Scott, J.A., and Jarvis, B. (1997). Observations on the microbial ecology of traditional alcoholic cider storage vats. *Food Microbiol.*, 14(4), 353–361.

Technavio (2020). Global Wood Packaging Market 2019–2023 [Online]. Available: https://www.technavio.com/report/global-wood-packaging-market-industry-analysis [Accessed 23 March 2020].

Vivas, N. (2002). *Manuel de la tonnellerie*, 2nd edition. Editions Ferret (ed.), Bordeaux, France.

2

Papers and Boards

Frédéric Debeaufort

Institute of Technology, University of Burgundy, Dijon, France

2.1. Introduction

Paper and paper-based materials are the oldest and most versatile packaging materials available on the market today. Paper and cardboard are the leading materials used in packaging, in France and worldwide. This lead is also true in terms of turnover. Worldwide, half of the paper produced is used as packaging material. Globally, about 420 million tons of paper-based packaging are produced, nearly 90 million tons in Europe in 2018 (PaperIndustryWorld 2020). In France, no less than 4.5 million tons are produced for transformation into the most varied packaging systems, affecting all industrial and commercial sectors – which represents a turnover of more than 2 billion euros. Fiber-based solutions, cartons, wrapping paper and corrugated cardboard are doing well in the packaging sector – particularly in high-end products and transport packaging – thanks to their moderate costs and, above all, their sustainable and ecological image. Paper and cardboard packaging dominates, however, in terms of volume, plastics are the most significant. Today, this major industrial sector revolves around three families of products: paper packaging, flat (or compact, as opposed to corrugated cardboard) and corrugated paper. While there may be common applications, these types of materials are mostly still used for secondary packaging (Cofepac 2020).

For a color version of all figures in this book, see www.iste.co.uk/debeaufort/packaging.zip.

Packaging Materials and Processing for Food, Pharmaceuticals and Cosmetics, coordinated by Frédéric DEBEAUFORT, Kata GALIĆ, Mia KUREK, Nasreddine BENBETTAIEB and Mario ŠČETAR.

The primary purpose of packaging is to protect and preserve products. To ensure this protection, it is imperative to find the optimal combination between the different packaging materials. Depending on the product, plastic or cardboard is often used. It is also not uncommon to find multi-material packaging containing at least two types of material, including paper or cardboard. In most cases, we know of corrugated boxes which are packed with filling and padding materials depending on the product and content. These cartons and packaging materials are particularly prevalent in e-commerce, which also concerns the food industry. More and more customers are ordering their food online, a quick, convenient and, above all, sustainable solution with corrugated packaging materials.

The main advantages of using papers and boards for commercial packaging include: excellent print quality for most boards, very good mechanical protection for products, relatively low production and processing costs and ease of recycling. Paper is made from cellulose fibers found in wood and grasses, making it renewable and sustainable. The process of making virgin paper involves cutting down trees to create virgin pulp, as well as using chemicals such as bleach. Paper and cardboard can be recycled up to seven times before being mixed with new fibers or virgin paper pulp. It is therefore important to recycle paper and cardboard to reduce its environmental impact.

Even if the invention of writing paper is attributed to the Egyptians 3 millennia BCE, it was not paper in the modern sense of the word – that is, obtained from vegetable fibers – but papyrus. It was in the year 105 CE that paper was discovered in China, during the reign of Emperor Hoti. Ts'ai Lun, the Minister of Agriculture, designed a paper pulp made from old rags, bark and crushed vegetable fibers dispersed in water (Cartonnage-Paulet 2020). At the beginning of the 20th century with the development, in particular, of trade and transport, the production and distribution of goods and food products required better protection of articles. A cheap and easy solution was to use boxes and thick paper bags. The transition from the sale of bulk products to the sale of packaged products also required that packaging offer not only protection but also a means of communication through its display (marketing). It has become more important to enable consumers to recognize products in-store. Paper-based materials lend themselves very well to printing and display and therefore responded to this demand from consumers and manufacturers very early on. Traditional thick paper was no longer sufficient. This led to the development of the first machines for making flat and corrugated cardboard in the USA in the 1920s. Cardboard, however, was a French invention, in 1751, attributed to a pupil of René-Antoine Ferchault de Réaumur, with Edward Healey and Edward Allen obtaining a patent, in 1856, having invented the first cardboard machine in the UK (Cartonnage-Paulet 2020).

A leader in the packaging world, paper and cardboard form the basis of a diversified range of packaging products, from simple folding paper, to cardboard cases, including large bags, the "American box", corrugated boxes and laminated packaging materials.

Wrapping papers can be unbleached or bleached, rubbed, primed, coated or combined with other materials and take a variety of forms: kraft bags for wrapping fruit and vegetables, crystal or parchment paper, technical and specialist papers (covers for yoghurt jars, sheet metal dividers or coffee bags, parchment paper, etc.). Paper is the base raw material for the manufacture of all carton types (CartonPro 2020). The flat cardboard generally has an interior wood or gray color and can be coated or uncoated. It is a material perfectly suited to automated production lines, capable of performing a wide range of different tasks, for example, acting as a barrier to grease, humidity and the air. Its uses, as it is or in combination with other materials, are very diverse: pastry boxes, cases for perfume, boxes for cigarettes, tubes or barrels for heavy products, blister packs or displays for distribution, bricks for milk, etc. Kraft papers, primarily, are the essential components of corrugated cardboard, a "sandwich" packaging material formed by assembling several sheets of these papers. Unbleached or white, light grammage or high density, resistant to bursting, humidity or compression, these papers constitute a wide range, which allows the manufacture of high-resistance industrial packaging, transport pallets, displays for mass distribution or microgroove packaging for consumer products. The highest usage rates of waste paper as raw material are found in the paper and cardboard packaging sector. Thus, constantly demonstrating innovation, developing new applications by combining with other materials, cardboard paper, made from renewable raw materials (wood and recovered paper), demonstrates daily that it is the packaging material of the future.

2.2. Pulp and cellulose extraction for paper-based materials

Paper is made from cellulose fibers extracted from trees, recycled paper and plant fibers such as cereal straw. Today, about 97% of the world's paper and board is made from wood pulp, of which more than 80% comes from spruce, firs and pines (softwood). Hardwoods, such as birch, aspen and others found in temperate climates (mainly Europe), are also used as raw material for corrugated board or for printing and writing papers. On the other hand, eucalyptus, native to Australia and New Zealand, has been successfully cultivated in countries with hot climates (Portugal, for Europe) as a raw material for a high-quality paste; this allows the manufacture of a wide range of papers, including specialist papers. Softwoods, however, offer longer fibers (on average 3 mm versus 1 mm for hardwoods). These long cellulose fibers are therefore preferred for paper and cardboard requiring higher mechanical strength characteristics. Wood composition and fiber characteristics affect the quality of the pulp (cellulose fibers) and the yield of extraction (Table 2.1).

Composition (%) and fiber characteristics	Softwood	Hardwood
Cellulose	42–44	43–47
Lignin	25–30	17–26
Hemicellulose hexosanes	15	5–8
Hemicellulose pentosanes	10–15	15–35
Waxes, resins, fats	1–10	0.5–2
Fiber length (mm)	2–3.5	0.9–1.2
Fiber diameter (µm)	20–50	20–30

Table 2.1. *Wood composition and fiber characteristics (Otenio* et al. *2004; Iggesund 2020)*

Chemically pure cellulose consists of long molecules made up of glucose units linked by β-(1→4) bonds. Glucose units are made up of carbon, hydrogen and oxygen atoms. These molecules are held together side by side by hydrogen bonds to form sheets, which in turn are stacked in tight layers to form "microfibrils". The microfibrils group together in bundles, with groups of these bundles forming the cellulose fiber, the basis of the pulp. Several processes are used to extract these cellulose fibers, depending on the origin of the raw material, the nature of the wood and the recycling of waste paper and cardboard. Pulp production can be integrated with paper production, or the pulp can be produced in a separate pulp mill. Pulp is produced using chemical (lignin dissolution-based process) and/or mechanical processes (fiber separation using either a shredder or a refiner), coupled, or not, with heat treatments.

Depending on the process under consideration and the origin of the raw material, the wood that arrives at a pulp mill can take different forms. It can be received in the form of round branches of wood (short logs, bolts) with its bark, or in the form of bark-peeled wood chips of about 1 to 3 cm. If bolts, branches or logs (trunks) of round wood are used, they must be debarked before any other operation. Debarking is usually done by tumbling the logs in large steel drums where wash water can be applied. These debarked logs are then shredded in a chipper if the pulping process requires chemical digestion. The chips are then sorted by size, cleaned and temporarily stored for the subsequent manufacture of pulp.

2.2.1. *Mechanical and thermomechanical pulping*

There are many mechanical processes, which can be classified into four main categories: stone ground wood (SGW), pressure ground wood (PGW), refined mechanical pulp (RMP) and temperature treated mechanical pulp (TMP).

Mechanical processes that use a rotating abrasive wheel transform wood logs into fibers, either at atmospheric pressure (SGW) or under pressure (PGW). The friction between the grinding wheel and the wood causes a local rise in temperature, reaching the glass-transition point of the lignin (about 175°C), which separates from the cellulose fibers. At this temperature, it is possible to considerably reduce inter-fiber cohesion forces, which enables dissociation between the cellulose fibers. Once the fibers are separated, the pulp is removed from the surface of the grinding wheel by rinsers and is collected in a tank before being purified (Petit-Conil 1999). As mechanical grinding produces pulps that are not very resistant, mechanical chip treatments have been developed in an attempt to produce longer, more resistant fibers.

Other mechanical processes are for wood chips, which use disk refiners without chemical wood treatment, and without heat treatment (RMP) or with heat treatment using steam (TMP) before being transformed into fibers (Petit-Conil 1999). The RMP process is based on the use of disks with special rotating blades and grooves to separate the fibers of the initial structure of the wood chip. As they pass through the blades of the disks, the chips undergo repeated cycles of compression and expansion. These cycles carried out in the presence of water cause energy absorption, leading to the mechanical breakdown of the bonds that hold the fibers together. This process is more energy-consuming than the SGW process, but the quality of the fiber is much better. The mechanical properties of the pastes can be improved by subjecting the chips – before mechanical treatment – to heat treatment, with steam, generally between 110 and 130°C, in the TMP process; often referred to as the thermomechanical process. The first stage of defibration under pressure (2–3 bar) makes it possible to obtain a coarse pulp and therefore requires a second mechanical treatment called refining. The latter is carried out under pressure or at atmospheric pressure.

2.2.2. *Chemi-mechanical pulping*

Mechanical processes on chips, using refiners, but with prior chemical treatment of the wood, are known as chemi-mechanical pulping (CMP) or chemi-thermomechanical pulping (CTMP) processes. Therefore, mechanical, thermal and chemical treatments are combined to improve pulp quality for use in higher grades of paper (Bajpai 2012).

The chemical treatment promotes defibration and makes the lignin more hydrophilic. After mechanical treatment, the fibers are longer, more flexible and therefore have better characteristics (reduced opacity). The name given to the paper pulp – CTMP or CMP – and the efficacy of pulp extraction (size of the yield) depends on the significance of the chemical treatment (temperature, pressure and duration). CTMP paste is generally produced at 120 to 130°C for 15 to 30 minutes; for CMP paste, the temperature is higher (150 to 170°C) for a period of time that may be longer than one hour. The yield of CTMP paste is then close to 90%, while that of the CMP paste is between 80 and 90%. The chemicals used are mainly sulfur-containing, such as sodium sulfite (1 to 5% compared to wood) used alone or with soda (pH = 9–10). Chemical treatment produces a higher-quality pulp than TMP, but the energy and water demands are higher.

2.2.3. *Chemical pulping*

Chemical pulps represent the majority of the world's pulp production to date. They are used in almost all grades of paper and board. The main purpose of this process is to dissolve and extract lignin from the cell wall and thus produce perfectly individualized fibers without mechanical damage. As most of the components of non-fibrous wood are removed with this technology, yields are typically in the range of 40–55%.

In the manufacture of chemical pulp, the wood chips and the chemicals in aqueous solution are cooked together in a pressure vessel (digester), which can operate batchwise or continuously. In batch cooking, the digester is filled with chips through an upper opening, digestion chemicals are added and the contents cooked at high temperatures (150–175°C) and pressures for long periods of time (several hours). Once cooking is complete, the delignified pulp is discharged from the digester and stored in a temporary retention tank. The sequence is then repeated. During continuous digestion, the previously steamed chips are fed into the digester at a continuous rate. Chips and chemicals are mixed together in the impregnation zone at the top of the digester and then passed through the upper cooking zone, the lower cooking zone and the wash zone, before being blown into the blowing tank.

The kraft or sulfate process is the alkaline process. It is the most widely used in the world for many lignocellulosic materials (wood, annual plants). It is based on the use of soda and sodium sulfate. The chemical treatment of the chips is carried out at 170–175°C for a period of 2–5 hours, depending on the origin of the wood. The cooking is carried out in a vertical reactor where the chips are immersed in various cooking liquors. During cooking, sodium sulfate is hydrolyzed into soda, NaHS and H_2S. The various sulfur compounds present react with the lignin to give more soluble thiolignins. Soda also plays a delignification role, which is associated with that of sulfate and its derivatives.

The acid or sulfite pulp process is only used for softwoods. It is based on the reaction between lignin and calcium, sodium, ammonium or magnesium sulfite. Ammonium bisulfite is preferred for softwoods with a high resin content. Sodium bisulfite is interesting because it dissolves resins and has the advantage of containing a higher content of sulfur dioxide. Therefore, it is used for woods rich in phenolic compounds (such as flavones) which can combine with lignin and thus become insoluble. Calcium bisulfite promotes the formation of insoluble resins that are difficult to remove from pulp.

2.2.4. *Recycled paper pulping and non-wood pulping*

In the recycled pulp (known as secondary fiber) process, pulp from previously manufactured products (e.g. cardboard, writing paper) is recovered by hydration and agitation (Bajpai 2018). Secondary fibers include any fibrous material that has undergone a manufacturing process and that is recycled as a raw material. However, secondary fibers have less strength and binding potential than virgin fibers. The fibrous material is introduced into a large reservoir, or pulper, and mixed by a rotor. The pulper may contain hot water or pulping chemicals to help dissolve the paper matrix. Debris and impurities are removed by wire filtration (debris accumulates on the wires) before being eliminated (EPA 2001).

Non-wood pulp is made from fibers other than wood. Non-woody fibers used for papermaking include straws and grasses (e.g. flax, rice), bagasse (sugarcane), hemp, flax, ramie, kenaf, cotton and fibers extracted from leaves. The pulping of these fibers can be carried out by mechanical means at high temperatures or by using the kraft or modified soda process. The production of non-wood pulp remains uncommon (EPA 2001).

2.2.5. *Pulp bleaching*

Paper pulps are either bleached or thinned, depending on the intended end product. Since all the components of wood are present and the cellulose is white in color, the brown color of these pulps comes mainly from the colored groups (chromophores) of the lignin. In addition, the mechanical preparation conditions are suitable for coloring: high temperatures, atmospheric oxidation, mechanical action, etc. The purpose of bleaching is not to extract the lignin but to modify its chemical structure by eliminating the chromophores.

For mechanical pulps (SGW, PGW, RMP and TMP), the chemicals commonly used are oxidizing or reducing agents and, in this case, are hydrogen peroxide and sodium hydrosulfite or dithionite.

Unbleached chemical or chemi-mechanical paper pulps (kraft) can be used in certain categories of paper or cardboard such as kraft packaging bags, however, in large part, it is necessary to bleach them for introduction into superior quality such as papers intended for printing. Unbleached pulp still contains a quantity of lignin. The bleaching of these pastes consists of removing this residual lignin completely, while trying to limit the degradation of the polysaccharides. This operation takes place in several stages, constituting a bleaching sequence. In general, a bleaching sequence consists of the alternation of oxidation stages of lignin followed by its extraction with alkaline agents. Each bleaching step is followed by washing to remove materials dissolved by the action of bleaching agents on the pulp and prevent the materials from consuming chemical reagents in the next step. The most effective oxidizing agent on lignin, which does not affect cellulose, is chlorine, but this is gradually being replaced by chlorine dioxide, which is more expensive but has much less impact on the environment.

In the case of recycled pulp, bleaching is associated with ink removal. The trituration and purification of the printed papers does not make it possible to obtain a pulp of white cellulosic fibers. Due to the presence of ink, the paste thus prepared has a gray appearance and its use in the manufacture of certain grades of printing paper will only be possible if residual ink is removed. The main technique used is the flotation process that is based on the differences in density of the fibers and ink particles to achieve their separation. The fibrous suspension is treated with chemical agents that facilitate the separation of the ink particles by foaming (flotation).

2.2.6. *Pulp properties and uses*

According to the process used for cellulose and pulp extraction, the pulp quality for paper-making varies (Table 2.2). The pulp characteristics determine the type of paper and thus the type of cardboard able to be produced (Bajpai 2010, 2012, 2018).

Compared to mechanical pulping, chemical pulping is more expensive and produces a lower yield (50% vs. 90% for TMP or RMP). It requires large amounts of chemicals, even if 99% are reusable. However, chemical pulp manufacture consumes less energy and the product has high stiffness and brightness and low opacity and yellowing.

Pulp type	Pulp grades (*and origin*)	Uses
Chemical pulps	Sulfite pulp (*softwood and hardwood*)	Printing papers
	Kraft pulp (*softwood and hardwood*)	Bleached-printing papers, paperboards, unbleached-heavy packaging papers
Chemically dissolved pulp	Dissolving pulp (*softwood and hardwood*)	Viscose, rayon, cellophane, acetate fibers and films
Semi-chemical pulp	Cold-caustic process (*softwood and hardwood*)	Newsprint and groundwood printing paper
	Neutral sulfite process (*hardwood*)	Newsprint and groundwood printing paper
Mechanical pulps	Stone groundwood pulp SGW (*softwood*)	Corrugating medium
	Refined mechanical pulp RMP (*softwood*)	Newsprint and groundwood printing paper
	Thermo-mechanical pulp TMP (*softwood*)	Newsprint and groundwood printing paper, paperboard front panels
	Chemi-mechanical pulp CTMP (*softwood*)	Printing papers

Table 2.2. *Pulp grades and uses (adapted from Bajpai 2012)*

2.3. Paper production and properties

2.3.1. *Paper sheet processing*

The production of paper can vary considerably depending on the type of paper produced and the raw materials used in the process. Nonetheless, most paper and paperboard production processes include the following key aspects: stock preparation, paper machine processes and optional finishing processes.

Stock preparation is the process in which different types of pulp fibers are mixed with water and various chemicals to form a fiber slurry. The cellulose (fiber) content of the slurry ranges usually from 0.3 to 7%, according to the thickness (50 μm to 1.2 mm) and grammage (12 to 600 $g.m^{-2}$) of the paper or paperboard expected (Riley 2012; Piergiovanni and Limbo 2016). Among the chemicals added to cellulose fibers are acids and bases for pH control; water-repellent sizing agents (binding agents); mechanical strength additives, either when dry (starches) or in the presence of humidity (polymers); fillers providing shine, luminosity or opacity (kaolins, TiO_2); and antifoaming agents. A cleaning step during the preparation of the stock aims to remove contaminants and impurities from the fibers (Castle *et al.* 1997; Zheng *et al.* 2001). The fiber slurry goes to the entrance of the paper machine, which could be in another factory or not.

The production of a sheet of paper is carried out using a "paper machine". The most common method for making paper or paperboard is the Fourdrinier technique (Riley 2012). Various stages transform a suspension of fibers dispersed in water into a continuous, homogeneous, resistant fibrous sheet that is adapted to the requirements of its use. The paper machine has three main parts (Figure 2.1). Formation of the sheet is carried out by draining the fibrous suspension, after it has been poured onto the fabric of the flat table. The wet sheet obtained, containing about 80% water, is compressed between two or more rollers. Pressing reduces the amount of water to about 40%. The compacting of the sheet and the multiplication of the inter-fiber contacts allows better cohesion of the fibers and therefore better mechanical strength of the sheet thus formed. The sheet then enters the drying section. Drying paper is an energy-intensive process. It is usually carried out by conduction, where the paper is in contact with a metal cylinder heated by steam. Sometimes, convection can be used. In this case, hot air is blown onto the sheet. Infrared radiation ramps can also be used for heating the sheet by radiative energy. The most common process, however, consists of passing the continuous sheet of wet paper from cylinder to cylinder until its humidity is about 6%. The sheet is kept in contact with the dryers by rolls made of cotton felt or artificial fabrics. It then leaves the "dryer" section to undergo, if necessary, one or more finishing treatments (Garcia Cerrillos and Mangin 2007).

There are different finishing treatments, depending on the type of paper and its intended use. These treatments can be summarized in three processes: sizing, coating and calendering or supercalendering (Bajpai 2010, 2012).

– Sizing gives the surface of the paper resistance to liquids, a property necessary for writing or printing. The most common types of sizing are pigments and starches, although animal glue and glycerol can also be used (for art and specialty papers or banknotes). Sizing presses are located after the drying step; rollers deposit size onto the wet paper, and this is followed by a drying step (Figure 2.1).

– Coating involves covering one or both sides of the paper with a "layer", generally composed of mineral pigments such as calcium carbonate, talc or kaolin. This process aims to improve the surface of the paper by increasing its impermeability against liquids, oils, gases and chemicals, increasing its adhesion characteristics and by reducing friction wear. Coatings can be classified as aqueous, organic solvent-based or extrusion. Aqueous coatings, used for base papers, contain water soluble binders. They are applied in liquid form and are followed by a drying step. Common aqueous binders are casein, starch, proteins, acrylics and polyvinyl acetates. Solvent coatings are used in situations where the binders are not soluble in water and are also used with specialty papers. With solvent casting, the drying is faster and less energy-consuming, but fire risks exist. High-solids and extrusion coatings are used for specialty papers where resistance to chemicals, gases or liquids is required. High-solids coatings are applied (monomers) in a molten or liquid state and then polymerized by heat, UV or electron beam treatments.

– Once the chemical processes are complete, physical processes, such as calendering (or supercalendering, at higher pressure), cockling and embossing, can be used to create the desired surface texture on the paper. Calendering involves passing the paper through a stack of several steam-heated cast iron cylinders. The finished paper is usually heavily sized and has the characteristics rather associated with high-quality paper. This process aims to improve the surface of the paper, making it more efficient for use in printing or providing more specific surface properties.

Over the past few decades, the paper industry has gradually evolved towards ever-larger and ever-faster paper machines. A modern paper machine produces a 10 m-wide and, on average, 120 km-long sheet every hour. A total of 90% of the production cost of the sheet of paper is due to the hot pressing and drying operations.

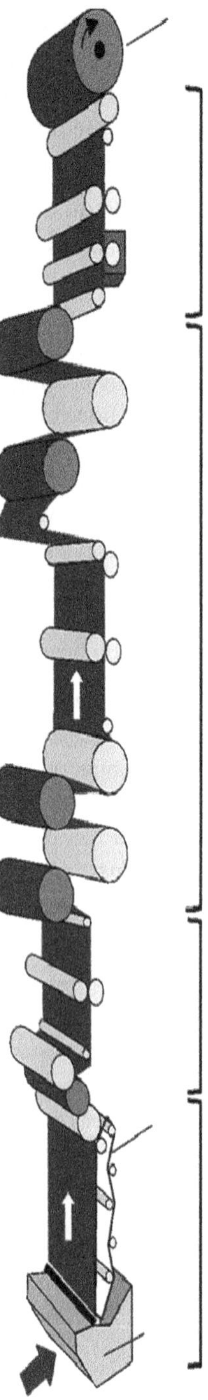

Figure 2.1. *Paper sheet processing by the Fourdrinier method*

2.3.2. *Types of paper for packaging*

The following list of paper types gives a general description of different paper products available for packaging purposes.

Greaseproof papers and parchment paper are derived from chemical pulps and packed tightly to prevent absorption of fluids. The paper very often has a special polymer-based coating to further enhance grease resistance. Poly-coated kraft paper is different from parchment paper, but it can also be used to resist grease penetration (Riley 2012).

Glassine paper (or parchment paper) has a glassy, smooth surface, is transparent and has a high density. It is obtained following a supercalender, during which it is carefully moistened with water and then dried on a cylinder at high temperature and pressure. This results in a very strong cohesion between the fibers, as well as almost zero porosity. The addition of titanium dioxide makes the paper opaque, and it is frequently plasticized to increase its toughness (Robertson 2013).

Wax papers provide a barrier against the penetration of liquids and vapors. Wax can be applied to many types of paper, including greaseproof papers and glassine. When the wax is applied during the finishing step, the barrier properties obtained are significantly superior to those obtained by adding wax to the slurry (Robertson 2013).

Kraft paper is one of the strongest papers available. It is used when maximum strength is needed. Kraft paper can be used for paper bags in the grocery store, void fill, cushioning, interleaving and a variety of other uses. The long fibers in kraft paper make it strong but poor for printing.

Bleached kraft paper is ideal for marketing when the appearance of the packed product counts. Kraft paper is necessarily coated and smoothed prior to being printed on.

Tissue paper is a term commonly used to refer to a lighter, liquid absorbent paper. It is often used as void fill packaging in baskets and smaller objects.

Label paper is similar to book/printable paper, often coated to allow for better printing quality. It can be coated on one or both sides and is used primarily for the labeling of other kinds of packaging.

Pouch papers are virgin kraft papers treated with plasticizers to make them more pliable; common uses include wrapping for soap and butter.

Containerboard paper is a kraft board made for the liners of corrugated board. Containerboard is a strong board that is composed of long cellulose fibers, mainly coming from pulp produced by chemical processing of softwoods.

The properties of the main papers used for food packaging are summarized in Table 2.3.

Type	Source	Grammage ($g.m^{-2}$)	Applications
Kraft	Chemical pulp (sulfite process, pine fibers), bleached or not	70–300	High mechanical resistance, bags, corrugated board, laminates
Sulfite	Chemical pulp (sulfite process, pine and eucalyptus fibers)	35–300	Printing, bags, labels, laminates
Greaseproof	Highly beaten pulp	70–150	Fat resistant, for confectionary and butter or cheese
Glassine	Highly beaten and calendered pulp	40–150	Oil and fat resistant, glassy, smooth surface and almost transparent
Vegetable parchment	Chemical pulp treated with sulfuric acid	12–75	Oil, fat and moisture resistant (stronger when wet)
Tissue	Different sources, including recycled, usually bleached	20–50	For cushioning, as food absorbent paper, napkins

Table 2.3. *Characteristics and applications of packaging papers*

2.3.3. *Paper-based bags, sacks and spiral wound containers*

Several forms of packaging can be obtained from paper: bags, sacks, wound containers and laminated brick packaging for beverages (laminates are detailed in Chapter 6).

Paper bags cover a wide range of uses, from grocery bags, often available for fruits and vegetables, to those that are artistically designed. Three main types are used: flat, satchel, strip window and self-opening satchel (SOS); these are illustrated in Figure 2.2. Flat bags are the simplest, two-dimensional in shape. The gussets present in satchel bags allow easier handling and opening for filling. Some can have a window allowing the content to be displayed. The windows are made, increasingly, of PLA (polylactic acid biopolymer) instead of polypropylene, to keep the bag sustainable and biodegradable. They are very often used for baguettes, for example. In SOS bags, which can be sealed, the internal side is laminated with a plastic film to prevent water, grease and gas transfers, and is thus more impermeable. They are used, for instance, for roasted meats, such as chicken.

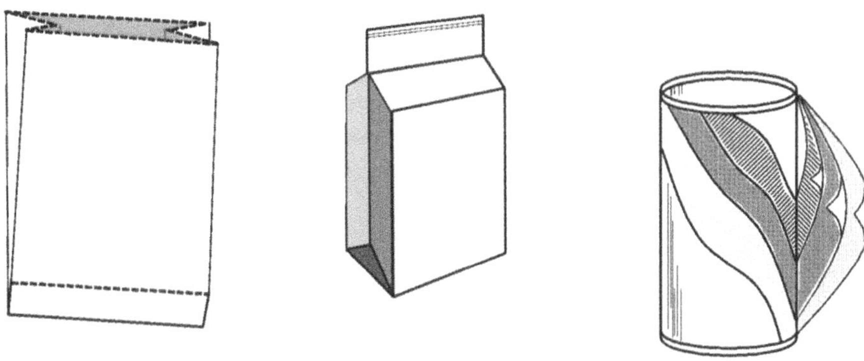

Figure 2.2. *Example of wound container and different paper-based bags*

Sacks are mainly designed and used for heavy contents (more than 5 kg) and are usually made of kraft papers with several plies, very often including a plastic ply inside to reduce water penetration. These are used for flour and other powdered foods and ingredients, for animal feed, etc. There is a wide range of sack designs, open mouth sacks, gusseted or not, etc.

Spiral wound containers are very often composite packaging that consists of at least four plies of paper and paper laminates wound together (Figure 2.2). The body is mainly composed of kraft paper plies with the internal and external surface made of plastic or aluminum ply and printed paper, respectively. The plies are glued together. Spiral wound containers are sealed (seamed) at the ends to provide quasi-hermetic packaging. These are very often used for snacks such as Pringles® or for fragile biscuits.

2.4. Paperboards and containerboards

2.4.1. *Paperboard processing*

Paper is qualified as paperboard or cardboard when the grammage is greater than 250 $g.m^{-2}$.

Cardboard is obtained either by the Fourdrinier method (identical to that for making paper sheets) or by the vat and cylinder method (Smook 1992).

The manufacture of cardboard using the Fourdrinier method differs from that of paper by the successive deposits of several layers of slurry, of different composition, onto the draining table (Figure 2.3), making it possible to obtain the desired qualities of the paperboard (mechanical resistance, folding, etc.).

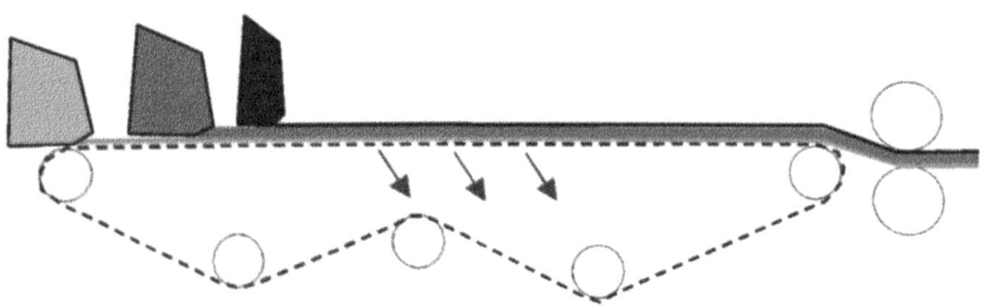

Figure 2.3. *Three-ply paperboard Fourdrinier table*

The vat–cylinder method involves injecting the slurry between two forming wires before it is drained under vacuum. This fiber sheet is transferred onto the underside of continuously moving felt. The felt then passes over the next vat where a second layer is added to the first, and so on, until the final specification is achieved (Figure 2.4). Different pulp fibers, for example, chemical, mechanical and recycled, can be added via the individual vats to build up a multilayer structure.

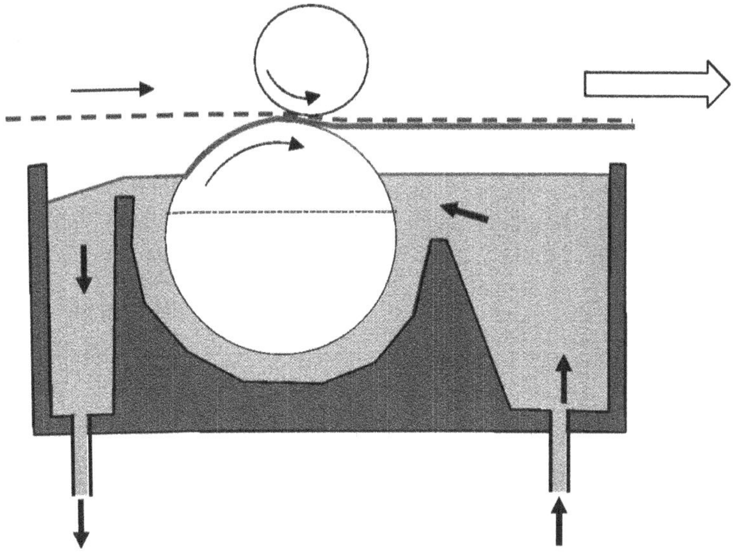

Figure 2.4. *Vat–cylinder paperboard method*

2.4.2. *Paperboard types and properties*

The paperboard/cardboard sheets are mainly intended for the manufacture of containers, obtained by folding, with a thickness between 300 and 1100 μm. They are cut and grooved in order to be folded and are often delivered flat, to be mounted prior to the filling operation.

Several types of cartons are used for food packaging:

Homogeneous coated and/or sulfate bleached board (SBB) is made exclusively from bleached chemical pulp (Figure 2.5). It is usually coated (one side or both) and has medium density, with excellent printing properties that meet the requirements of the packaging industry (Iggesund 2020). It can be die-cut, grooved, hot-stamped and embossed. Pure and hygienic, it is ideal for packaging food products that are sensitive to taste and aroma such as chocolate, cheese, coffee, tea, reheatable foods and frozen foods (Robertson 2013). It can also be used for rolled materials used to make bricks for liquid products (see Chapter 6).

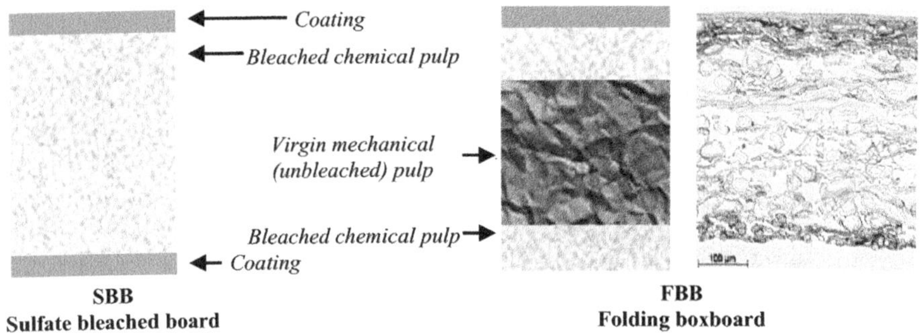

Figure 2.5. *Structure of SBB and FBB multi-ply paperboard (adapted from Iggesund 2020)*

Wood interior paperboard or folding box board (FBB) is best suited for folding. FBBs are made from bleached or unbleached mechanical pulp, sandwiched between two layers of chemical pulp (Iggesund 2020). The chemical pulp front is bleached and coated. The back of the sheet is often cream colored, and the back layer can be thicker or coated with kaolin; the product is then often referred to as White Back Folding Box Board (FBB). Combining mechanical pulp on the inside with outer layers of chemical pulp increases the stiffness and strength of the board, despite its low density.

Coated solid unbleached board (SUB) or sulfate unbleached solid board (SUS) is made from 100% virgin, unbleached chemical pulp. It consists of two to three layers of mineral or synthetic pigment coating on top and one layer on the back to facilitate printing. It is also known as coated unbleached kraft (CUK). Recycled fibers are sometimes used to replace unbleached chemical pulp (Emblem and Emblem 2012; Robertson 2013).

White lined chip board (WLC) is typically made from recycled fibers. It usually has two or three layers of coating on the printing surface, consisting of bleached chemical pulp. The second layer is made from bleached chemical or mechanical pulp, while the middle layers (usually three in total) are made from unbleached recycled pulp from mixed waste paper or waste cardboard. WLC is used for dry products, breakfast cereals and frozen or chilled products (Robertson 2013). The use of fibers from recycled paper greatly reduces the suitability of the cardboard for contact with food, due to the risk of migration of printing products (mineral oils).

A number of steps are involved in converting the cardboard into a compatible material for food packaging. Coatings and laminates are applied to obtain specific barrier properties. The wax deposit provides a very effective water barrier, used, for example, in the manufacture of yogurt pots. The glassine coating allows better resistance to grease, and the polymer deposits by extrusion make it more impermeable and facilitate heat-sealing. During folding, the cardboard is subjected to complex stresses (IggeSund 2020). Paperboard shaping capacity depends on its rigidity, its resistance to delamination of folds and its deformability (stretching), as shown in Figure 2.6.

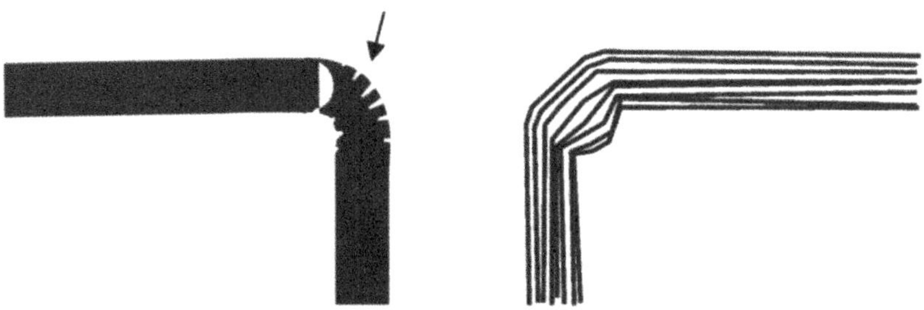

Figure 2.6. *Folding capacity of monoply and multi-ply board*

The possible uses of paperboard as packaging are numerous but depend on the type of paperboard (Table 2.4). Primary packaging is in direct contact with the product, and secondary packaging contains the primary package. The most common example of such a structure is breakfast cereals, with a plastic bag inside a WLC folding carton (Kiviranta 2008, pp. 55–75). Indeed, where primary packaging based on paperboard is concerned, higher purity is required for consumer safety because it is in direct contact with food.

Carton styles follow the ECMA (European Carton Makers Association) requirements that state: the glue flap should never be on the front face; the carton should open squarely when erected from the flat state to be shaped prior to use; no end with a glue flap at the base before being shaped definitively; unglued paperboards must be supplied flat; the fiber direction of the board should be at 90° to the major crease; and so on. The carton folding is designed according to the dimensions required for the product to pack, the type of opening, the method of filling (vertical or horizontal), the gluing method (hot melt or water-based adhesive), the mechanical resistance required, the side exposed to consumer, etc. (Figure 2.7). The creasing steps are of key importance in the carton-making technology, and thus the capacity of folding is crucial.

Product	Requirements	Type of paperboard
Direct contact with food	Purity, cleanliness, runnability	FBB
Frozen food	Strength, barrier, purity, cleanliness, runnability	SBS, SUS
Indirect contact with food	Runnability	WLC
Confectionery	Printability, purity, cleanliness, odor and taint free	FBB, SBS, SBB
Beverage carriers	Strength	SUS
Cosmetics	Printability and appearance	FBB, SBS
Tobacco products	Printability, runnability, cleanliness, odor and taint free	FBB, SBS
Pharmaceuticals	Identification, runnability	FBB, WLC
Milk, juices	Runnability, cleanliness, purity and strength	LPB

Table 2.4. *Applications and required properties of paperboards*

Board types are defined according to the standard DIN 19303, which is a combination of two letters and one figure. The first letter describes the type of board surface treatment: A refers to cast-coated, B to pigment-coated and U to uncoated. The second letters give information about the main origin of the pulp used for the board: Z is related to bleached chemical pulp; N to unbleached chemical pulp; C to mechanical pulp; T to secondary(recycled)-fiber pulp with white, cream or brown reverse; and D to secondary-fiber pulp with gray reverse. For the D grades, the figure describes the density of the bulk of the paper from more than 1.45 cm^3/g (grade D1) to less than 1.3 cm^3/g (grade D3). For grades different from D, the figure defines the color of the reverse side: 1 for white, 2 for cream and 3 for brown.

The different packaging types obtained from paperboard are bottle wrap carton, trays and carton containers (boxes). All of these container types and folding design styles were defined and encoded by the ECMA (Poças 2020). This describes the shape, dimensions (height/depth, width and length), type of paper, etc. (Figure 2.7).

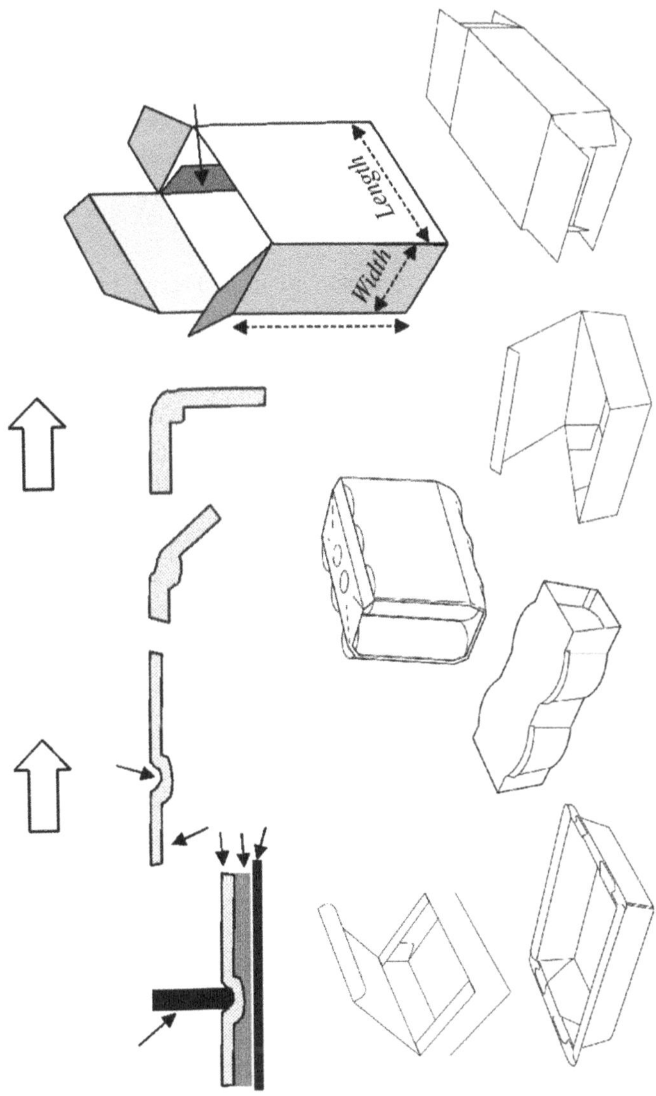

Figure 2.7. *Process of paperboard folding and examples of carton design (adapted from Alexir Packaging 2020; Poças 2020)*

2.5. Corrugated paperboard

2.5.1. *Corrugated board history and economy*

Invented in 1856 to make hats, corrugated paper – ancestor of corrugated cardboard – was the result of the work of two Britons, Edward Healey and Edward Allen. At the time, they had a simple, manual corrugating machine. The first use of this material as a packaging tool was in the USA in 1871, to wrap fragile objects such as bottles. At this time, they only used single-sided cardboard. Soon after, the American Oliver Long came up with the idea of combining corrugated paper with a paperboard to make the cardboard more solid (double face).

The main applications of corrugated cardboard are in packaging. Often criticized and considered unnecessary, it is nevertheless essential and constitutes a key factor of the sustainable development of shipping and delivery of goods. Indeed, thanks to this material, almost all products reach the user in a perfect state of preservation. The development of e-commerce was mainly possible thanks to the tertiary packaging made of corrugated cardboard, which allows product protection, its transport and its easy and hygienic delivery. In figures, in 2018, the European corrugated cardboard industry was producing about 48 billion m^2 of corrugated board every year (almost enough to cover the area of Denmark). More than 40% of corrugated packaging is used to package food products, and it protects more than 75% of European goods (FEFCO 2020). At European level, this concerns about 400 companies and almost 100,000 employees.

2.5.2. *Corrugated board processing and characteristics*

A sandwich material, corrugated cardboard consists of assembly by the gluing of flat cover papers (liners) held equidistantly by corrugated-shaped fluting papers (medium) whose profile thickness varies from approximately 1 mm to 8 mm and whose average weight is 575 $g.m^2$. Fluting/corrugating machines are very technical and can produce up to 450 meters of corrugated board per minute (Figure 2.8). The principle assembly between the layers of fluted paper and paperboard, or rigid paper, is done by gluing the flutes. A corrugator is a set of machines designed to bring together three, five or seven sheets of paper to form single-, double- or triple-wall board in a continuous process. The sheets of paper are unrolled, heated, fluted and then glued on top (Carton Ondulé de France 2020). Three sheets of paper (containerboard) are unrolled simultaneously on the corrugator: the inner liner, the outer liner and, in the middle, the medium (also called fluting).

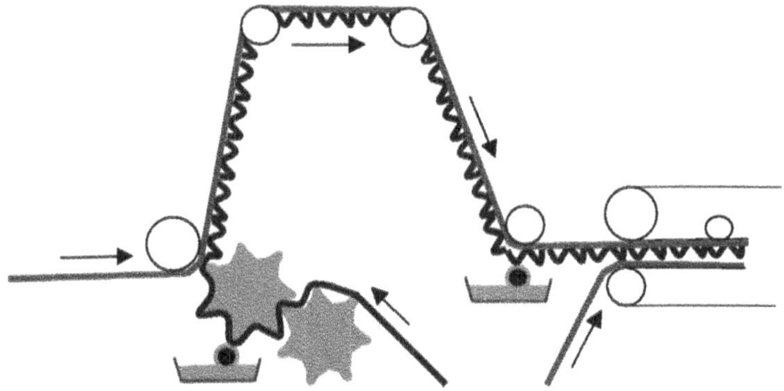

Figure 2.8. *Corrugating machine for corrugated board (adapted from Watkins 2012)*

The medium sheet is given a "wave". This corrugation is obtained by pressing the sheet, under heat, through grooved metal rolls. This process gives the paper its fluted shape in the single facer. Starch is applied to the tips of the flutes on one side; the outer liner and the inner liner are glued to the tips of the corrugated medium sheet of paper, from the outer and the inner side, respectively. The fluted sheet is then placed on the liner (flat paperboard) to form a single-sided corrugated cardboard. The settings at their disposal make it possible to vary the corrugations of the flutes, according to intended use. To form a double face, the tops of the flutes opposite the liner are glued and then deposited on a liner again. This operation can be repeated as many times as the number of layers desired to form double-, triple-layer cardboard, etc. (Table 2.5). Converting corrugated board includes all the processes of transformation from a flat board to a finished product (mainly packaging). These processes consist of printing, die-cutting, folding and gluing.

Liners contribute to the mechanical strength of the packaging and serve as a communication and/or information medium (Table 2.5). The flutes ensure the rigidity of the packaging as well as maximum elasticity since they serve as shock absorbers in the event of impact (Carton Ondulé de France 2020). To offer maximum use, corrugated cardboard comes in many shapes, sizes, thicknesses, mechanical strengths, surface treatments (waterproofing, etc.). Number of layers, height and width of flutes and number of flutes per meter are factors to take into account for conception and uses of corrugated containers (Table 2.5) The FEFCO code defines three main classes for corrugated board. Class I is a single wall and is dedicated to containment, class II (single wall) is for loading and class III is double wall for loading and additional protection (against moisture, shocks, etc.). The FEFCO code also gives information on the closure made by gluing, taping, stitching or belting.

Flute type	Flute height (mm)	Flutes per meter	Stacking strength	Puncture resistance	Cushion	Flat crush	Surface printing ability
A	4.5–4.7	105–125	Best	Good	Best	Poor	Poor
C	3.5–3.7	120–145	Good	Best	Good	Fair	Fair
B	2.1–2.9	150–185	Fair	Fair	Fair	Good	Good
E	1.1–1.2	290–320	Poor	Poor	Poor	Fair	Very good
F	0.7–0.9	400–440	Poor	Poor	Poor	Poor	Excellent

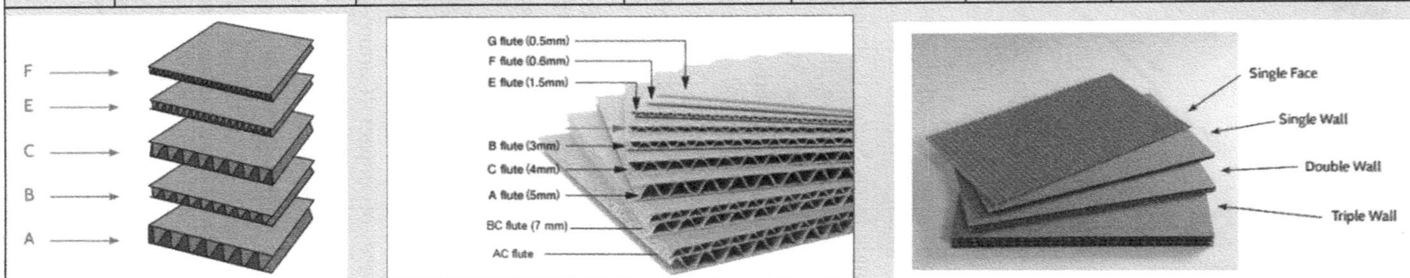

Table 2.5. *Corrugated board type, flute characteristics and mechanical properties (adapted from Watkins 2012; Carton Ondulé de France 2020)*

Corrugated board specifications include thickness, grammage, bursting/puncture resistance, edge crush resistance and stacking strength resistance, flat crash resistance (flute surface resistance to pressure), compression resistance of the box, bending resistance and water absorption resistance. All of these are regulated due to transportation concerns.

2.5.3. *Corrugated board sustainability*

A ton of corrugated cardboard emits 538 kg of CO_2 equivalent over its entire lifecycle, reflecting the efforts of corrugated cardboard manufacturers to minimize its environmental impact. In the case of aluminum and plastic, these emissions are four to six times higher. Corrugated cardboard is therefore part of an almost closed material loop that optimizes the use of natural resources. Corrugated cardboard is made from cellulose fibers, 89% of which are recycled from paper and cardboard; the remaining 11% are new fibers injected into the circuit to maintain the quality and technical characteristics of the finished product. These new fibers come from sawmill waste and forest by-products. Cardboard has the highest recycling rate, reaching 95% in France. This rate is among the highest in Europe, taking into account all packaging.

2.6. Regulations and safety concerns of papers and boards

Food contact materials (FCM) are defined by the European Food Safety Authority (EFSA 2015) as "all materials and articles intended to come into contact with food". FCM are estimated to be the largest source of food contamination. Apart from plastics, the most commonly used FCM are made of printed paper and board. FCM are thus routinely involved throughout the production, processing, packaging, transport, storage, preparation and serving of food and beverages (EFSA 2015; European Union 2015). Regulation (EC) 1935/2004 outlines 17 FCM groups including plastics, paper and boards, glass, as well as printing inks, coatings and adhesives used in product finishing. Unlike their plastic counterparts, these are not covered by a specific European regulation (Poças and Hogg 2007). The Council of Europe has compiled resolutions for several non-plastic FCM including paper and boards. These policy statements are meant to serve as guidance in cases where no specific regulation is adopted for a particular FCM group. Resolution ResAp 2002-1 (Council of Europe 2009) is completed by five Technical Documents, the first of which contains a list of additives that may be used in the manufacture of paper and board materials and articles intended to come into contact with foodstuffs (Van Bossuyt *et al.* 2016). Over 6,000 unique substances can be used in the manufacture of printed paper and board FCM, mainly related to coating processes, printing and gluing. The vast majority of them are considered non-evaluated, pointing out an important knowledge gap in the safety evaluation (migration, toxicity level, etc.) of these substances.

When safety is ensured, avoiding the chemical contamination of potential toxic compounds due to their transfer from the packaging to the food, any other transfer should be considered and accepted. Food quality is obtained mainly by avoiding changes to organoleptic properties and the loss of nutritive compounds. These organoleptic changes may be produced by the interaction between the packaging and the food, adding new off-odors (Vera *et al.* 2020). This type of contamination may come simply from the composition of the material used or may be derived from the degradation products after the manufacturing process, like irradiation or high-temperature exposure, all of them producing a negative effect on the quality of the food (Czerny and Buettner 2009; Osorio *et al.* 2019).

2.7. References

Alexir Packaging (2020). Alexir Corporated [Online]. Available: http://www.alexir.co.uk/.

Bajpai, P. (2010). *Environmentally Friendly Production of Pulp and Paper*. John Wiley & Sons, Inc., Hoboken.

Bajpai, P. (2012). *Biotechnology for Pulp and Paper Processing: More Comprehensive Coverage of All Aspects of Biotechnology in the Pulp and Paper Industries*. Springer, Berlin.

Bajpai, P. (2018). *Biotechnology for Pulp and Paper Processing: Up-to-Date with Current and Emerging Technologies*. Springer, Berlin.

Cartonnage-Paulet (2020). Histoire du carton [Online]. Available: http://www.cartonnages-paulet.fr/le-carton/histoire-du-carton.html [Accessed 29 June 2020].

Carton Ondulé de France (2020). Fédération Carton Ondulé de France [Online]. Available: https://www.cartononduledefrance.org/.

CartonPro (2020). Histoire du carton [Online]. Available: https://www.carton.pro/page.html?chapter=1&id=13 [Accessed 29 June 2020].

Castle, L., Damant, A.P., Honeybone, C.A., Johns, S.M., Jickells, S.M., Sharman, M., and Gilbert, J. (1997). Migration studies from paper and board food packaging materials. Part 2. Survey for residues of dialkylamino benzophenone UV-cure ink photoinitiators. *Food Additives and Contaminants*, 14(1), 45–52.

Cofepac (Comité français de l'emballage papier carton) (2020). L'emballage papier-carton : les produits [Online]. Available: http://www.cofepac.org/all_page.asp?page=37&lg=fr [Accessed 1 August 2020].

Council of Europe (2009). Policy statement concerning paper and board materials and articles intended to come into contact with foodstuffs. Resolution ResAP(2002)1 [Online]. Available: https://rm.coe.int/16804e4794 [Accessed 10 November 2020].

Czerny, M. and Buettner, A. (2009). Odor-active compounds in cardboard. *J. Agric. Food Chem.*, 57(21), 9979–9984.

Deutsches Institut für Normung (2011). Paperboard – Terms and grades. DIN 19303 [Online]. Available: https://www.din.de/en.

ECMA (European Carton Makers Association) (2020). European Carton Makers Association [Online]. Available: https://www.ecma.org/ [Accessed 2 August 2020].

EFSA (European Food Safety Agency) (2015). Food contact materials [Online]. Available: http://www.efsa.europa.eu/en/topics/topic/foodcontactmaterials.

Emblem, A. and Emblem, H. (2012). *Packaging Technology, Fundamentals, Materials and Processes*. Woodhead Publishing Limited, Oxford.

EPA (Environmental Protection Agency of the United States) (2001). Available and emerging technologies for reducing greenhouse gas emissions from the pulp and paper manufacturing industry [Online]. Available: https://www.epa.gov/sites/production/files/2015-12/documents/pulpandpaper.pdf [Accessed 30 July 2020].

EPA (Environmental Protection Agency of the United States) (2008). Pulp, paper and paperboard effluent guidelines [Online]. Available: https://www.epa.gov/eg/pulp-paper-and-paperboard-effluent-guidelines [Accessed 30 July 2020].

European Commission (2004). Materials and articles intended to come into contact with food and repealing Directives 80/590/EEC and 89/109/EEC. Regulation 1935/2004 [Online]. Available: https://eur-lex.europa.eu/legal-content/EN/ALL/?uri=CELEX%3A32004R1935.

European Commission (2011). Plastic materials and articles intended to come into contact with food. Commission Regulation 10/2011 [Online]. Available: https://eur-lex.europa.eu/legal-content/EN/ALL/?uri=celex%3A32011R0010.

FEFCO (2020). European Federation of Corrugated Board Manufacturers [Online]. Available: https://www.fefco.org/.

Garcia Cerrillos, B. and Mangin, N. (2007). Papiers et cartons au contact des denrées alimentaires. *Les Techniques de l'Ingénieur*, F 1315, 1–8.

Iggesund (2020). Iggesund Holmen Group [Online]. Available: https://www.iggesund.com/.

Kiviranta, A. (2008). *Paper and Board Grades*. Fapet Oy, Helsinki.

Osorio, J., Aznar, M., and Nerin, C. (2019). Identification of key odorant compounds in starch-based polymers intended for food contact materials. *Food Chem.*, 285, 39–45.

Otenio, D., Escabasse, J.-Y., and Podd, B. (2004). *Packaging Materials 6: Paper and Board for Food Packaging Applications*. ILSI, Brussels.

PaperIndustryWorld (2020). What is the role of fibre based packaging in the global packaging market? [Online]. Available: https://www.paperindustryworld.com/what-is-the-role-of-fibre-based-packaging/ [Accessed 28 June 2020].

Petit-Conil, A. (1999). Procédés papetiers – fabrication des pâtes. *Les Techniques de l'Ingénieur*, J 6900 V1, 2–12.

Piergiovanni, L. and Limbo, S. (2016). *Food Packaging Materials*. Springer International Publishing, London.

Poças, M.-F. (2020). FITNess. Online course "paper and paperboard" in Fitness Erasmus Project. Food packaging open courseware for higher education and staff of companies [Online]. Available: https://fitness.agroparistech.fr/fitness/lectures/.

Poças, M.F. and Hogg, T. (2007). Exposure assessment of chemicals from packaging materials in foods: A review. *Trends Food Sci. Tech.*, 18, 219–230.

Pulpandpaper-Technology (2020). Pulping [Online]. Available: https://www.pulpandpaper-technology.com/ [Accessed 30 July 2020].

Riley, A (2012). Paper and paperboard packaging. In *Packaging Technology, Fundamentals, Materials and Processes*, Emblem, A. and Emblem, H. (eds). Woodhead Publishing Limited, Oxford.

Robertson, G.L. (2013). *Food Packaging – Principles and Practice*. CRC Press, Boca Raton.

Smook, G.A. (1992). *Handbook for Pulp and Paper Technologists*. Angus Wilde Publications, Inc, Vancouver.

Van Bossuyt, M., Van Hoeck, E., Vanhaecke, T., Rogiers V., and Mertens, B. (2016). Printed paper and board food contact materials as a potential source of food contamination. *Regul. Toxicol. Pharmacol.*, 81, 10–19.

Vera, P., Canellas, E., and Nerin, C. (2020). Compounds responsible for off-odors in several samples composed by polypropylene, polyethylene, paper and cardboard used as food packaging materials. *Food Chem.*, 309, 125792.

Watkins, T. (2012). Corrugated board packaging. In *Packaging Technology, Fundamentals, Materials and Processes*, Emblem, A. and Emblem, H. (eds). Woodhead Publishing Limited, Oxford.

Zheng, M.-H., Bao, Z.-C., Zhang, B., and Xu, X.-B. (2001). Polychlorinated dibenzo-pdioxins and dibenzofurans in paper making from a pulp mill in China. *Chemosphere*, 44(6), 1335–1337.

3

Glass Packaging

Frédéric Debeaufort

Institute of Technology, University of Burgundy, Dijon, France

3.1. Introduction

Each glass bottle we open embodies a tradition that dates back several millennia. Glass is one of the oldest materials and is closely linked to historical and cultural techniques. Glass is used in crafts, architecture and for the preservation and packaging of precious foods. However, each glass bottle also represents a decision for the future, because glass protects the environment and preserves our natural resources. Around 1,500 BCE, the Egyptians manufactured the first hollow glass containers, which then served as containers for ointments or oils (Henderson 2013).

The benefits of glass packaging are clear: it is sustainable, inert, fully and infinitely recyclable, reusable and refillable and safe for storing food and drinks. Glass is always a safe choice to use for packaging products, particularly for beverages and liquids in the food and pharmaceutical industry. However, when glass is used to make jars or vials, and not bottles, it is most commonly used to contain cosmetics and perfumes. The impermeable and inert nature of glass makes it a safe form of protective packaging. In addition to being an excellent barrier against the external environment, glass containers do not interact with or alter the taste, odor or composition of the products they contain. However, its brittleness and heavy weight remain real disadvantages for the carriage, transportation and worker safety involved

---

For a color version of all figures in this book, see www.iste.co.uk/debeaufort/packaging.zip.

Packaging Materials and Processing for Food, Pharmaceuticals and Cosmetics, coordinated by Frédéric DEBEAUFORT, Kata GALIĆ, Mia KUREK, Nasreddine BENBETTAIEB and Mario ŠČETAR.

in glass packaging. Glass packaging provides optimum long-term conservation of the original qualities and virtues of bioproducts. Seventy-six percent of European citizens recognize that glass is the most environmentally friendly packaging material (FEVE 2020). This increased by 50% in the last decade. Glass is clearly considered the best packaging material for beauty, fragrance and medical products, with a 2018 study finding that more than three in four consumers prefer it over plastic/PET or metal. Indeed, although the use of plastic in packaging has largely increased in the last three decades, consumer research consistently indicates that consumers attach a high-quality perception to glass-packaged products (Balzarotti *et al.* 2015).

The world production of glass containers and bottles has a linear growth of about 8% every five years, increasing from about 50 million tons in 2015 to almost 61 million tons in 2020. It represents about 21.5 million tons in Europe and about 78 billion units. Glass is the leading packaging material for spirits, wines and beer, while it is increasingly gaining a share in the food, water and dairy sectors. It is the second leading packaging material in Europe in terms of volume and one of the most promising growing sectors of the packaging industry. A total of 50,000 direct jobs were involved in the glass packaging sector in 2019, and up to 250,000 indirect jobs (VerreOnLine 2020).

3.2. Composition and properties of glass

3.2.1. *Glass composition*

Glass can be defined either as a hard, fragile (brittle) and transparent material, mainly based on silicate, or as a non-crystalline solid exhibiting the glass-transition phenomenon. This definition is common, and it was also used by scientists until the 19th century. Before the 20th century, in fact, silicate glasses (soda-lime glasses) were practically the only transparent materials that we knew how to produce industrially. Glass has been defined by the ASTM (2010) as "an amorphous, inorganic product of fusion that has been cooled to a rigid condition without crystallising". Although glass is often regarded as a synthetic material, it was formed naturally from common elements in the Earth's crust (VerreOnLine 2020).

The two main types of glass container used in food packaging are bottles (70%) and jars (30%), whereas in the pharmaceutical industry, vials, bottles/flasks and ampoules dominate. The food industry uses mostly clear glass (85%), whereas the cosmetic and pharmaceutical industries prefer opaque or amber glass, in order to provide better protection against light. The addition of metal oxides is mainly responsible for these differences in transparency and color. Today's glass containers are lighter as well as stronger. Glass packaging weight was reduced by 25–50% over the last 50 years, after formulas and processes were optimized (Robertson 2013).

The raw materials for glassware come from mines or quarries and must be melted or chemically reduced to their oxides at temperatures above 1,500°C. The main ingredient of glass is silica, derived from sand, flint or quartz. Silica is melted at very high temperatures (>1,700°C) to form molten silica glass. For most glass, silica is combined with other raw materials in various proportions. To lower the fusion temperature (down to 850°C) and viscosity of molten silica, some alkali fluxes (commonly sodium and potassium carbonates) are added. Calcium and magnesium carbonates (limestone and dolomite) prevent glass from dissolving in water and are also considered as structure stabilizers. By adding about 25% of sodium oxide to silica, the melting point is reduced from 1,723 to 850°C, however such glasses are easily soluble in water. The addition of lime (calcium oxide or CaO), supplied by limestone, renders the glass insoluble again. However, when added in too high an amount, it induces a devitrification, which corresponds to a loss of amorphous structure that could shift it to crystalline.

To give glass specific physical properties such as clarity and brilliance, additives can be added. For instance, alumina increases hardness and durability and boron allows the formation of borosilicate glass, which reduces the leaching of sodium (which is loosely combined with the silicon) and is of great interest for uses in pharmaceutical and chemical industries.

The composition of glass is not a single one, but a mixture, whose composition is adapted to the target use, the color and properties required. It is the combination of alkali oxides, metal oxides and silica dioxides, associated with additives. Its formula could be expressed as $aX_2O + bYO + 6SiO_2$, where a and b are the number of molecules, X is an atom of an alkali metal (i.e. Na, K, etc.) and Y is an atom of a bivalent metal (i.e. Ca, Pb, etc.). The soda-lime glass is mainly composed of $Na_2O + CaO + 6SiO_2$. Potash-lead glass is made with K_2O, PbO and $6SiO_2$. The optimum composition for soda-lime glass is about 75% silica, 10% lime and 15% soda, but even this is liable to devitrification during certain mechanical forming operations to be satisfactory. Bottle glass often contains about 2% alumina (aluminum oxide, or Al_2O_3). The typical composition of a food glass container is given in Table 3.1 (Piergiovanni and Limbo 2016). Borosilicate glass, also called Pyrex®, is mostly used for pharmaceuticals and chemistry/laboratory usages, or for culinary purposes. The mixture in borosilicate glass binds all the chemicals more tightly together, leading to a much higher melting point and a higher chemical and mechanical resistance. Borosilicate glass has a higher proportion of silicone dioxide than soda-lime glass (80.6% in borosilicate glass and 69% in soda-lime glass), 13%

of boron oxide B_2O_3 (maximum 1% in soda-lime glass), but a much lower content of sodium oxide Na_2O (4%) and aluminum oxide Al_2O_3 (2.4%). The composition of lead glass is similar to soda-lime glass, but lead oxide replaces calcium oxide and it is mainly used for homeware crystal, such as luxury Champagne or Cognac glasses and carafes or wine decanters (Gul *et al.* 2018).

Components	Content (% weight)
SiO_2 from sand (99.8%)	71–75
Na_2O mainly from soda ash (Na_2CO_3 89%) and cullet (67%)	12–14
CaO mainly from aragonite ($CaCO_3$ 49%) and cullet (49%)	9–10.5
Al_2O_3 mainly from feldspar (7.6%) and cullet (9.2%)	0.15–1.3
Fe_2O_3 from sand (0.1%)	0.04–0.1
TiO_2	< 0.01
SO_3	0.2–0.3
K_2O	0.03–0.3
MgO	0.2–4

Table 3.1. *Composition (% of weight) of a typical soda-lime glass for food packaging (adapted from Boyd* et al.*1994; Seward and Vascott 2005)*

Color or UV barrier agents	Metal oxides
Colorless, UV absorbing	CeO_2, TiO_2
Blue	Co_3O_4, $Cu_2O + CuO$
Purple	Mn_2O_3, NiO
Green	Cr_2O_3, $Fe_2O_3 + Cr_2O_3 + CuO$
Brown	MnO, $MnO + Fe_2O_3$, $TiO_2 + Fe_2O_3$, $MnO + CeO_2$
Amber	Na_2S
Yellow	CdS, $CeO_2 + TiO_2$
Orange	CdS + Se
Red	CdS + Se, Au, Cu, Sb_2S_3
Black	Co_3O_4 (+ Mn, Ni, Fe, Cu, Cr oxides)

Table 3.2. *Metal oxides used to color a typical soda-lime glass for food packaging (adapted from Robertson 2013)*

Due to the absence of a crystalline structure (fully amorphous), glass is homogeneous and free from any stresses and it is optically isotropic. The optical properties of glass relate to the degree of penetration of light and the subsequent effect of that transmission as a function of wavelength (Dias *et al.* 2010; Robertson 2013). In order to control the transmission of light through glass packaging, bottles, flasks, bulbs, etc. are colorful. The absorption of light, not only in the visible but also in the UV and IR regions of the spectrum, can thus be modulated by the addition of metal oxides. Most transition-metal oxides (e.g. cobalt, nickel, chromium, iron, etc.) absorb at specific wavelengths and allow the tuning of light transmissions, as well as giving glass a specific color (Figure 3.1). Their thermal stability is also compatible with the glass-making process (Table 3.2). The presence of iron oxides in glass produces a green color, owing to the absorption bands in the UV and IR regions, particularly appreciated for beer bottles. The three main colors of glass used to produce containers for food and pharmaceutics are flint or clear, amber or brown and green.

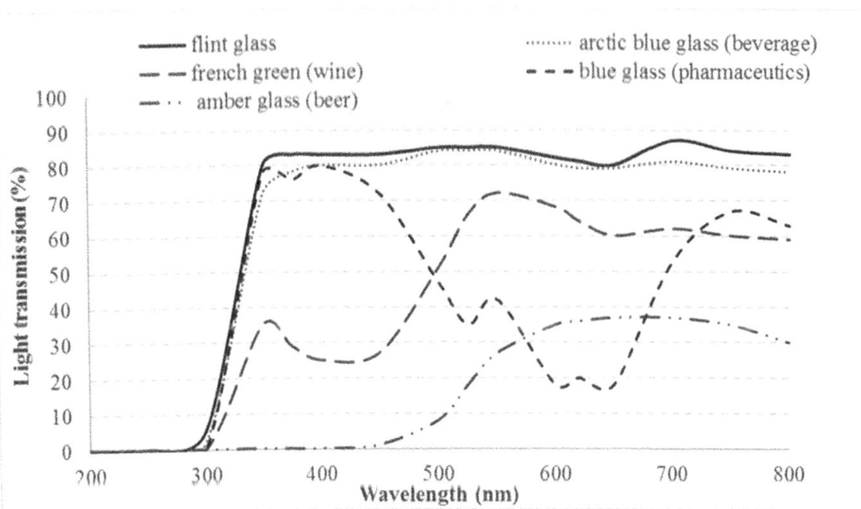

Figure 3.1. *Light transmission though glass packaging for food and pharmaceutical applications (adapted from Dias* et al. *2010; Robertson 2013)*

3.2.2. *Chemical and functional properties of glass*

Although made of natural raw materials, glass is often considered a synthetic material. Constituted mainly of silica, it is often associated with other compounds, for example, with sodium or potassium carbonates, calcium and magnesium carbonates. These carbonates are mainly used to modify the manufacturing conditions, in particular, the melting temperature, as well as influence its physical properties (Robertson 2013). Likewise, lead is often added to bring clarity and shininess to the glass but may induce greater flexibility and fragility. Aluminum oxides, on the other hand, increase the hardness and durability of glass. Boron, added up to 6% in borosilicate glasses, prevents the leaching of sodium silicates present in glass when in contact with acid solutions.

Glass can be corroded in three ways:

– etching corresponds to an attack by alkaline compounds, which gradually destroys the silica network, which then releases the other constituents of the glass. This is what happens in dishwashers in the presence of detergents, for example, and what induces the whitening of flint glass containers;

– leaching corresponds to an acid attack on the glass, and the hydrogen ions are exchanged with the alkalis or with the other positively charged ions;

– finally, weathering occurs under very extended storage conditions inducing surface blooming. Weatherization refers to a milky, hazy or crystalline (snowflake) residue (mainly refers to humidity-induced deterioration).

90% of the glass produced for packaging is soda-lime glass. Used industrially to make small bottles or flasks, this glass does not need an exceptional chemical resistance, nor particular thermal resistance. In the opposite case, where a thermal resistance and a high resistance to acids are expected, borosilicate glass such as Pyrex® is used.

Glass is neither a liquid nor a solid. It exists in a glassy state for the expected uses. The glassy state is a state where the molecules are organized randomly but with enough cohesion to obtain the mechanical rigidity expected for a solid. It is actually an extremely viscous liquid that behaves like a solid and has its properties. During its manufacture, the cooling of glass paste results in a reversible change in viscosity until it reaches a value so high that the glass becomes rigid and brittle (Lehman 2001).

The completely amorphous structure of glass, therefore without a crystalline zone, makes it a homogeneous material, which can be optically isotropic (Piergiovanni and Limbo 2016). This isotropy allows it to be transparent with a very high degree of light transmission. The spectrum of light transmission through glass depends on the reflection on its surface and its optical absorbance. It ranges from 150 nm in the UV to almost 6,000 nm in the infrared. The addition of metal oxides, such as cobalt, nickel, iron or chromium, increases the absorption of light on certain wavelengths, in particular, the UV range. The chemical composition of glass allows it to keep its optical properties in many circumstances. Apart from degradation by chemical action (etchings, leaching or weathering) in the presence of acids or alkalis, the glass loses its transparency when it is bombarded by high-energy radiation over long periods.

At room temperature, glass appears to be a solid but is, in fact, an undercooled liquid, which can flow so slowly that it is not visible on a human timescale. Its basic structure is a tetrahedron of silica and oxygen, which brings stiffness and brittleness. From a mechanical point of view, glass is a very brittle material; it breaks under the effect of tensile stress or bending stress (Gul *et al.* 2018). The ultimate tensile or bending stress of a glass fracture determines when the container, or the bottle, will break. The rupture of a glass bottle can be favored by micro-cracks, which occur when the glass is not handled with care, in particular, on industrial packaging lines (Robertson 2013). The mechanical resistance of the glass can be improved by modifying its chemical composition, either by adding compounds which increase its hardness, or by adding compounds which increase its flexibility. The mechanical strength of glass is the measure of the fracture resistance capacity when a force is applied. Glass deforms elastically until it breaks. This means that the deformation is linear with the stress applied. Young's modulus, which characterizes this elastic behavior, is of the order of 70 to 100 GPa for packaging glass (Nielsen *et al.* 2009).

The analysis of a glass fracture or break is an important element to consider for packaging. The mechanical resistance of glass packaging can be questioned mainly in four situations of use. The first case is related to internal pressure, for example, when packaging carbonated or fermented drinks. This also occurs during heat treatment operations such as the sterilization of milk bottles or jars. The second reason for failure may be due to the excessive vertical pressure, for example, when stacking pallets (Gul *et al.* 2018). The third case, which is the most often observed, is mainly due to an impact or shock during the rapid movement of glass containers throughout the filling process or transport and storage operations. Finally, the last reason for breakage is due to the weakening of certain parts of the bottle, influenced by its shape and design (angular shape, thinness, etc.), or by the presence of micro-defects (scratches, microcracks, abrasions, etc.).

The ability of glass to withstand large or sudden changes in temperature depends on its thermal resistance (Piergiovanni and Limbo 2016). Indeed, during the manufacture of glass, in particular, during cooling, tensile stresses are set up on the outer surfaces, whereas compressional stresses occur on the inner surfaces. A very rapid rise in temperature therefore induces surface compression as well as internal tensions. In general, very rapid cooling is more damaging than sudden heating. The evaluation of the thermal resistance of a glass package is generally determined empirically. The glass container is successively immersed in boiling water and then in ice water, and thereafter the number of cracks or microcracks is determined.

To sum up its properties, glass absorbs, refracts and transmits light. Its amorphous structure (having no definite crystalline parts) makes it easy to fuse and mold as many times as possible. There is no sharp melting point. Glass does not react with water and other atmospheric agents – but is affected by alkalis. The characteristics of glass, both physical and chemical, vary considerably with the addition of other materials. For instance, if B_2O_3 is added, glass becomes alkali resistant. Glass is easy to clean, is resistant to stains and can be easily colored. Glass has a considerable compressive strength up to certain point, but does not deform. Indeed, deformation at low range induces breaks and it is thus a highly brittle packaging. Its thermal conductivity is low (Gul *et al.* 2018).

Glass containers have several characteristics that make them ideal for food, beverage and pharmaceutical packaging. They are impervious to moisture, gases, odors and microorganisms. They are inert and do not react with or migrate into food products. They have filling speeds comparable to those of cans in industrial packing lines. They are suitable for heat processing (sterilization) when hermetically sealed. They are transparent to microwaves. They are reusable and recyclable. Glass can be molded into a variety of shapes and colors. Glass bottles and containers are transparent and display the contents. They are rigid and allow stacking without container damage. They can be directly printed on or use paper labels. They are perceived to add value to the product by the customer.

However, the main disadvantages of glass as a packaging container are also numerous. Glass has higher weight and hence, transportation costs more than other types of packaging containers. It has a lower resistance than other materials to fractures, scratches and thermal shock. Due to its brittleness, potentially serious hazards arise from glass splinters or fragments in food. Glass permeability to UV light favors food oxidation. However, this latter problem can be overcome by incorporating various oxides, sulfides or selenides to color glass and block out the incident UV radiation.

3.3. Glass manufacturing

Glass manufacturing takes place in several stages: some are common and some are specific to give the final shape to the packaging. The process can be summarized in five steps (Figure 3.2). The first step consists of mixing the raw materials and melting them to obtain a glass paste. The shaping (forming) step will depend on the object that we want to obtain. Then, an annealing step is necessary to stabilize the structure, before applying surface treatments (Shelby 2005).

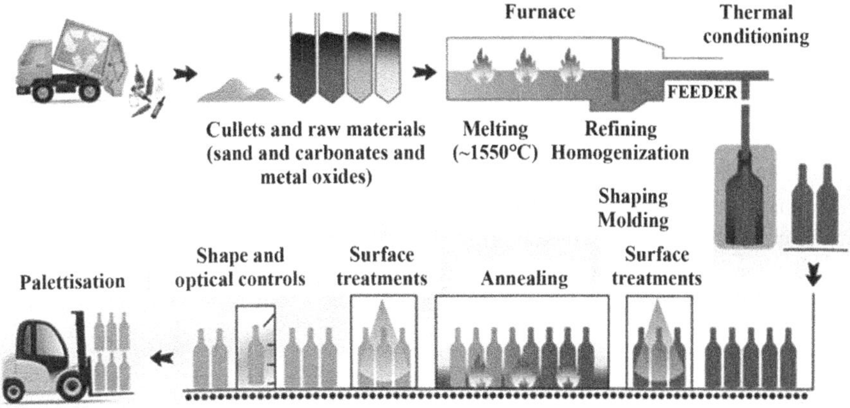

Figure 3.2. *Glass bottle or container process (adapted from VerreAvenir 2020)*

3.3.1. *Collection of raw materials and mixing*

The collection of raw materials depends on the type of the glass. Soda-lime glass preparation needs chalk (CaO), soda ash (Na_2O) and clean sand (SiO_2), whereas for borosilate glass, boron is added. Up to 50% of cullets, in the form of waste glass or broken glass, are added to increase the fusibility as well as to prevent the loss of alkali through volatilization. For flint glass, decolorizers are added to eliminate, for instance, the yellowish tint of ferric oxides or the greenish tint of ferrous oxides. Cobalt oxide, manganese oxide and nickel oxide are the most used decolorizers. On the other hand, when glass has to be colored to stop or decrease visible and UV light transmission, some metal oxides are added (see section 3.2.1).

3.3.2. *Melting in furnace*

The glass is produced in a furnace made of refractory materials, resistant to both the high temperatures involved and to the erosion caused by the glass melting bed. Obviously, these refractories must minimize heat loss to the outside.

The elaboration of the glass consists in transforming the vitrifiable mixture into a homogeneous liquid mass of a suitable viscosity to be shaped. It includes several steps. The first step is the fusion of raw materials, where the vitrifiable mixture is transformed into a liquid mass without unmolten ingredients (Shelby 2005). Second, the chemical homogenization of the liquid mass is carried out by controlling the convection currents in the molten glass and the refining corresponds to the release of gas bubbles at the end of melting. Then, during the thermal conditioning, in the part of the oven called the work basin, the temperature required to feed the glass to the feeder openings is obtained. This temperature is much lower than the refining temperature.

The melting temperature is about 1,550°C and the output varies between 50 and 500 tons of glass per day depending on the oven. Thanks to the progress made and the use of cullets, the consumption of energy per ton of glass produced has decreased significantly from 1,600 kWh in the 1970s to about 1,100 kWh today.

3.3.3. *Shaping/molding*

The glass is brought from the work basin to the machine through refractory channels called feeders. Upon exit from the work basin, the glass is still hot enough to be shaped by the machines. The feeders allow the glass to be conditioned at the temperature corresponding to the correct working viscosity. The temperature of the parisons or gobs (drop of semi-liquid glass paste) distributed to the machines must be precise, to the nearest degree, but it is also advisable to have within this parison an almost perfect thermal homogeneity (within a few degrees). This temperature regulation of the feeders is imperative to obtain good manufacturing of products. The gobs (200 per minute) are distributed to the shaping sections of the machine by slides. Their temperature, depending on the weight and the shaping technique, is between 1,100 and 1,230°C (Verallia 2020; VerreOnLine 2020).

Several manufacturing processes (shaping) for bottles, flasks and jars are used, depending on the type of packaging required. All operate in pretty much the same way. The glass-forming machine has several sections (up to 12 sections). Each section has several molds used successively:

– one ring mold, which will definitively fix the shape and the dimensions of the packaging ring (bottle neck or jar screw, for example);

– one blank mold for the formation of the packaging preform/blank, which will be transferred to the finishing mold;

– one finishing mold in which, by blowing the blank/preform, the article takes its final form.

Preform formation can be obtained either by blowing the parison after the formation of the ring (blow–blow process, Figure 3.3) or by pressing the parison with a punch (press-and-blow process, Figure 3.4).

The press-and-blow process was first developed for the manufacture of wide-opening packaging, and provided to be superior in terms of the distribution of glass, whereas the blow–blow process is used for articles with narrow openings and seems to be essential for obtaining light packaging.

For each article, specific molds are developed and are very expensive, which partly explains the mass production of glass packaging. Different materials are used to build these molds: "gray" cast iron, sometimes treated on the surface; copper-based alloys (their thermal conductivity is superior to cast iron and is advantageous for better temperature control during production at high speed) and finally, nickel alloys of a high hardness for less wear (VerreOnLine 2020).

3.3.3.1. *Blow–blow process*

The principle of the blow–blow process is presented in Figure 3.3.

The parison arrives at the center of the funnel and enters the blank mold (loading). The rough bottom is placed on the funnel and closes the mold. Compressed air is injected through it. The glass is pushed down into the ring mold and onto the punch (compression). To warm up the unblocking, the punch goes down and the rough bottom is removed, as well as the funnel. The blank bottom is placed in the low position on the mold. During these operations, the imprint in the glass of the punch heats up: it is the "cracking" of the uncork. When the roughing bottom is in the base position, air is "blown" through the base of the punch and comes to "pierce" the glass and press it against the walls of the roughing mode (drilling). The blank is held by the ring mold and is transferred into the finishing mold. The blank warms up (cracking) and lengthens under its own weight before being inflated. The inflation head is positioned on top of the finishing mold and compressed air is blown into the blank to give it its final shape. The finishing mold opens, and the bottle is extracted by pliers. The bottle is held by the grippers in a flow of ventilation air and then it is released and evacuated by a conveyor to the annealing arch (Misaglass 2020; VerreOnLine 2020).

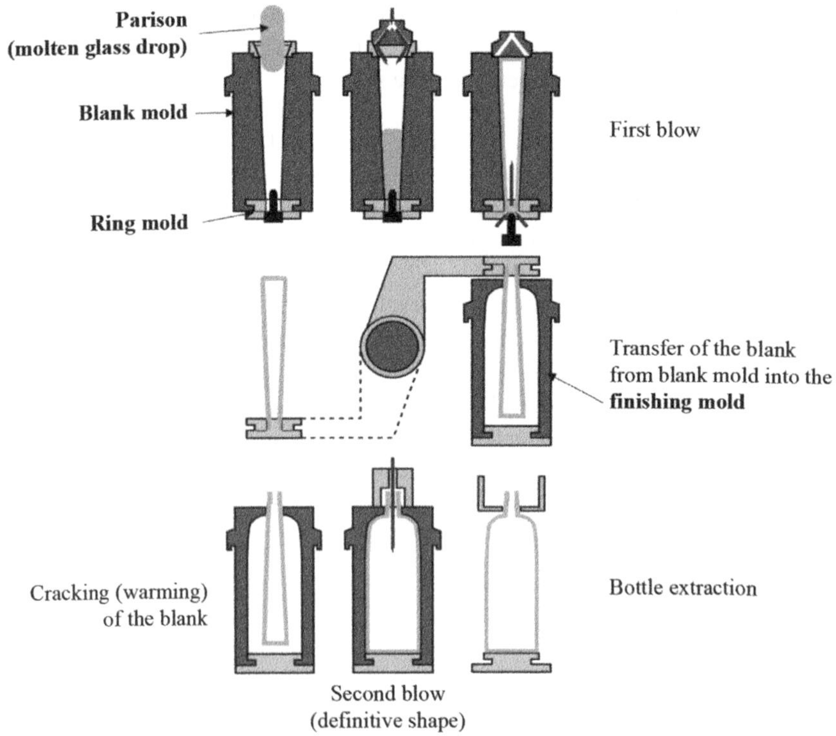

Figure 3.3. *Blow–blow processing of glass bottles (adapted from Wikibooks 2020)*

3.3.3.2. *Press-and-blow process*

The principle of the press-and-blow process is presented in Figure 3.4.

This is the only process that allows us to make wide opening items. It is sometimes used to produce light bottles (blown press for narrow opening). Here, the blank is pressed instead of blown, which provides better control of the distribution of the glass and the possibility of increased lightening. Manufacturing follows the following stages. The parison arrives at the center of the funnel in the blank mold (loading). The roughing base is put in place on the roughing mold, and the punch begins to rise (start of pressing). The punch rises completely and presses the glass, which fills the space between the punch and the mold (main step of pressing). The blank is then transferred to the finishing mold. This process has some drawbacks. The punches, which are very hot, must not release micro-particles, which would increase the risk of a subsequent breakage of the manufactured articles and food contamination. This therefore implies the use of materials resulting from the latest advances in metallurgy, as well as controlling the temperatures of the punches. The operating costs are higher, and the production rates are often lower for the narrow opening process (Misaglass 2020; VerreOnLine 2020).

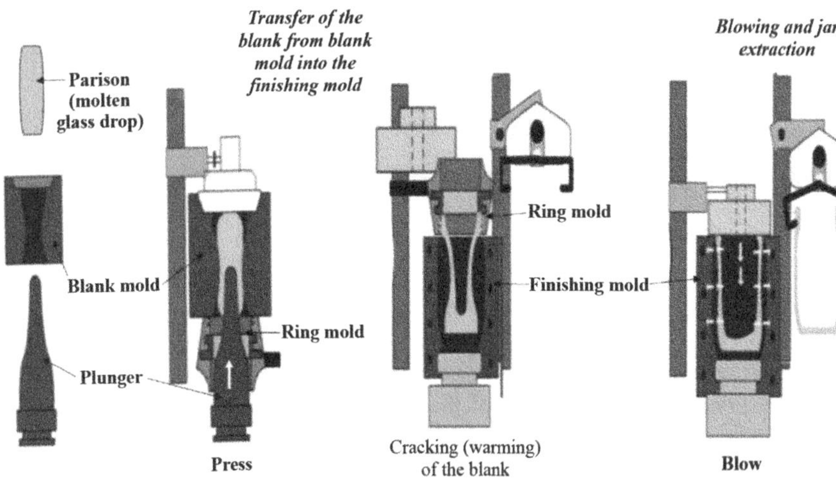

Figure 3.4. *Press-and-blow processing (adapted from Misaglass 2020)*

3.3.3.3. *The manufacture of glass packaging from drawn tubes for pharmaceuticals*

From these tubes, ampoules, pill boxes and vials are made for the pharmaceutical industry. Specialized machines, generally vertical and with a carousel, heat the tubes with torches to form the tips of the ampoules, the necks or the bottoms of the bottles and pill boxes. Production rates reach 5,000 items per hour. Glass tubes are produced in two ways: the Danner process, generally for tubes with a diameter above 25 mm, or the Vello process, for tubes with a smaller diameter (VerreOnLine 2020).

3.3.3.4. *Manufacturing processes for cups*

This type of packaging is widely used for the condiment market, including mustard, sauces, pickles, etc. Two main processes are used.

In the "pressed process", the parison falls into a mold, then a punch descends to squeeze the glass and form the article. The pressing table includes 12 to 36 molds. The punch control is either hydraulic or pneumatic. Multi-parent systems exist (with up to four cavities filled simultaneously). The mold is generally made of stainless steel for the manufacturing of cups.

In the Hartford 28 pressed-blown-rolled process, the preform is pressed and then rotated in the finishing mold during inflation, which makes it possible to have no trace of the mold. The article, after manufacture, is cut and flame finished at the right height. This process is often preferred for high-quality or high-value articles such a perfume glass vials (VerreOnLine 2020).

3.3.4. *Annealing and surface treatments*

After manufacture, the glass packaging which has been brutally cooled (faster cooled outside than inside) presents stresses and tensions within the walls of the bottles or jars. The purpose of annealing is to reduce these tensions to a level which no longer presents a risk of breakage when the articles are used. In the annealing arch, the temperature first rises above the annealing temperature (with the annealing point corresponding to a relaxation of the stresses in 15 minutes and depending on the composition of the glass: approximately 560°C for the classical soda-lime glass). All the stresses created during manufacturing are canceled. Then, a slow and controlled cooling down to a temperature below the strain point (corresponding to a relaxation of the stresses in four hours) makes it possible to obtain articles whose outer skin has very high compression allowances. These articles therefore have better resistance, both to thermal shock and to the resistance of internal pressure (Robertson 2013; VerreOnLine 2020).

The mechanical resistance depends, above all, on whether or not there is a presence of surface defects, which the glassmaker will seek to avoid by using protective surface treatments. They are of two types: hot or cold surface treatments (Misaglass 2020).

The "hot" treatment (560°C) is deposited in the vapor phase, between the manufacturing machine and the annealing arch. Its purpose is to coat the outer surface of the glass packaging in order to increase its hardness and to allow a good application of the second treatment, which is deposited after the annealing arch. The hot treatments consist of tin oxide (SnO_2) or titanium (TiO_2) generally deposited from tin or titanium tetrachloride (a thickness of about 5 nm). For returnable or reusable items, a type of surface treatment (with TiO_2) is a better choice (although more expensive). In fact, during washing with soda, the surface treatment lasts longer and increases the number of re-uses (Shelby 2005; Emblem and Emblem 2012).

The "cold" spray treatment, for the most effective products, consists of depositing a lubricant (anti-scratch), facilitating slipping during the packaging processes and reducing the formation of scratches. This lubricant is generally based on polyethylene (thickness of about 1 micron). Some conditioners request a "fleeting" treatment based on oleic acid being deposited in the vapor phase. In bottling, various treatments based on ester wax and fatty acid ester derived from polyoxyethylene glycol are also used (Shelby 2005; Emblem and Emblem 2012).

3.4. Bottle design and terminology

Glass containers are engineered along very specific design principles. Container size, orifice diameter, neck length and other attributes, as well as the ratios between such morphological characteristics, are determined largely by basic principles, such as the frequency of access, the amount of contents removed with each access and the type of content (bulk solid, liquid, semi-liquid, etc.). Most bottle shapes are closely associated and identified with a certain product (trademark identification or region/indication origin products) or products as "form follows function" to a large degree in bottle shapes and styles. Wine, beer, sodas and mineral water are prime examples of products which are very closely identified with distinct bottle shapes that are rarely used for other products (Emblem and Emblem 2012).

The different parts of a bottle respond to functional properties and are described in Figure 3.5. The neck is the narrowest part of the bottle, where the opening for pouring is. When the neck has a smooth tube inside (finish), it is designed to welcome a natural or synthetic cork stopper. When the finish neck is screwed or finished by a flat sealing surface with a ring, the bottle may be closed by screw caps or with a capsule, such as for beer bottles or the wire of a Champagne bottle. The neck ring (or transfer bead) has a horizontal ridge at the base of the finish, used for transferring the bottle from one part of the production process to the next. It is also called a "crown". In the past, it was used for anchoring the cork to the bottle with strings or for a better grip for sealing wax. The bottle shoulder is the link between the neck and the body. The shoulder is important for pouring and settling. The slope of the shoulder is one factor in determining how quickly a product will be dispensed when the bottle is inverted. It is the shape of the shoulder that very often distinguishes the different types of wine origin. The body is the biggest part of the bottle, that fixes the final volume of liquid contained within. The main part of the bottle is where the sidewalls are usually (but not always) vertical. The body supports the label, which should be as large as possible. The bottom is the entire lower part of the bottle below the sidewalls. The bottom includes the heel, base and push-up. It allows the bottle to keep a vertical position. The heel joins the sidewall to the bottom-bearing surface and may have a small recessed spot that serves as a registration device for labeling and decorating equipment. The bottom may have letters and symbols molded into it that indicates the number of the mold cavity that produced the container and the manufacturer. The manufacturer symbol is called a "punt mark". The bottom may also have a small projection that serves as a registration device for labeling and decorating equipment. The device can also take the form of a small recess along the heel of the container. The push-up is an inward dome in the center of the base. The resulting ring around the outside of the bottom ensures stability, providing an even bearing surface, which the bottle rests on. The push-up is exaggerated on some bottles, such as wine bottles. This is a punt, that is more or less hollowed out, useful for allowing the wine to be served elegantly by the sommelier (wine waiter). Sometimes a special feature is a mold in the glass bottom, for positioning labels in fast labeling operations. The push-up allows bottles to be given a better resistance to internal pressure, like for carbonated beverages, sparkling wines (Champagne) and beers (SHA 2020).

In jars and wide-opening glass containers, the anatomy is almost the same as for bottles, except the neck and shoulder are smaller; the finish is adapted to specific closers (very often screw type). The mouth (finished open part) at the top of the jar permits an overflow capacity where the maximum volume of the jar can be filled or emptied (including solids) as fast and easily as possible. The base ring is usually given a stippled finish in the mold to mask scratches that occur during handling. Concentrating abrasions on the high points of this raised pattern prevent the container from being weakened. The base is inside and underneath the heel and surrounds the push-up.

The standard capacity of a wine bottle is 0.75 liters, but for beer, there is no specific standard capacity. Beer bottle volume varies according to consumer habits. For instance, in France, Spain, Germany or Belgium, it is mainly 25 or 33 cl, whereas pint (56.8 cl UK or 47.3 cl US) or half pint bottles are preferred by English, Irish and American consumers. Every type of wine has its own bottle (Figure 3.6.), and the volume can also vary (Table 3.3) (SHA 2020). The most common shape for wine is that of Bordeaux wine bottles, which have straight sides and distinct shoulders. This typical shape widely inspires the bottle design of many wines from the new world (South America, California, South Africa). Instead of name indications related to the village, or "chateau" in Bordeaux, the other wines that are bottled refer to the grape variety, such as Cabernet Sauvignon, Merlot, Malbec, Sauvignon, etc. (Rosa 2019). Usually, the color of the bottle is dark green or brown for red wine and light green or transparent for white wine.

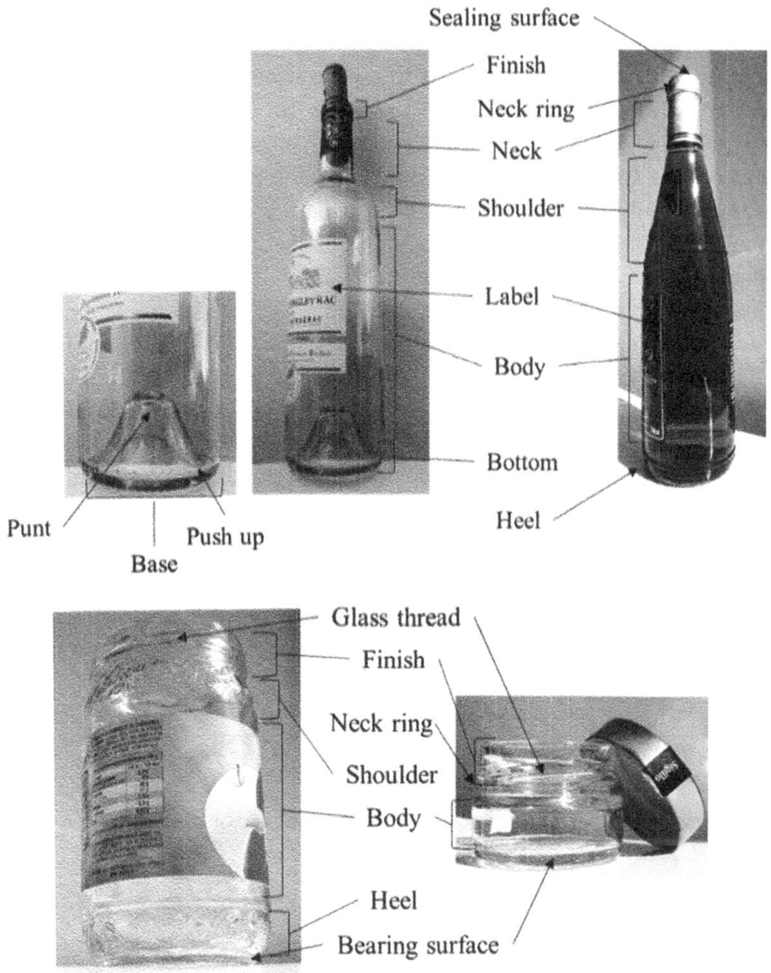

Figure 3.5. *Bottles and jar characteristics*

The other famous wine bottle is the one used for Burgundy wines or Albeisa (Italian wine). Champagne and sparkling wine (Crémant, Prosecco, etc.) require a particular bottle because it must resist the high pressure of the sparkling wines (6 atm). The shape is similar to the Burgundy bottle, but is also heavier and thicker, with a hollow bottom. The color is usually green, both light and dark. In Alsace, the bottles also look like the Burgundy bottle, but with a short neck, whereas Tokaji bottles have very long necks. Port wines, Chianti and marsala have very characteristic bottles. The typical bottles for port, madeira and sherry have a bulb in the neck, which is intended to trap excess sediment when pouring. The marsala bottle is very similar to the port bottle, but it is taller and thinner and the color is typically dark brown or black. The typical bottle for the white wines from the Côtes de Provence (France) or for Verdicchio wine (Italy) is a mix between an amphora vessel and a bowling pin. The glass for this type of bottle is typically transparent (light green for Verdicchio). The Bocksbeutel type is completely different from the other wine bottle shapes; it is a flattened ellipsoid. The glass is dark green in color. It is used for the red wines from the Franconia region, some Portuguese wines and any Orvieto wines. This shape is protected under the European Union. It is often copied and preferred for some spirits, such as Cognac or Armagnac or grappa. Some bottles have a specific volume and shape, like for Tokaji, whose volume is 0.5 liters, or for Clavelin, whose volume is 0.62 liters. This bottle is short, stocky and heavily built. The Clavelin is the only bottle legally authorized for Vin Jaune. For special events, and particularly for Champagne and Bordeaux, bigger bottles are also used, from 0.2 liters (piccolo) to 18 liters (Melchior) corresponding to 24 standard bottles (Rosa 2019; Winerylovers 2020).

Bottle name	Number of 0.75-liter bottles	Volume (liters)
Piccolo (for marketing, flight)		0.2
Chopine (Champagne, red wine)	1/3	0.25
Fillette (Champagne)	1/2	0.375
Tokaji (Hungarian wine)	2/3	0.5
Clavelin (only Vin Jaune-Jura)		0.62
Magnum (Champagne and Bordeaux)	2	1.5
Double Magnum (only Bordeaux)	4	3
Jeroboam (Champagne and other sparkling wines)	4	3
Jeroboam (only Bordeaux)	6	4.5
Methuselah (Champagne and other sparkling wines)	8	6
Methuselah (only Bordeaux)	12	9
Salmanazar	12	9
Balthazar	16	12
Nebuchadnezzar	20	15
Melchior	24	18

Table 3.3. *Wine bottle volumes and related names*

Figure 3.6. *Variety of bottle design (adapted from WineryLovers 2020)*

3.5. Lifecycle, recycling and safety concerns

Glass is the only packaging material generally recognized as safe (GRAS) by the FDA (GPI 2020). Made from natural raw materials, glass is impermeable and nonporous, which protects its contents from degradation and tampering. In addition, glass is chemically inert, which preserves its contents from moisture and oxygen, and ensures that the freshness and flavor of its contents is maintained along with the product's shelf life. With GRAS classification, glass containers require no leaching or extraction testing. The FDA requires all other packaging options to predict the amount of migrating foreign substances that will be consumed and sets limits on the amount consumed when products are packaged in anything other than glass. The only limitation in terms of risks for the consumer, apart from risks related to splinters or breaks, is the migration of heavy metals used, such as oxides, in glass containers. These metals are lead, chromium and cadmium, whose maximum migrations limits are defined in standards and European Community directives (ISO 7086-1:2019; 84/500/EEC, 1935/2004a, 2023/2006). Dealing with lead and cadmium, the specific migration limits (SML) are from 0.8 to 2 and from 0.07 to 0.2, respectively. These SML values depend on the type of glass container, shape, volume surfaces, etc. (Table 3.4).

Category of glass and ceramic packaging	Lead	Cadmium
Category 1: non-fillable objects and fillable objects, whose internal depth, measured between the lowest point and the horizontal plane passing through the upper edge, is lower than or equal to 25 mm. Specific migration limits **in mg/dm²**	**0.8**	**0.07**
Category 2: fillable objects other than those covered by categories 1 and 3. Specific migration limits **in mg/L**	**4**	**0.3**
Category 3: cooking utensils; packaging and containers of storage with a capacity greater than 3 liters (all glass or ceramic objects intended for use in the oven, including the microwave, which include labeling information (logo or user instructions for the consumer)), providing for their use in a conventional oven or in a microwave; and objects for which consumers can reasonably be expected to use in a microwave oven: these include fillable cookware such as mugs, bowls and plates. Specific migration limits in **mg/L**	**1.5**	**0.1**
Oral contact (concerns any object decorated externally on 20 mm width measured from the outer edge). Specific migration limits in **mg/item**	**2**	**0.2**

Table 3.4. *Specific migration limits of lead and cadmium of glass containers*

Glass packaging containers (jars and bottles) can be infinitely reused up to breakage if glass packaging is well cleaned and/or sanitized.

The European glass packaging manufacturing industry is truly a model of the circular economy, by recycling used bottles and jars to make new ones. Once produced, glass is one of the rare materials that can be recycled 100% and endlessly, without losing quality. Recycled glass should not be considered as waste, but as a precious resource to reduce the removal of virgin raw materials. In the EU, over 70% of all post-consumer glass packaging is recycled. A plant in Reims (France) uses up to 90% cullet in glass manufacturing because of the area's high-quality recovery and recycling methods of glass bottles (Lonsway 2006). Glass represented 23% of the packaging waste by mass (Tallentire and Steubing 2020). The primary production of glass included virgin material and energy inputs, water consumption, emissions to air and water, and waste production, based on European averages. One ton of recycled glass represents a reduction of virgin raw materials by 0.66 tons of sand and 0.1 tons of limestone, as well as an energy saving of 1.46 MWh and avoiding the release of 0.46 tons of CO_2 (GPI 2020). Glass packaging waste management is governed by the overall directive 94/62/EC (European Commission 1994, 2004b), which is related to all packaging material waste (Tencati *et al.* 2016).

A recent survey of 8,000 people in 11 European countries showed a great disparity in recycling behavior depending on the age or the region concerned. The oldest people seem to be the most aware of the infinite recyclability of glass. The 60+ age group claims to recycle almost all of their packaging and outperforms the 18–29 age group. It seems that the lifestyle of this young generation may be a cause of their lack of understanding of the infinitely recyclable nature of glass. It appears that women are less aware of the environmental properties of glass and men use more glass-packaged products than women. Despite this, glass packaging remains the material of choice for its environmental qualities: in Switzerland, 93% claim to recycle glass, followed by Italy (91%), Germany (89%), Spain (77 %) and France (74%) (Klaiman *et al.* 2016).

Glass recycling is organized in a loop and involves many players. Local authorities organize glass collection points and install special containers and bins available to the public. Consumers sort their glass packaging after use and place them in glass containers or tubs. Collectors empty containers and bins and transport the glass to processing centers. Glass caterers clear the glass of all the impurities it contains and transform it into cullet. Glassmakers use cullet prepared by caterers to recreate glass packaging. The conditioners fill the new packaging and put it back into the distribution channel. In Europe, they contribute financially to the systems put in place to collect and recycle packaging (VerreAvenir 2020).

The quantity, as well as the quality, of the glass collected depends on the recycling chain. By putting the packaging used in glass tubs or containers, the consumer contributes to the fact that used glass is recycled 100% and infinitely. Only glass packaging can be recycled, that is, food bottles, jars and pots. Jars with a metal frame and/or removable seals are also recyclable. Pharmaceutical bottles are recyclable, as well as perfume bottles and glass cosmetic jars. Culinary glass (transparent Pyrex® dishes) should not be placed in the containers because they are made from transparent ceramic and not glass. These objects have a higher melting temperature than glass and deteriorate the quality of food glass packaging production. The initial sorting by consumers is a fundamental step in the efficiency and quality of recycled glass.

After collection, the glass undergoes a series of sorting in the treatment center. A first sorting eliminates ferrous (steel) or non-ferrous metals (aluminum, etc.) by magnetic detection. Optical laser sorting identifies and extracts infusible materials (ceramics, Pyrex®). Blowing allows the extraction of the light elements: the caps, the paper (labels). The glass is then transformed into cullet, calibrated at the request of the glassmakers, to recreate the glass packaging. The only limit to recycling glass is color. Depending on the country (France, Italy and Spain), the collection is organized in a mixture and is made up of glass of different colors: green, brown, colorless. With this glass, you can only recreate colored glass. With a complementary sorting (used in Sweden, Germany and other northern countries), the de-mixing allows the separation of colored glass from colorless glass, which also allows the production of new packages with a clear tint (Lehmann *et al.* 2020).

3.6. References

ASTM (American Society for Testing and Materials) (2010). Standard terminology of glass and glass products. ASTM Standard C162–05.

Balzarotti, S., Maviglia, B., Biassoni, F., and Ciceri, M.R. (2015). Glass vs. plastic: Affective judgments of food packages after visual and haptic exploration. *Procedia Manufacturing*, 3, 2251–2258.

Boyd, D.C., Danielson, P.S., and Thompson, D.A. (1994). Glass. In *Kirk-Othmer Encyclopedia of Chemical Technology*, 4th edition, Kroschwitz, J. (ed.). John Wiley & Sons, New York.

Dias, D.A., Smith, T.A., Ghiggino, K.P., and Scollary, G.R. (2010). Wine bottle colour and oxidative spoilage. Project UM 0902, University of Melbourne [Online]. Available: https://www.researchgate.net/publication/236230297_Wine_Bottle_Colour_and_Oxidative_Spoilage [Accessed 1 April 2020].

Emblem, A. and Emblem, H. (2012). *Packaging Technology, Fundamentals, Materials and Processes*. Woodhead Publishing Limited, Oxford.

European Commission (1984). The approximation of the laws of the Member States relating to ceramic articles intended to come into contact with foodstuffs. Council Directive 84/500/EEC [Online]. Available: https://eur-lex.europa.eu/legal-content/EN/ALL/?uri=CELEX%3A31984L0500.

European Commission (1994). Packaging and packaging waste. European Parliament and Council Directive 94/62/EC [Online]. Available: https://eur-lex.europa.eu/legal-content/en/TXT/?uri=CELEX:31994L0062.

European Commission (2004a). Materials and articles intended to come into contact with food and repealing Directives 80/590/EEC and 89/109/EEC. Regulation 1935/2004 [Online]. Available: https://eur-lex.europa.eu/legal-content/EN/ALL/?uri=CELEX%3A32004R1935.

European Commission (2014b). Amending Directive 94/62/EC (1994). on packaging and packaging waste. Directive 2004/12/EC. [Online]. Available: https://eur-lex.europa.eu/legal-content/EN/ALL/?uri=CELEX:32004L0012.

European Commission (2006). Good manufacturing practice for materials and articles intended to come into contact with food. Commission Regulation 2023/2006 [Online]. Available: https://eur-lex.europa.eu/legal-content/EN/TXT/?uri=celex%3A32006R2023.

FEVE (The European Container Glass Federation) (2020). The European glass container Industry [Online]. Available: https://feve.org/about-glass/european-glass-container-industry/ [Accessed 30 March 2020].

GPI (Glass Packaging Institute) (2020). Lifecycle assessment [Online]. Available: https://www.gpi.org/life-cycle-assessment [Accessed 30 March 2020].

Gul, K., Wani H.M., Singh, P., Wani I.A., and Wani A.A. (2018). Testing glass as a food packaging material. In *Food Packaging Materials – Testing & Quality Assurance*, Singh, P., Wani, A.A., and Langowski, H.-C. (eds). CRC Press, Boca Raton.

Henderson, J. (2013). *Ancient Glass*. Cambridge University Press, Cambridge.

ISO (International Organization for Standardization) (2019). Glass hollowware in contact with food – Release of lead and cadmium. ISO 7086-1:2019 [Online]. Available: https://www.iso.org/standard/67562.html.

Klaiman, K., Ortega, D.L., and Garnache, C. (2016). Consumer preferences and demand for packaging material and recyclability. *Resources, Conservation and Recycling*, 115, 1–8.

Lehman, R. (2001). Method of recycling batches of mixed color cullet into amber, green or flint glass with selected properties. US Patent US 6,230,521 B1.

Lehmann, S., Fisher, M., Rosin, A., Gerdes, T., and Frenkel W. (2020). The feasibility of CO_2-laser-induced breakdown spectroscopy for fast lead determination in glass cullet. *Int. J. Appl. Glass Sci.*, 11, 369–379.

Lonsway, M. (2006). Glass packaging a clear choice. *Food Quality and Safety from Farm to Fork* [Online]. Available: https://www.foodqualityandsafety.com/article/glass-packaging-a-clear-choice/.

Misaglass (2020). Understand the production process of glass bottle manufacturer. Available: http://www.misaglass.com/glass-bottle-manufacturer-production-process/ [Accessed 1 April 2020].

Nielsen, J.H., Olesen, J.F., and Stang, H. (2009). The fracture process of tempered soda-lime-silica glass. *Exp. Mechanics*, 49, 855–870.

Piergiovanni, L. and Limbo, S. (2016). *Food Packaging Materials*. Springer International Publishing, London.

Robertson, G.L. (2013). *Food Packaging – Principles and Practice*. CRC Press, Boca Raton.

Rosa, S. (2019). *Guide Hachette des vins 2019*. Hachette Livre, Vanves.

Seward, T.P. and Vascott, T. (eds) (2005). *High Temperature Glass Melt Property Database for Process Modeling*. John Wiley & Sons, New York.

SHA (Society of Historical Archeology) (2020). Historic glass bottle identification & information website [Online]. Available: https://sha.org/bottle/ [Accessed 30 March 2020].

Shelby, J.E. (2005). *Introduction to Glass Science and Technology*, 2nd edition. The Royal Society of Chemistry, Cambridge.

Tallentire, C.W. and Steubing, B. (2020). The environmental benefits of improving packaging waste collection in Europe. *Waste Manag.*, 103, 426–436.

Tencati, A., Pogutz, S., Moda, B., Brambilla, M., and Cacia, C. (2016). Prevention policies addressing packaging and packaging waste: Some emerging trends. *Waste Manag.*, 56, 35–45.

Verallia (2020). A propos du verre : processus de fabrication [Online]. Available: https://www.verallia.com/a-propos-du-verre/processus-de-fabrication [Accessed 1 April 2020].

VerreAvenir (2020). Les différentes étapes de la fabrication du verre d'emballage [Online]. Available: http://www.verre-avenir.fr/Le-verre-d-emballage/Les-differentes-etapes-de-la-fabrication-du-verre-d-emballage [Accessed 3 April 2020].

VerreOnLine (2020). Le verre creux : la fabrication et la transformation [Online]. Available: http://www.verreonline.fr/v_creu/fabr_vc3.php [Accessed 1 April 2020].

Wikibooks (2020). Fichier: Glass forming process blow-blow.svg [Online]. https://fr.m.wikibooks.org/wiki/Fichier:Glass_Forming_Process_Blow-Blow.svg [Accessed 3 April 2020].

WineryLovers (2020). Wine bottle shapes: Why are they so different? *WineryLovers* [Online]. Available: http://winerylovers.club/wine-bottle-shapes/ [Accessed 30 March 2020].

4

Metal Packaging

Frédéric Debeaufort
Institute of Technology, University of Burgundy, Dijon, France

4.1. Introduction

Metal packaging for food products is part of our daily universe. While cans seem to be mainly dedicated to canned foods and beverages, this packaging is naturally associated with the ideas of robustness and durability. The range of products packaged in light metal packaging is very extensive. The main sectors that use this packaging are canned products, drinks, dairy products, non-canned food products (using metal closures, metallized films, laminates, etc.), as well as non-food products, such as chemical and industrial products (paints, varnishes, lubricants, etc.), cosmetics, biological products and medicines, and clogging in general. Convenient for transport and storage, metal packaging is also in use thanks to the development of easy opening devices. Finally, and this is an important point today, metal boxes are easily recyclable and are indeed largely recycled. These strong advantages of metal boxes are added to a primary condition, which is the good compatibility of foods, in their diversity, and packaging materials. The properties of the base materials, most usually steel or aluminum, and their shaping methods make it possible to ensure gas, light and microorganism impermeability in a very reliable manner (Pelletier 2000; Robertson 2013). This is necessary for the prolonged storage of canned food for up to several years.

Tinplate is one of the oldest packaging materials and was originally used for round, square and rectangular boxes (Emblem and Emblem 2012). The first

For a color version of all figures in this book, see www.iste.co.uk/debeaufort/packaging.zip.

Packaging Materials and Processing for Food, Pharmaceuticals and Cosmetics, coordinated by Frédéric DEBEAUFORT, Kata GALIĆ, Mia KUREK, Nasreddine BENBETTAIEB and Mario ŠČETAR.

commercial manufacture of tin cans began in England in 1699 and in France in 1720, where it was used for household utensils, such as plates. Around the middle of the 18th century, the Dutch navy first used food cooked in fat and packaged in tin cans (Thorne 1986). In the early 1800s, Napoleon Bonaparte offered a prize for the first person capable of developing a means of preserving food under all conditions and over long periods for his navy (Emblem and Emblem 2012). Thus, the first tin cans produced industrially for foodstuffs were proposed and patented in August 1810 by Peter Durant, based on the invention of Nicolas Appert, who developed the sterilization process for a product in its packaging (in glass bottles, then in metal) for long-term preservation.

According to the North American Packaging Alliance, frozen products require almost 70% more energy than foods stored in metal. Tins are very light and can reduce CO_2 emissions by up to 60% during transport compared to heavier materials (such as glass). The sustainability of metal packaging is therefore one of the main factors driving the growth of the global metal packaging market. Metal packaging materials such as aluminum and steel are the two most used raw materials for packaging because of the systems dedicated to their recycling and recovery. Steel and aluminum can be easily separated from other waste and can be 100% recycled without any loss of material quality. Steel is generally separated using magnets, while aluminum is separated using eddy current technology. The metal boxes are light, unbreakable and offer mechanical protection and strong barriers for the conservation and transport of food products. Thus, the durability of metal packaging is one of its main advantages, which consistently stimulates market growth. The metal packaging market in 2018 was nearly USD 24 billion, with an annual growth rate estimated at more than 3% for the next five years. However, in the early 2010s, the market seemed to be saturated (Parisi 2013, p. 420). The market is moderately concentrated, with some very large players such as Amcor, Ball, Crown, Sonoco and Ardagh Group (Technavio 2020). A third of companies in the metal packaging sector are American. One of the main market trends that explains this growth is the advent of resealable boxes. Resealable boxes are integrated with a flat plastic opening mechanism. They can be easily opened and closed with a simple rotary movement, which allows them to be opened and closed effortlessly. With the growing demand for resealable boxes, several suppliers of metal packaging offer this functionality.

4.2. Metal packaging types

In Europe, almost 100 billion units of metal packaging are produced each year for the beverage, food, health and beauty, household and industrial markets. This means that, on average, every EU citizen consumes four units weekly (Metal Packaging Europe 2020). Many types and formats of metal packaging are available for the food industry, the best-known of which are cans. Some cans are illustrated in Figure 4.1.

Food cans: with a proven safety and sustainability record, food cans offer unrivalled performance throughout the supply chain, and this is true for virtually every product. The storage of food cans saves up to 70% of energy over the whole supply chain. Cans need significantly less secondary packaging compared to alternative packaging due to the natural strength of metal. Cans and food cans consist of either two or three separate components ("two-piece" or "three-piece"), composed of a cylinder (body) and one or two ends (lids). The lids are always attached (seamed) after the cans are filled with the foodstuff. The size of food cans ranges from very small (50 grams), up to catering pack size (5–20 kg). Tinplate is still preferred for this kind of packaging.

Beverage cans and bottles: canned beverages are a perfect fit for active, outdoor and on-the-go lifestyles. Beverage cans are lightweight, yet unbreakable. They are impact-resistant, puncture-resistant and can withstand extreme temperatures and pressure. Cans provide a total barrier against light, gas and oxygen, and thus preserve their contents perfectly in order to deliver safe products. They also chill faster than other packaging. The sound of opening a beverage can is a unique feature guaranteeing product freshness. Beverage cans are most often two-piece type packaging. Beverage cans and bottles are almost all made from aluminum.

Closures, lids and crown corks: tinplate closures combine the advantages of metal with other materials. Products are often sealed with tinplate screw caps to keep them completely airtight, preventing the entry of air and guaranteeing taste neutrality. In the food industry, tinplate closures are particularly popular for products such as baby food, processed fruits and vegetables, jams, dairy products and delicacy sauces. The versatile aluminum screw cap is an all-rounder: it fits on all bottles, resealable or non-resealable, and covers all beverages such as alcoholic, water and juice. Thanks to the thread, the containers can be opened and closed as often as desired. Screw caps made of aluminum, often called aluminum roll-on closures, are rolled onto the bottle thread when closing and only then do they get their thread. Like all metal closures, they are 100% airtight and can be printed to suit the content. Closures and corks are detailed in Chapter 9.

Drums and pails: drums and pails are, in essence, large three-piece steel or stainless steel cans. They are not subjected to any other process after they are filled. Usually, for foodstuffs (concentrated juice, fruit flesh, oils and fats, for instance), they are aseptically filled. Drums usually contain large volumes, ranging from 100 to 250 liters, the most common being 220 liters, while pails usually refer to a volume of 5–25 liters.

Aerosols: aerosol cans are not very common in food applications (whipped cream, for example), but are used very often for cosmetics, body care products, insecticides and lubricants. Aerosol cans are either three-piece or two-piece cans.

Trays and foils: rigid and semi-rigid aluminum trays for food applications are made of rolled aluminum and aluminum alloys, with a thickness ranging from 70 to 300 μm. Aluminum foil is mainly used for wrapping food in catering, or at home.

Tubes: only tubes with contents needing minimal interaction with oxygen or light are metal-based, and others are plastic-based. Some food pastes (tomato paste, mayonnaise, harissa, spread or concentrated milk) are packaged in metal tubes, but they are more commonly used for cosmetic and health care products (toothpaste, sunscreen, etc.). Metallic tubes are extruded from a slug of metal (usually aluminum).

Figure 4.1. *Illustration of the variety of types of food packaging from metal-based materials*

4.3. Composition and properties: metals and alloys

Three main classes of metallic materials for food packaging applications are currently available on the market: aluminum, coated plates (tinplate, tin-free steel, polymer-coated and steels) and stainless steel plates. Minor metals are also used in the manufacture of metallic packaging for food (nickel, copper, titanium). For instance, copper is used as a wire electrode during the welding of tinplate containers, but does not become part of the finished product (Robertson 2013).

4.3.1. *Steel-based (iron-based) and tin-plated steel materials*

Tin, steel and chromium steel are used as composite materials in the form of tinplate and electrolytically chromium-coated steel (ECCS), the latter being somewhat unhelpfully referred to as tin-free steel (TFS).

The material used will be different depending on whether it is for:

– drums (or other industrial packaging such as jerry cans and vats): they are more than 0.49 mm thick and are made from commercial sheet metal;

– light packaging (cans, cans, buckets, etc., with a capacity of less than 40 liters or 50 kg): their thickness is generally less than 0.49 mm. They are made from tin or black iron. For certain applications, however, it is possible to seek to replace the tinplate with iron without tin (chromed iron, for example) or stainless steel, which is considered a very inert material, which is only used for the packaging of very corrosive products and/or very expensive goods (Kleniewski 1995).

Iron, steel and stainless steel are often confused. Iron is a chemical element used in the composition of the other two metals. Steel and stainless steel are produced mainly from iron, but contain other components and are differentiated by their carbon content and that of other metals. Iron ore, coal, lime and ferroalloys are needed to make steel: aluminum, chromium, manganese, silicon, titanium, vanadium, etc. These give steel specific characteristics that are suited to its various uses: architecture, machine tools, food packaging, etc. Manufacturing also requires a lot of water for cooling the installations, as the melting temperature exceeds 1,500°C. Steel can be made in two types of factories: in a blast furnace, from iron ore and coke (almost pure carbon extracted from coal), or in an electric furnace, from recovered steel (recycled steel). The use of 20%–40% of recycled steel enables saving about 60%–70% of the energy and up to 80% of the water and air necessary to produce steel for can production, compared to raw matter (Selke 1994). The composition of the steel ranges from 0.03% to 2% carbon, which is the most important alloying element, strongly affecting the strength of the can. The properties of steel are obtained either by adjusting the carbon content (the less carbon, the more deformable, but not very resistant, the steel), by the application of a subtle metallurgy, combining control of the chemical composition at the steelworks and the steel manufacturing process (thermomechanical rolling treatments), or by a possible thin metallic or organic coating (A3M 2020). Usually, the higher the carbon in steel, the higher the manganese content, in order to maintain weldability properties. The aluminum and silicon contents of the steel vary according to the type of calming used (processing step consisting of preventing the bubbling of the molten steel in the vat). Phosphorus and nitrogen play the role of stiffening agents: they are, however, not present in steel unless they have been intentionally added in order to make tinplate for special uses, such as packaging that must withstand strong internal pressures (Selke 1994; Robertson 2013). In the presence of corrosive media, the phosphorus content must be reduced, as this element reduces the corrosion resistance. In general, the chemical composition of the steel has an impact on the tinplate's corrosion resistance to acidic food products. Thus, the fewer traces of non-metals in the steel, the better the natural resistance to corrosion.

Four types of steel are defined for the manufacture of tin cans for food uses. Tinplate steel, intended for highly corrosive products, must be low in both non-metal and residual metallic elements, and it is type L (low metalloid) steel (Kleniewski 1995). For moderately corrosive food products such as vegetables or meat, MR type steel could be used. MR steel has a higher content of carbon, aluminum, phosporus and copper because the risk of residual elements able to migrate into the food is reduced and the latter are efficient in preventing corrosion. The other two grades of steel are between these two (Table 4.1).

Type	Properties	Applications	Composition (maximum content %)						
			C	Al	Si	Mn	P	S	Cu
L	High purity	High corrosion resistance, acidic foods	0.13	0.08	0.01	0.60	0.01	0.05	0.06
D	Non-aging steel	Dedicated for severe drawing operations	0.12	0.06	0.02	0.60	0.02	0.05	0.20
N	Nitrogenized steel for higher rigidity	For high-strength containers, such as drums or can ends	0.13	0.06	0.01	0.60	0.01	0.05	0.06
MR	Most widely used for tinplate	For foods that are not too corrosive such as meat or vegetables	0.13	0.08	0.01	0.60	0.02	0.05	0.20

Table 4.1. *Properties and composition of the maingrades of steel used for the production oftinplate (adapted from Robertson 2013)*

Tinplate remains the main raw material used to make packaging and metal boxes. It is made of mild steel with a carbon content of less than 0.08%, which is covered with a layer of tin. This "sandwich" alloy of steel and tin has a good hardness and guarantees a high resistance to corrosion. It is a non-toxic alloy, which is why it is used massively in food packaging (Boite Aluminium 2020).

Tinplate should readily lend itself to shaping and must offer sufficient strength during the intended use. The steel will, therefore, be chosen to be more or less stiff, depending on the case (the stiffness corresponds to the resistance to deformation by bending). Suitability to deformation is therefore the main criterion for choosing steel, according to its mechanical characteristics. The stiffness is assessed by measuring the depth of penetration of an indenter under a large load compared to the penetration made by a preload and expressed in Rockwell units. Single reduction tinplate is produced from a strip of cold rolled steel which is annealed and surface-hardened. The final thickness is between 0.1 and 0.49 mm, and the hardness varies from class T50 to T65 on the Rockwell scale. Double reduction tinplate is obtained after first cold rolling and annealing, then undergoing a second cold thickness reduction of approximately 30%. The steel obtained is generally stiffer than that in simple reduction, which makes it possible to use thinner iron without losing the strength of the packaging. Their mechanical characteristics are anisotropic, and it is necessary to take this into account for the direction of rolling. The thicknesses of double reduction tinplate that is usually made ranges from 0.14 to 0.29 mm and is preferred for cans (Kleniewski 1995; Roberston 2013; ISO 11949 2016).

We are all familiar with tin cans, which are used for holding many types of food and beverages; these containers are actually manufactured from sheet steel that is coated with a thin layer of tin known as tinplate. Tinplating mainly aims to prevent the corrosion of steel from the internal contents (food), or the external environment (a moist atmosphere). Tinplating can produce a whitish-grey color that is preferable when a dull or matte appearance is desired. It can also produce a shiny, metallic look when more luster is preferred. Tin offers a decent level of conductivity, making tin plating useful in the manufacturing of various electronic components. Tin is also FDA and EC regulations approved for use in the food service industry (Sharretts-Platting 2020).

Today, tinplating is obtained by electrolytic deposition. This process replaced the traditional hot tinning technique (used until the 1980s), which consisted of dipping the steel sheets into a bath of molten tin. Tinning by either method is preceded by a thorough cleaning (descaling (also known as pickling) and degreasing, followed by washing) to prepare the surface. Electrolytic tinplate is obtained either by alkaline tinning from tetravalent tin baths, or by acidic tinning from bivalent tin baths. The deposit of tin acquires its shine only after the coating is melted by a heat treatment at a temperature higher than the melting point of tin (usually between 260 and 280°C). The tinning rate is expressed in grams per square meter (g/m^2) and, single-sided, varies from 0.4 to 1.25 (Klienewski 1995). After the tinplating step, the coating is passivated, and then lightly oiled (Robertson 2013). Post-treatment is usually not required at the conclusion of the tinplating process. Passivation, which is the application of a light coating of protective material, may be used in special plating applications to provide additional corrosion protection or enhance the reactive properties of the tin. A heat treatment may also be used to prevent hydrogen embrittlement, which is a weakening of the metal caused by exposure to hydrogen. Research into new high-performance materials, while reducing the level of tin, has resulted in a new generation of LTS (low tin steel) iron with a low tin coating. The materials undergo an original development of the coating, for example, by optimizing the deposition of chromium and chromium oxide from the passivation, or even by using a layer of nickel. The schematic structure of tinplate and tin-free steel is given in Figure 4.2.

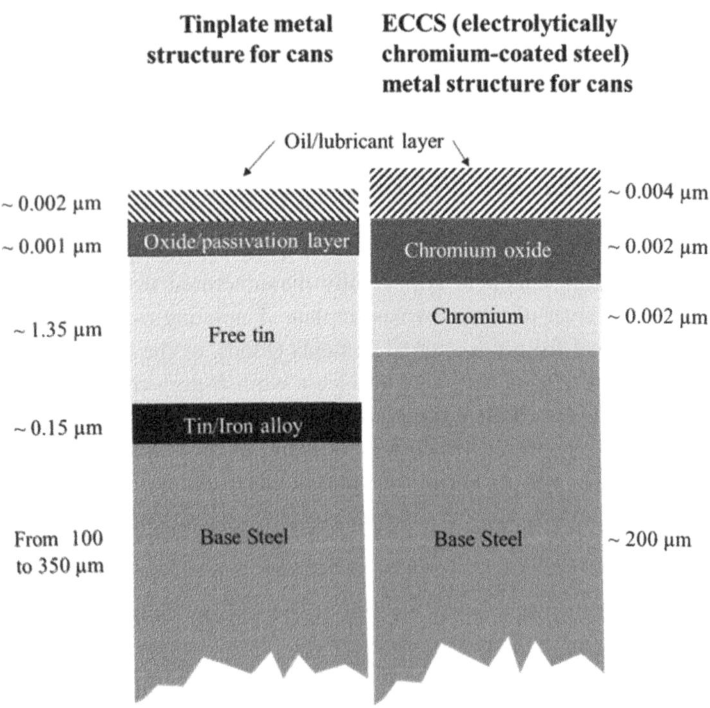

Figure 4.2. *Representation of the multi-layered structure of tinplate and tin-free steel (ECCS) (adapted from Robertson 2013)*

4.3.2. *Tin-free steel or ECCS (electrolytic chromium oxide-coated steel) materials*

Chromed iron (tin-free steel or TFS, standardized under the name of ECCS: electrolytic chromium- or chromium oxide-coated steel) is obtained from a steel sheet or strip which, this time, is coated with a film of chromium and chromium oxides, much less than a micrometer thick. This process is very similar to the electro tinning, but uses chromium ions instead of tin ions. Requirements for single and double cold-reduced electrolytic chromium-/chromium oxide-coated steel (ECCS) in the form of sheets or coils are specified in ISO 11950 (2016). Single cold-reduced ECCS is generally specified in nominal thicknesses that are multiples of 0.005 mm, from 0.150 mm up to and including 0.600 mm. Double cold-reduced ECCS is generally specified in nominal thicknesses that are multiples of 0.005 mm, from 0.100 mm up to and including 0.360 mm. The structure of ECCS is described in Figure 4.2.

4.3.3. *Stainless steel for food packaging materials*

Some metals, such as stainless steel, are inert, whereas others are non-corrosive if they are well-coated. This is an important property because it ensures that containers will last for a long time and will be easy to clean (Chemseal Inc. 2020). Metals such as stainless steel do not corrode easily. Reactive chromium forms a passive coating which creates a surface barrier to corrosion. This barrier is resistant and protects the surfaces of metal containers. Indeed, an addition of metallic chromium of at least 10.5% to the steel makes it possible to achieve this performance (Kleniewski 1995). There are many advantages to using stainless steel in the processing and preservation of food and beverages. The number one advantage of stainless steel is its resistance to corrosion. Stainless steel also has very good durability and heat resistance. Due to the corrosion resistance, its surface hardness and its quality surface finish (smooth surface), it is easy to clean and maintain, which is essential for reusable packaging. However, despite its many advantages, stainless steel remains little-used because of its uncompetitive price for single-use food packaging, despite the fact that it is 100%, and almost infinitely, recyclable. However, stainless steel is gaining market shares in the transport of food products in large quantities or volumes from 50 to 10,000 liters (drums, tanks, etc.).

There are approximately 100 grades of stainless steel suitable for packaging materials. These metal alloys come in five main groups, which include precipitation-hardening, duplex, martensitic, ferritic and austenitic stainless steel. 304 stainless steel has a metallic composition of up to 24% chromium. This metallic alloy may be made of manganese, carbon or nickel. This alloy has a property of perfect anti-corrosion that makes it ideal for food products. 316 or 318 stainless steel is almost the same as the 304 stainless steel metallic alloy, but is preferred in the food industry for both packaging and processing. It has an addition of up to 3% of molybdenum. Molybdenum reduces chloride-related corrosion and adds extra strength, which makes it possible to store mild salts and even some meat products (Chemseal Inc. 2020).

4.3.4. *Aluminum-based materials*

Compared to tinplate and ECCS, aluminum is a lighter, thinner, more ductile (deformable) material, but cannot be welded easily. From the principle of reducing waste and also the cost of the raw material, over the past three decades, a considerable effort has been made to reduce the thickness of this metal. This reduction was possible thanks to the development of new aluminum alloys and the optimization of the manufacturing processes for aluminum foils and cans (Robertson 2013). Coors Company pioneered the use of aluminum cans in 1959. From this start, there has been a steady growth in the production of sodas, energy drinks, sparkling waters and, increasingly, craft beers. Aluminum cans became the most sustainable beverage packaging and are infinitely recyclable (The Aluminum Association 2010). Aluminum foil is used in food containers, bins, bottle caps, soft packages for liquids or bulk food stuffs, capsules (such as the coffee capsule for Nespresso®), bottle stops, doypacks, medication tablet blister sheets, tubes, bottles, cans and many other types of containers.

The aluminum used to manufacture light metal packaging is used in a relatively pure form. Only manganese and magnesium are added to improve its resistance properties. Aluminum is the third most abundant element on earth in bauxites (soils or minerals rich in aluminum in various forms). Aluminum oxide (alumina) is extracted from the minerals present in bauxite before the start of the smelting process (Page 2012). This process requires considerable electrical energy to produce aluminum on a commercial scale. The primary aluminum and the aluminum from the recycling channels are then mixed and moved into an oven to purify the liquid metal. It is then filtered (through a ceramic filter) to be cleaned before being cast into ingots, before being rolled into sheets. The faces of the ingot are pre-ground (scalped) to remove surface imperfections and to provide clean surfaces for rolling. The metal is then reheated and hot rolled to an intermediate thickness. It is then left to cool in the air for two to three days, at which point it is completely soft. The final cold rolling takes place to obtain the defined thickness and hardness without the need for an additional heat treatment.

Aluminum packaging uses different types of alloys which are processed by hot rolling and then cold rolling into strips or coils. The composition of the alloy varies according to the manufacturing or shaping processes of the finished products. The production techniques for aluminum packaging call upon three technological disciplines: shaping, surface treatment and coatings (Kleniewski 1995). Most uses of aluminum for food packaging require special properties (resistance to acids, for example) that pure metal cannot guarantee. Thus, other compounds are added to give these aluminum alloys resistance or improve the shaping characteristics (deformability, ductility) and modulate the resistance to corrosion. However, these alloys all contain a very high aluminum content, often greater than 90% (Table 4.2). Alloys are identified by four-digit numbers, where the first digit corresponds to the area of use, which depends on the elements added. Commercially pure (unalloyed) aluminum (e.g. 1xxx) contains different levels of impurities, Fe and Si being the most common. It is used for the manufacture of aluminum foil and extruded containers. The 3xxx alloy family contains manganese (Mn) and the 5xxx series contains 4% to 5% magnesium (Mg) and 0.5% Mn, producing a very rigid material, suitable for the manufacturing of cans. The 8xxx series covers a wider range of composition and applications (Reboul and Baroux 2011).

Applications	Type	Composition (maximum content %)							
		Mn	Mg	Fe	Cu	Cr	Zn	Ti	Si
Foils and flexible tubes	1050	0.05	0.05	0.40	0.05	0.05	0.03	0.03	0.25
Beverage cans, three-piece bodies and ends	3104	1.4	1.30	0.70	0.25	-	0.25	0.10	0.60
Two-piece bodies and easy-opening ends	5042	0.50	4.00	0.35	0.15	0.10	0.25	0.10	0.20
Easy-opening beverage cans and tabs	5182	0.50	5.00	0.35	0.15	0.10	0.25	0.10	0.20
Pilfer-proof caps	8011	0.20	0.05	1.00	0.10	0.05	0.10	0.08	0.90
Foils for laminations	8079	–	–	1.3	0.05	–	0.10	–	0.30

Table 4.2. *Properties and composition of the main grades of aluminum alloys (adapted from Robertson 2013)*

Copper greatly reduces the corrosion resistance of aluminum and leads to a higher rate of general corrosion. Zinc has little influence on corrosion resistance in most environments, tending to reduce the resistance of alloys to acidic media and increase their resistance to alkalis. Silicon also slightly decreases corrosion resistance, depending on its shape and location in the alloy microstructure. Iron also reduces corrosion resistance and is probably the most common cause of pitting in aluminum alloys, but a high iron content increases the burst strength of cans.

Titanium has little influence on the corrosion resistance of aluminum alloys. On the contrary, manganese, magnesium and chromium generally have a beneficial influence on the corrosion resistance of aluminum-based packaging (Robertson 2013).

4.4. Can manufacturing

The following two basic methods are mainly encountered in the manufacturing of cans: either the three-piece can (comprising of a cylindrical body rolled from a sheet of tinned steel with the side welded and two ends joined mechanically by crimping), or the two-piece box (comprising of a seamless cylindrical box body obtained by stamping and, after filling, the crimped cover). The box format is often characterized by the geometric relationship between diameter and height. This relationship can be described as (Figure 4.1):

– tall cans whose height is greater than the diameter (e.g. beer cans);

– short cans whose height is equal to or slightly less than the diameter (e.g. tuna cans);

– shallow cans whose height is significantly less than the diameter or the width (e.g. sardine cans);

– tube-type cans whose height gives the shape of a bottle (for instance, metal bottles for oils, aerosols for whipped cream, or tubes for tomato paste or mayonnaise).

4.4.1. *Three-piece can-making process*

Three-piece cans, which were the first to be developed in the middle of the 19th century, consist of a cylindrical body, rolled from a piece of flat metal with a longitudinal weld joint (usually formed by welding or soldering), together with two can ends, which are seamed onto each end of the body. Three-piece welded cans for food, drink and cosmetics and industrial cans, as well as aerosols, are only constructed from steel-based materials. Indeed, thin gage aluminum cannot be welded by this process. Most of these cans are made from tinplate, as ECCS or TFS is difficult to weld consistently without first removing the metallic coating (Page 2012). Three-piece cans may be made in almost any practical combination of height and diameter. This process is particularly suitable for making cans with mixed specifications, as it is relatively simple to change the specification of the cans being made (MPMA 2020).

The making process of the three-piece welded cans consists of 12 steps and is illustrated in Figure 4.3 (MPMA 2020).

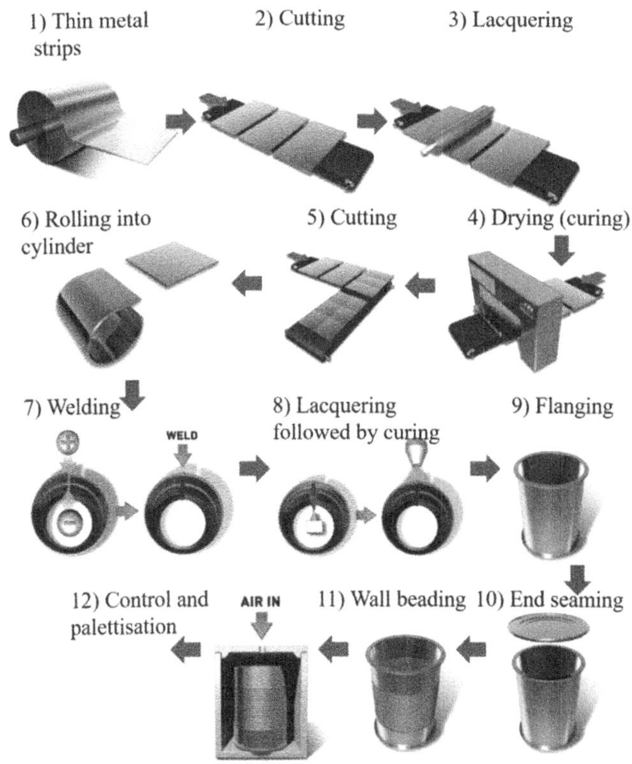

Figure 4.3. *Three-piece can manufacturing (adapted from MPMA 2020)*

1) Steel or tinplate strips arrive at the can manufacturing plant in large coils. 2) Strips are cut into large sheets. 3) Lacquer is applied to the side of the sheet that will become the internal surface. This special lacquer is used to protect the can from corrosion and from any possibility of interaction between the contents and the metal. 4) The lacquered sheets are dried and/or cured in an oven. 5) The large sheets are cut into small sheets, one for each can body. 6) The small sheets are rolled into cylinders (when an oval or an almost rectangular parallelepiped shape, for sardine cans, for instance, is required, the shaping occurs after welding). 7) The cylinder

edges are welded by squeezing them together while passing an electric current through them. This heats the metal sufficiently for a sound joint to be made. 8) The inside surface of the weld is sprayed with lacquer and then cured by blowing heated air onto the outside of the cans. 9) The cans are passed through a flanger where the top and bottom of the can are flanged outwards to accept the ends (or lids). 10) Plain ends (ends, lids or easy-opening ends) are seamed to the can bodies to close one end of every can. 11) The cans are passed through a beader where the walls of the cans have circumferential beads formed in them to give added strength. Indeed, forming circumferential beads in the can body wall significantly increases the hoop strength, to resist the implosion of the can during the further sterilization processes. 12) Each can is tested at each stage of manufacture, and at the final stage, cans pass through a pressure tester, which automatically rejects any cans with pinholes or fractures. The finished can bodies are then transferred to the warehouse to be automatically palletized before being dispatched to filling plants or food factories.

Two of the critical steps in this process relate to the welding and seaming. The seaming step is discussed in section 4.4.3. The area of the weld is left without coating or print to ensure that a sound weld is made. The two longitudinal edges dedicated to the weld overlap by approximately 0.4 mm (Page 2012). Electric resistance spot welding is often used where alternating current passing through the metal seam heats up the material. The tinplate is then sufficiently softened to be squeezed together to form a sound joint. Each peak of electric current creates a spot of weld. A series of overlapping spots is created to form a continuous weld along the cylindrical can body (Emblem and Emblem 2012). A stripe coating is sometimes applied over the weld area at this time. This welding process can only generate bodies of a circular cross section, therefore, where non-round – that is, square, rectangular, slightly tapered, etc. – bodies are required, the circular body must be reshaped after the completion of the welding process.

4.4.2. *Two-piece can manufacture*

Two-piece cans are made from a disk of metal which is reformed into a cylinder with an integral end. A loose end is seamed to this body to close the can. The operation of reforming sheet metal without changing its thickness is called "drawing" (DR). The operation of reforming a two-piece can into one of smaller diameter, and therefore with a greater height, without changing its thickness, is called "re-drawing" (DRD). The operation of thinning the walls of a two-piece can by passing through circular dies is called "ironing". Drawn and ironed cans are referred to as "DWI" or "D&I" cans. The DWI process is used for making cans where the height is greater than the diameter, and is particularly suited to making large volumes of cans of the same basic specifications.

The differences between DRD and DWI are related to the drawing steps, and also to the coating. Drawn and redrawn (DRD) cans are used for food (particularly processed fish products) and are made from steel or aluminum. In most cases, the metal is coated internally and externally and printed, if necessary, before the can-forming operation, as the presence of these materials on the metal surface provides some degree of lubrication to aid the forming. Lubricants, such as waxes, may need to be added to the outside surface to enhance the drawing steps. The DRD process may also be used to form taper wall flanged cans, as well as aluminum or steel taper thin wall trays, to which heat sealed foil lids are applied (Emblem and Emblem 2012). DWI cans are used for food, drink and aerosol cans and are normally made from uncoated aluminum or tinplate. However, DWI processed food cans are only made from tinplate, as thin wall aluminum cans do not have sufficient strength to withstand the external pressure imposed during the heat process cycle. In DWI manufacturing, the coil is uncoated because subsequent wall thinning (ironing) operations are so severe that the coating would be stripped off of the metal surface. Because of this, a water-soluble synthetic lubricant is applied to the coil before the cup blanking or drawing operation. The application of coatings and print (where appropriate) later in the process are an integral part of the DWI can-making process (Page 2012).

The manufacturing process of the two-piece draw and wall-ironed cans (DWI) can be organized into 14 main steps and illustrated in Figure 4.4 (MPMA 2020):

1) An aluminum or steel strip arrives at the can manufacturing plant in large coils. 2) The strip is lubricated with a thin film of liquid and then fed continuously through a cupping press, which blanks and draws thousands of shallow cups every minute. 3) Each cup is rammed through a series of tungsten carbide rings. This is the drawing and ironing process, which redraws the cup to a smaller diameter and thins the walls while increasing the height. 4) Trimmers remove the surplus irregular edge and cut each can to a precise height. The surplus metal is recycled. 5) The trimmed can bodies are passed through highly efficient washers and then dried. This removes all traces of lubricant in preparation for internal and external coating. 6) The clean cans are coated externally with a clear or pigmented base coat which forms a good surface for the printing inks. 7) The cans pass through a hot air oven to dry the lacquer. 8) The next step is a highly sophisticated printer or decorator which applies the print design in up to six colors, plus a varnish. 9) A coat of varnish is also applied to the base of each can by the rim coater. 10) The cans pass through a second oven, which dries the ink and varnish. 11) The inside of each can is sprayed with lacquer. This special lacquer is used to protect the can from corrosion and from

any possibility of interaction between the contents and the metal. 12) Lacquered internal surfaces are dried in an oven. 13) The cans are passed through a necker flanger, where the diameter of the wall is reduced (necked-in). The tops of the cans are flanged outwards to accept the ends after the cans have been filled. 14) Every can is tested at each stage of manufacture and at the final stage they pass through a light tester, which automatically rejects any cans with pinholes or fractures. The finished can bodies are then transferred to the palletization before being sent to the customers (the food industry or filling industry).

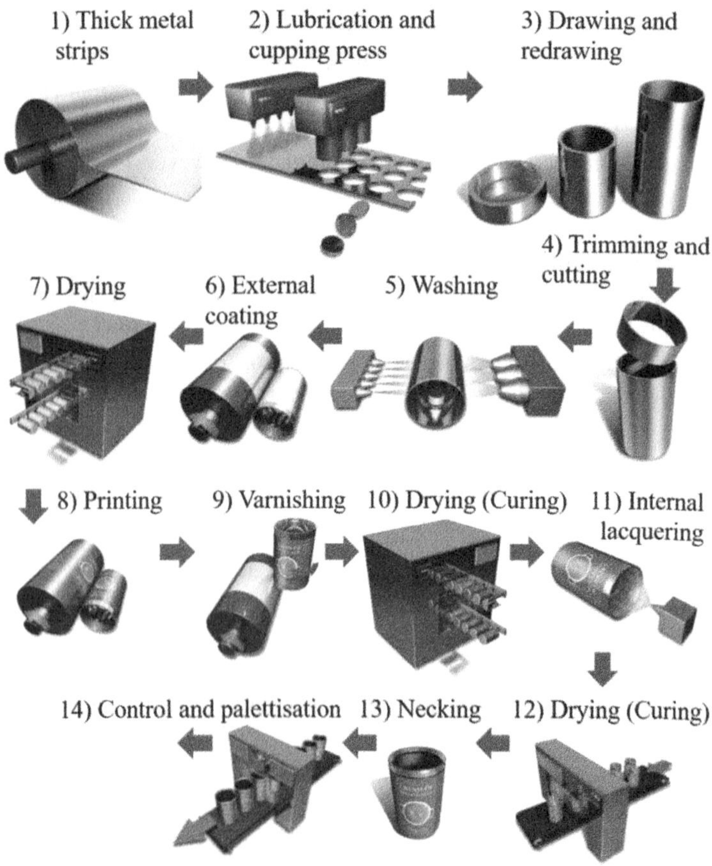

Figure 4.4. *Processing flowchart of the two-piece can making (DWI process, adapted from MPMA 2020)*

Tubes and aerosols made of aluminum (Page 2012) are a special case of two-piece metal packaging. Indeed, it is not possible to produce these tall and narrow packages (bottles, aerosols, tubes) in steel by a stamping-stretching process by impact. In this process, a very thick metal disk with a diameter equal to the finished product is cut before being stamped. The disk is then placed at the bottom of a die and a punch strikes (stamps) the cold metal at high speed. Many passages are made during stamping. The cold metal is expelled from the die and slides along the punch until the end reaches the desired length (Figure 4.5). The same process is used to form foldable tubes, but the base of the mold (drawing die) allows the formation of the nozzle of the tube. As the forming (deep drawing) process hardens the wall of the tube, it is usually necessary to soften it using a posterior heat treatment (annealing) so that the tube can be folded or rolled up after filling. The other manufacturing steps, including coating and printing, are similar to those used for DWI beverage cans. The rim, however, is replaced by a forming process to shape the top of the aerosol (a flange on which will be crimped an aerosol valve mechanism, a rolled edge to accept a crown end or a screw neck to accept a roll-on pilfer-proof cap).

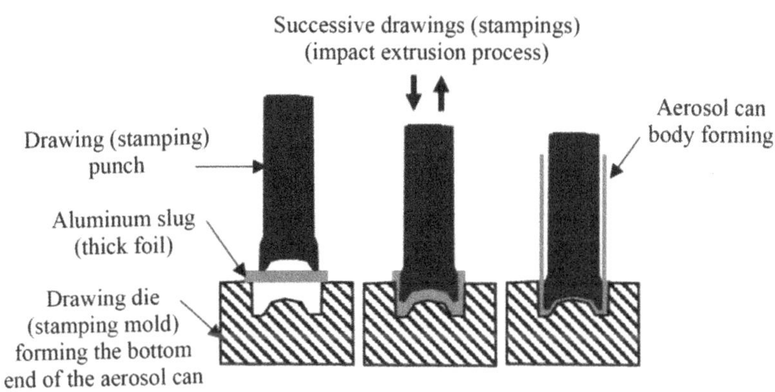

Figure 4.5. *Processing of two-piece impact cold extrusion process for aerosols in aluminum (adapted from Emblem and Emblem 2012)*

4.4.3. *Can end making and seaming onto can bodies*

The can ends or lids for mechanical double stitching are made of aluminum, tinplate or tin-free steel. Aluminum and TFS are always coated on both sides with organic lacquer or a laminated film while the metal is always in the form of a coil or flat sheet. The end or lid of the box is of a complex design, developed for optimal deformation behavior (Robertson 2013). This deformation is necessary during the seaming step and also very important during the subsequent operations of filling, seaming, heat treatment and storage (stacking). This depends on the thickness of the metal sheet, the precise contour of the expansion rings and the milling depth. It is important that the ends can deform under internal and external pressure, but reversibly. The edges of the lids are stamped on presses from a sheet of tin (generally of high hardness). After stamping, the edges are shaped to match the lower or upper edges of the body of the can. A coating or seal is then applied to the hollow of the cover (Figure 4.6). The sealant used is based on natural or synthetic rubber and is dissolved in water or a solvent. The purpose of the sealant allows the formation of a hermetic seal between the layers of crimped metal.

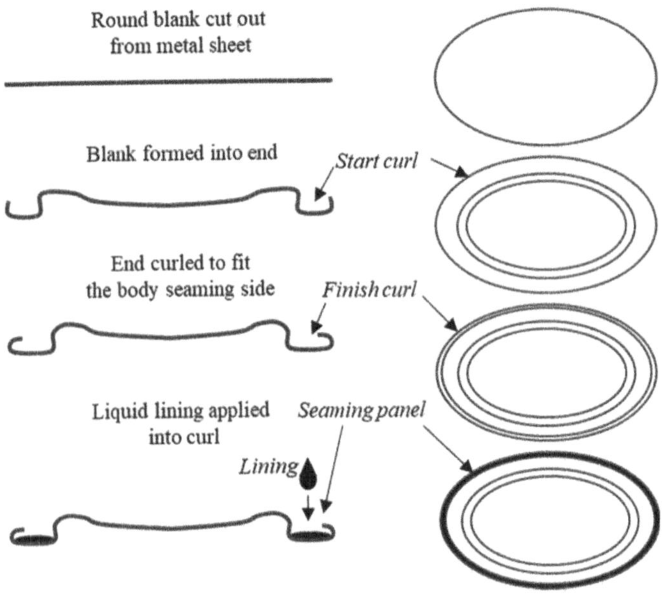

Figure 4.6. *Can end making (adapted from Page 2012; Robertson 2013)*

The seam operation in can manufacturing and can filling is a crucial step of the can quality and consumer safety. In the field of packaging, seaming is the operation that concerns metal cans (canned food, drink cans, etc.). It allows the attachment of the body (in general, cylindrical) to one or two ends (in general, circular). It is the mechanical assembly of two parts of rigid metallic sheets or foils without the use of welding, gluing or tapping: the two sheets are folded or rolled so that they fit together. This technique must guarantee the airtightness of the closed metal package. Neither the sealing compound, nor the interlocked can body and end, are able to form a hermetic seal; they must complement each other. The double seam must be correctly formed. The compound, notwithstanding its resilience and ability to fill voids in the double seam, cannot compensate for an improperly formed seam (AFDO 2011).

The double seam is uniformly used for food and beverage cans today. A double seam is a simple and proven way to provide a hermetic seal. This process is completed on as many as 500 billion cans annually. However, while the end goal is simple, the process can still appear to be very complex (Crown Inc. 2020). In the seaming process (Figure 4.7), it is first necessary to deposit or fit an end or a lid onto the body of the can. The body and the cover must be perfectly adjusted. The parts are put into compression on a mandrel, and then a knurling wheel (called the first pass) rolls the edge of the box up with the edge of the end or lid. Then, a second pass (second roll) wheel tightens this winding to ensure a perfect seal. Modern multi-head seamers reach rates of up to 1,000 boxes per minute.

The body flange and the curl of the end interlock during the double seaming operation to form a strong mechanical structure. Each double seam consists of three thicknesses of the can end and two thicknesses of the can body with an appropriate sealing compound distributed throughout the folded metal to form a hermetic seal (Figure 4.7). In the first stage of the seaming operation, the curl of the end is interlocked (sometimes referred to as engaged) with the flange of the can body. The actual interlocking is performed with a roll that has a specially contoured groove. The first operation seam should not be too loose or too tight, since there is no way to correct a faulty first operation seam during the remaining seaming steps. A good quality first operation seam has the body hook approximately parallel to the cover hook, the edge of the flange of the body (which becomes the body hook) well tucked down into the cover hook radius, and the curl of the cover adjacent to, if not actually touching, the body wall of the can. The second stage (second roll) has a flatter groove profile than the first operation roll. The flatter profile is designed to press the preformed hooks together, to iron out wrinkles in the cover hook, to distribute the sealing compound in the seam, and, specifically, to develop the double seam tightness. The tightness or compactness of the finished double seam is a function of the adjustment of the second roll, which has to be adjusted to tighten the finished double seam. If the correct roll profiles are not used, or the rolls are excessively worn, the desired seam structure and tightness cannot be achieved. During the double seaming operation, considerable pressure is exerted on the can end, the can body and the sealing compound. The compound should be enclosed by the double seam. The compression by the seaming rolls will cause the sealing compound to flow and fill voids in the seam, thereby blocking potential leakage paths.

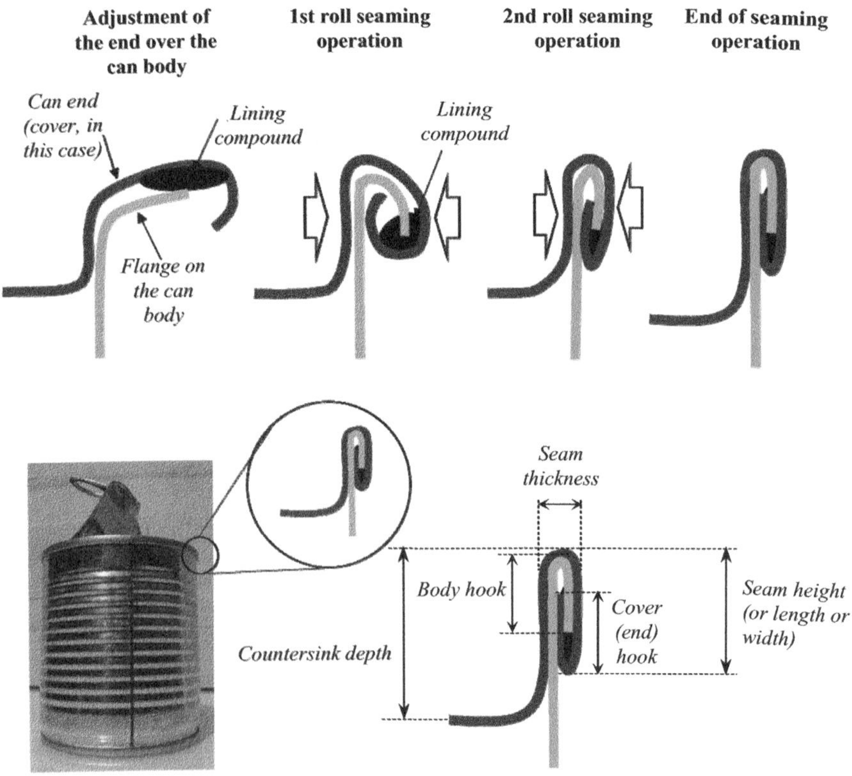

Figure 4.7. *Seaming process and seam characteristics (adapted from AFDO 2011; Robertson 2013)*

Seaming defects could occur and need to be well-controlled (AFDO 2011). Indeed, the quality of the seam is crucial for the preservation of the food and for the product and consumer safety. A smooth projection of a double seam below the bottom of a normal seam is identified as a droop. Vees are irregularities in the double seam due to insufficient and sometimes no overlap of the cover hook with the body hook, usually in small areas of the seam. The cover hook metal protrudes below the seam at the cover hook radius in a "V" shape. A sharp seam refers to a sharp edge at the top inside portion of the seam, at any point around it. The condition results from a portion of the cover being forced over the top of the seaming chuck lip during double seaming. The sharp seam is very often the first indication of a cut over. A cut over is a seam that is sharp enough to fracture the metal at the top inside portion of the seam. A jumped seam is a portion of a double seam that is adjacent to the juncture area of a soldered side seam can that is not rolled tight enough. It is caused by the seaming rolls jumping after passing over the lap. This situation does not occur with welded or two-piece cans. Seam bumps are most often found on the packer's end of two-piece cans and three-piece welded side seam cans. The defect is found in a relatively small area of the double seam, up to 32 millimeters long, where seam thickness suddenly increases from 0.08 millimeters or more. The increase in thickness is accompanied by a cover hook that is pulled away from the body wall and, when viewed in cross-section, a distorted body hook. A deadhead is an incomplete seam caused by the chuck spinning in the countersink of the end during the seaming operation. This seam defect is also known as a "spinner", "skidder" or "slip". A false seam is a seam, or portion of a seam, that is entirely unhooked and in which the folded cover hook is compressed against the folded body hook. The damaged end curls result when the end curl is flattened in one or more spots, causing the curl to fold back on itself. The can body buckling is a condition found directly under the finished seam where the can appears to be buckled or twisted. A cocked body occurs when the can body blank was out of square at the time of manufacturing, causing an unevenness at the lap or juncture. This is a can manufacturing defect that could result in double seam dimensions that vary excessively around the circumference of the can. A misassembly, sometimes referred to as a misplaced cover, is the result of the can body and end being improperly aligned in the closing machine. Therefore, the seam is completely disconnected partway around the can. A cut seam is a fractured double seam wherein the outer layer of the seam is fractured. This is a common problem with soldered side seams at the juncture area where the double seam crossed the side seam and there was excess solder at the side seam lap. Fractured embossed or debossed codes result when the metal of the end has been cut through at the code mark. Fractured embossed or debossed codes are often caused by misalignment.

4.5. Can surface treatments

Iron and steel cans were originally covered with tinplate to prevent oxidation. The steel boxes used today for light-colored fruit and fruit juices still use a tin coating without any additional organic coating. Indeed, the oxidation of tin delays the oxidative degradation of foods, such as fruit browning (Blunden and Wallace 2003). In boxes without an organic coating, the dissolution of tin (non-lacquered boxes) provides electrochemical protection to steel or iron (Coles and Kirwan 2011). However, most acidic foods and beverages will corrode the tin and, eventually, the iron, which will become accessible. Steel corrosion induces pitting of the metal and can also lead to the production of hydrogen, swelling and potential damage. In aluminum cans, a thin film of aluminum oxide (Al_2O_3) is formed when the aluminum is exposed to air or water. Although this coating does not flake, its solubility increases at low and high pH, and also at high concentrations of NaCl (Oldring and Nehring 2007). Once the box is sealed, the regeneration of the coating depends on the presence of oxygen in the box, which is generally limited. Without the oxide layer, aluminum corrosion occurs. Thus, without an organic coating, the shelf life of aluminum may be insufficient.

For steel and aluminum cans, an additional organic coating is necessary. The first organic can liners were made from porcelain wood oil and natural resins. Today, mainly synthetic resins (which started to be developed in the 1940s) are used, because they allow more flexibility for high-speed manufacturing (LaKind 2013).

The key function of the coating is to ensure that food or drink does not corrode the metal. Due to the consumer's expectation of a long shelf life, the coating must be strong enough to protect the integrity of the box during the manufacturing, filling and sterilization processes (Page 2010). Interactions between the food and the metal can affect the taste. Although organoleptic problems may seem secondary, unpleasant tastes can make the consumer reject the product (Huber *et al.* 2002). In Europe, the migration of components from packaging materials into contact with food should not cause organoleptic changes or unacceptable changes in the composition of the food. Coatings are also applied to improve the visual appearance of the box and the product (Lakind 2013). For example, the interactions between sulfur compounds in food (proteins, preservatives or pesticide residues) and metal can lead to the formation of spots of iron sulfide or tin sulfide, which induces rejection from the consumers (Robertson 2006).

Several types of can-coating resins are commercially available, including oleoresinous compounds (natural waxes, natural oil-based coatings derived from fossil gums) and synthetic resins (acrylic, epoxy, phenolic, polyester and vinyl resins). Each resin can be produced from several raw materials (natural substances or monomers) and the final polymer is generally mixed to obtain the desired attributes adapted to each can and lacquer combination. Resin mixtures are identified by a double chemical family name (e.g. epoxy-amino), the second resin having a concentration of less than 50% (LaKind 2013). Other agents are often added to create the desired properties (e.g. coatings with a dispersion of TiO_2 are white); release agents are added to coatings to improve the removal of meat and fish from boxes (Robertson 2013). The different coatings authorized and used for lacquering cans, as well as their functional properties, are given in Table 4.3. The safety of these coatings is an important aspect to take into consideration, and is highly regulated.

Lacquers	Epoxy	Acrylic	Phenolic	Vinyl organosol	Polyester laminates	Polyester coatings	Oleoresins
Applications	Most widely used coating. Universal golden coating for three-piece and shallow-drawn cans	Universal lacquer for beer and beverage cans. Side seam stripe in high solids form welded cans	Drums and pails	Drawn cans, Easy-open ends. Often used over an epoxy-phenolic base coat	Shallow drawn cans. Easy-open and standard ends	Internal and external coating for two-piece and three-piece cans and ends for meat, fish and vegetables	Initially used for all types of cans, but now only very limited use
Properties	High-MW epoxy resins cross-linked with phenolic resins or anhydride hardeners. Very good chemical resistance and fabricability	High-MW epoxy resins cross-linked with amino or acrylate resins. Waterborne coatings	Low cost, poor flexibility, excellent resistance to aggressive foods	PVC dispersed in an appropriate solvent and stabilized with low-MW epoxy resin	Extrusion coated or laminated film of polypropylene, polyester or polyamide. Good fabricability; superior corrosion resistance	Polyester resins cross-linked with phenolic or amino resins. Very good chemical resistance. Good fabricability	Synthetic modification of natural oils and fatty acids, golden-colored, inexpensive coating
Corrosion resistance	☺	😐	☺	☹	😐	☹	😐
Ease of fabrication	😐	☹	☹	☺	☺	☺	☺
Ease of application	☺	☺	😐	☹	☺	☺	☹

Properties	Epoxy	Acrylic	Phenolic	Vinyl organosol	Polyester laminates	Polyester coatings	Oleoresins
Universality of food applications	☺	☹	☹	😐	😐	☹	☹
Organoleptic properties and appearance	☺	😐	😐	☺	☺	😐	☹

Table 4.4. *Properties and composition of can lacquers (adapted from LaKind 2013; Robertson 2013)*

The classes and characteristics of the resins summarized in Table 4.4 are described as follows:

Acrylics: these resins are not intended for applications where they come into contact with food because they can impart undesirable tastes due to the possible migration of one of their components, ethyl acrylate. Acrylic varnishes are too brittle to resist the manufacture of boxes, in particular, the process for boxes drawn by stamping. Thus, acrylic resins are mainly used for the external coating of boxes (Robertson 2016).

Epoxy: the use of epoxy coatings in the manufacture of cans began in the 1950s. These resins (bisphenol A diglycidyl ethers) are produced by condensation. Compared to the oleoresins that preceded the use of epoxies, the epoxy resins provided significantly greater flexibility. Currently, epoxy-based resins account for 95% of the coating of aluminum and steel boxes. Only a very small percentage of epoxy resins do not use BPA as the starting monomer. Its flexibility, good adhesion to metal surfaces, chemical resistance and compatibility with most foods and drinks explain its success compared to other resins (Lakind 2013). Despite the excellent mechanical properties of epoxy resins (Robertson 2013), they resist the drawing and redrawing processes very poorly. Epoxy resins do not impart flavor to food and retain their appearance.

Oleoresins: oleoresins are derived from the fusion of gums and natural rosins mixed with drying oils. Until 1965, oleoresins were the only coatings used in cans, before being replaced by epoxy resins (Oldring and Nehring 2007). Oleoresins are used primarily for fruit drinks, fruits and vegetables (Robertson 2006). Oleoresins are less resistant to corrosion, but can withstand the various methods of can manufacturing despite poor adhesion to the metal and a long hardening time. Due to its low resistance to corrosion, the use of oleoresin is limited to non-aggressive foods (e.g. dry beans), but it can induce color and taste changes when it is in the presence of sulfur compounds (Robertson 2013). The bis-phenol A crisis has renewed interest in these resins in recent years.

Phenolic: phenolic resins are obtained by condensation of one or more phenols with one or more aldehydes. They are very resistant to corrosion, but have limited uses due to their low deformability, which limits their suitability for high production rates (Oldring and Nehring 2007). Instead, phenolic compounds are used as a cross-linking agent with other resins to improve the corrosion resistance of very aggressive foods. The percentage of incorporation in mixtures must remain low because phenolic compounds impart flavor and odor to certain foods (LaKind 2013).

Polyester: polyester resins are produced by the condensation of an acid with one or more alcohols or epoxides, followed by copolymerization with one or more cross-linking agents (Robertson 2006). Polyester resins fail with aggressive or acidic foods (Oldring and Nehring 2007) due to the hydrolytic attack of the ester bond at an acidic pH. This severely limits their use, although the flexibility range is the widest and adapts to all manufacturing processes. Polyester resins do not impart taste or smell to food and beverages, but can absorb aromas or small molecular weight compounds. A polyester multilayer system (PET) is only widely used in Japan (Hanlon *et al.* 1998). These laminate coatings are compatible with all processes that do not require welding (three-piece boxes). The laminates are easily applied and have a low capacity for absorbing aromas compared to other polyesters.

Vinyl: vinyl resins are generally used in mixture with alkyd, epoxy and phenolic resins because they do not resist corrosion, but are flexible. They are generally applied to an underlying epoxy coating because they do not adhere well to the metal substrate. Even though they do not give taste to food or drink, they are little used, because they do not resist thermal sterilization treatments (Robertson 2013). In the form of an organosol (dispersion of PVC in a solvent and mixed with an epoxy resin), vinyl resins allow improved chemical resistance, thermal stability and adhesion (Robertson 2006).

4.6. Food–can interactions

European Regulation 1935/2004 (or "framework regulation") of the European Parliament and of the Council of 27 October 2004 defines the general requirements which apply to materials and articles intended to come into contact directly or indirectly with foodstuffs, food products and beverages placed on the community market, to ensure a high level of consumer protection. This regulation obviously applies to metallic packaging.

In the absence of a harmonized regulation applicable to metals and alloys in the EU and subject to respecting the principle of mutual recognition, certain metals and alloys are covered by the regulations of each country (Table 4.5).

Uncoated stainless steels and articles made exclusively of stainless steel must comply with the criteria for chemical composition, but criteria are not provided in terms of release limits. Other steels, uncoated aluminum and aluminum alloys, tin and tinplate and objects made of these metal alloys must only comply with the chemical composition, but criteria are also not provided in terms of release limits. The end products concerned are packaging (tins for canned food, beverage cans, capsules, aerosols, covers, thin foil, cups, trays, tubes, staples) or reusable items (household items, household appliances, industrial equipment in the food industry). When organic coatings are applied to aluminum, steels or their alloys, the recommendations must comply with Article 3 of EU Regulation 1935/2004 (European Commission 2004), modified for bis-phenol A (BPA) in epoxy resins in 1895/2005 (European Commission 2005) and 2018/213 (European Commission 2018) and for plastics (polyester coatings) in 10/2011 (European Commission 2011), fully described in Chapter 5, which is dedicated to plastics.

Country	Regulation	Substitutive recommendations
Belgium		Practical guide of COE on metals
Finland	KTM decision no. 268/92	
France	Decrees of 28/06/1912, 15/11/1945, 13/01/1976 and 27/08/1987	DGCCRF guidance website
Greece	Decision no. 232/98	
Hungary	Regulation no. 17/1999 VI.16	
Italy	Ministerial decrees (DM) of 21/03/1973, 18/02/1984, 01/06/1988, 13/07/1995, 18/04/2000 and 21/12/2010	
The Netherlands	Packaging and consumer products edition 40-07/2014 Chapter 4	
Sweden	Code of 09/06/2003 Appendix 3	

Table 4.5. *National regulations regarding metals for packaging (adapted from Fitness Erasmus project 2020)*

4.7. References

A3M (Alliance des minerais, minéraux et métaux) (2020). [Online]. Available: http://www.a3m-asso.fr/wp-content/uploads/2017/02/comment-fabrique-acier.pdf.

AFDO (Association of Food and Drug Officials) (2011). A guide to can defects and basic components of double seam containers. Guide, AFDO, York, PA.

Blunden, S. and Wallace, T. (2003). Tin in canned food: A review and understanding of occurrence and effect. *Food Chem. Toxicol.*, 41, 1651–1662.

Boite Aluminium (2020). Les boites métal durable [Online]. Available: https://www.boitealuminium.com/boite-en-fer-blanc/ [Accessed 1 July 2020].

Chemseal Inc. (2020). The benefits of stainless-steel containers to the food industry [Online]. Available: https://www.chemsealinc.net/blog/the-benefits-of-stainless-steel-containers-to-the-food-industry/ [Accessed 6 July 2020].

Coles, R. and Kirwan, M.J. (2011). *Food and Beverage Packaging Technology*, 2nd edition, John Wiley & Sons Ltd, Hoboken, NJ.

Crown Inc. (2020). All about cans: Double seaming is simply made [Online]. Available: https://www.crowncork.com/news/all-about-cans/double-seaming-made-simple-part-1 [Accessed 1 July 2020].

Emblem, A. and Emblem, H. (2012). *Packaging Technology, Fundamentals, Materials and Processes*. Woodhead Publishing Limited, Oxford.

European Commission (2004). Materials and articles intended to come into contact with food and repealing Directives 80/590/EEC and 89/109/EEC. Regulation 1935/2004 [Online]. Available: https://eur-lex.europa.eu/legal-content/EN/ALL/?uri=CELEX%3A32004R1935.

European Commission (2005). The restriction of use of certain epoxy derivatives in materials and articles intended to come into contact with food. Commission Regulation 1895/2005 [Online]. Available: https://eur-lex.europa.eu/legal-content/EN/ALL/?uri=celex:32005R1895.

European Commission (2011). Plastic materials and articles intended to come into contact with food. Commission Regulation 10/2011 [Online]. Available: https://eur-lex.europa.eu/legal-content/EN/ALL/?uri=CELEX%3A32011R0010.

European Commission (2018). The use of bisphenol A in varnishes and coatings intended to come into contact with food and amending Regulation (EU) No 10/2011 as regards the use of that substance in plastic food contact materials. Commission Regulation 2018/213 [Online]. Available: https://eur-lex.europa.eu/eli/reg/2018/213/oj.

Fitness Erasmus Project (2020). Food packaging open courseware for higher education and staff of companies [Online]. Available: https://fitness.agroparistech.fr/fitness/lectures/.

Hanlon, J., Kelsey, R., and Forcinio, H. (1998). *Handbook of Package Engineering*, 3rd edition. Technomic Publishing Company Inc., Lancaster, PA.

Huber, M., Ruiz, J., and Chastellain, F. (2002). Off-flavour release from packaging materials and its prevention: A foods company's approach. *Food Additives & Contaminants*, 19, 221–228.

International Standard Organization (2016a). ISO 11949/2016 Cold-reduced tinmill products – Electrolytic tinplate [Online]. Available: https://www.iso.org/standard/.

International Standard Organization (2016b). ISO 11950/2016. Requirements for single and double cold-reduced electrolytic chromium/chromium oxide-coated steel (ECCS) in the form of sheets or coils [Online]. Available: https://www.iso.org/standard/.

Kleniewski, A. (1995). Emballages métalliques. *Les Techniques de l'Ingénieur*, A9760, 1–24 [Online]. Available: https://www.techniques-ingenieur.fr/base-documentaire/genie-industriel-th6/conception-d-emballage-42133210/emballages-metalliques-a9760/.

LaKind, J.S. (2013). Can coatings for foods and beverages: Issues and options. *Int. J. Technol. Policy Manag.*, 13, 80–95.

Metal Packaging Europe (2020). The best of metal [Online]. Available: https://www.metalpackagingeurope.org/metal-packaging [Accessed 1 July 2020].

MPMA (Metal Packaging Manufacturer Association) (2020). How cans are made [Online]. Available: https://www.mpma.org.uk/information/how-cans-are-made/ [Accessed 1 July 2020].

Oldring, P.K.T. and Nehring, U. (2007). Packaging materials – 7. Metal packaging for foodstuffs. *ILSI Europe Report Series*, 1–44.

Page, B. (2012) Rigid metal packaging. In *Packaging Technology: Fundamentals, Materials and Processes*. Emblem, A. and Emblem, H. (eds). Woodhead Publishing Limited, Oxford.

Parisi, S. (2013). *Food Industry and Packaging Materials: Performance-oriented Guidelines for Users*. Smithers Group Company Shawbury, Shrewsbury.

Pelletier, Y. (2000). Revêtements intérieurs pour emballages métalliques. *Les Techniques de l'Ingénieur*, F1310, 1 – 12 [Online]. Available: https://www.techniques-ingenieur.fr/.

Reboul, M.C. and Baroux, B. (2011). Metallurgical aspects of corrosion resistance of aluminium alloys. *Materials and Corrosion*, 62, 215–233.

Robertson, G.L. (2006). *Food Packaging: Principles and Practice*, 2nd edition. CRC Press, Boca Raton, FL.

Robertson, G.L. (2013) *Food Packaging – Principles and Practice*, 3rd edition. CRC Press, Boca Raton, FL.

Selke, S. (1994). *Packaging and the Environment: Alternatives, Trends and Solutions*. CRC Press, Boca-Raton, FL.

Sharretts Plating (2020). Surface treatments [Online]. Available: https://www.sharrettsplating.com/blog/the-tin-plating-process-a-step-by-step-guide/ [Accessed 1 July 2020].

Technavio (2020). Global Wood Packaging Market 2019–2023 [Online]. Available: https://www.technavio.com/report/global-metal-packaging-market-analysis-share-2018?tnplus [Accessed 8 June 2020].

The Aluminum Association (2020). The aluminum advantage [Online]. Available: https://www.aluminum.org/product-markets/aluminum-cans [Accessed 4 July 2020].

Thorne, S. (1986). *The History of Food Preservation*. Parthenon Publishing Group, London.

5

Plastics

Mario Ščetar

Faculty of Food Technology and Biotechnology, University of Zagreb, Croatia

5.1. Introduction

Due to ease of processing, low costs and available raw materials for their production, plastics are considered to be one of the most commonly used packaging materials. Plastics are widely used in almost all industrial segments, such as construction and civil engineering, automotive, electronics, aerospace, the prevention and control of corrosion and finally, the packaging industry. Plastics are materials with extensive possibilities when it comes to their properties; thus, in this chapter, the focus will be on their use as packaging materials in different areas such as food, pharmaceuticals and cosmetics.

The raw material used for the production of plastics is oil, which has been treated by refining processes. Therefore, it is important to emphasize that the price of plastic as a finalized material is related to the market price of oil.

According to Grand View Research (2020), the global plastics market was estimated to be worth $568.9 billion in 2019, and it is expected to grow with a compound annual growth rate (CAGR) of 3.2% from 2020 to 2027. In terms of revenue, polyethylene (PE) dominated the market with a share of 25.7% in 2019. This high share is attributed to the growing demand for packaging, which includes plastic foils and bags, as well as containers and bottles.

For a color version of all figures in this book, see www.iste.co.uk/debeaufort/packaging.zip.

Packaging Materials and Processing for Food, Pharmaceuticals and Cosmetics, coordinated by Frédéric DEBEAUFORT, Kata GALIĆ, Mia KUREK, Nasreddine BENBETTAIEB and Mario ŠČETAR.

5.2. Plastic materials and processes

Plastic polymers are synthetic substances and materials made of macromolecules (polymer molecules). Synthetic polymers are composed of the most common structural units – monomers. Plastics are formed by chemical reaction – polymerization (initiated by heat, pressure and catalysts) that involves joining several small molecules together into a chain, in order to form a large molecule. Molecules are then merely arranged in succession. The product retains the same composition as the starting materials, and no secondary products are eliminated. If the starting material consists of identical molecules, the product is known as a homopolymer, while if it consists of different molecules, it is known as a copolymer.

Plastics are divided into two basic types: thermosets and thermoplastics. Thermoset plastics are characterized as materials that, during a curing process (heating and shaping), sustain a permanent change (cross-linking) to their molecular structure. This means that once cured, they cannot be re-melted. Thermoplastics are materials that soften on heating but their chemical structure remains unchanged when either heated or shaped. Thermoplastic materials (e.g. polyolefins) are commonly used for food packaging manufacturing (Vujković *et al.* 2007).

Commercial thermoplastics are classified according to their performance as a "commodity" (low performance, like monofilms: polyethylene (PE), polypropylene (PP), polystyrene (PS) and polyvinyl chloride (PVC)). Others are classified as "engineering" materials (polycarbonate (PC), polyamide (PA) and polyethylene terephthalate (PET)), or "advanced" materials (highest performance, like liquid crystal polymers (LCP), polyphenylene sulfide (PPS) and polyether ether ketone (PEEK)). Commercial laminates, which may contain up to as many as eight different layers, are designed for a particular product. Laminations of different materials may be formed by various processes (bonding with a wet adhesive, dry bonding of layers with a thermoplastic adhesive, hot melt laminating where one or both layers exhibit thermoplastic properties, as well as special extrusion techniques). Such structured plastic films may be further bonded to papers or metal foils to produce more complex laminates (see Chapter 6).

Increased barrier properties are obtained via incorporation of high barrier layers such as ethylene vinyl alcohol (EVOH), polyvinyl alcohol (PVAL), polyvinylidene chloride (PVDC) and polyamide (PA).

The polymeric material is defined as a polymer with added low molecular weight compounds (additives) in order to ease processing (Hahladakis *et al.* 2018).

Depending on the desired application, additives often vary. Examples of additives used are given in Table 5.1.

Most widely used production processes for plastic are continuous and offer high performance in all stages of production operations. Two well-known processes include: extrusion and calendering. Different processes correspond to the development or modification of certain procedures with the introduction of additional techniques, or by improving a stage of the original process. Which procedure will be used essentially depends on the materials to be produced.

Additives	Application
Antistatics	Prevent electrostatic charging of the packaging
Colorants	Soluble/insoluble (pigments), inorganic/organic, used to color plastics
Flame retardants	Modify the combustion behavior of plastics
Fillers (glass fiber, chalk, graphite, carbon black)	Extend the plastics and reduce their cost, improve the qualities of the plastic
Lubricants	Facilitate the processing of plastics
Plasticizers	Modify certain properties of a plastic: increase resilience, improve flexibility, durability and stretchability of polymeric films
Stabilizers	Protect against atmospheric oxidation and UV radiation

Table 5.1. *Plastics additives with processing properties and application (Vujković* et al. *2007)*

Extrusion is the best-known process for the production of plastic (see Figure 5.1).

Thermoplastic pellets drop from the hopper onto the turning screw. Plastic pellets melt as the turning screw pushes them forward by the heaters. Then, molten polymer is forced through the shaping die, under pressure, to form the final product (extrudate), as shown in Figure 5.1. This process is used in the production of films, coatings and tubes (Parisi 2013).

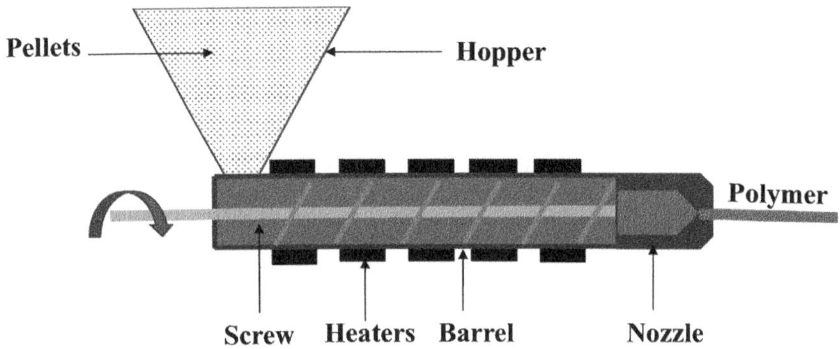

Figure 5.1. *Extrusion of plastics*

Calendering is a continuous process of producing infinitely long films, according to the principle of passing the softened polymeric material between roller systems with adjustable spacing. Feedstock is passed through a series of rolls to reduce thickness to the desired gage. Calendering requires expensive equipment but results in high production rates. The process is important to achieve a good surface finish and high gage accuracy.

A typical calendering process consists of five main components: plasticizing unit, calender, cooling unit, accumulator and windup station. An example of a three-roll calendering system is shown in Figure 5.2. Typical materials produced by calendering processes are rubber or rubbery thermoplastics, such as plasticized polyvinyl chloride (PVC), acrylonitrile butadiene styrene (ABS) and polystyrene (PS), although also some grades of polyethylene (PE) and polypropylene (PP) may also be used.

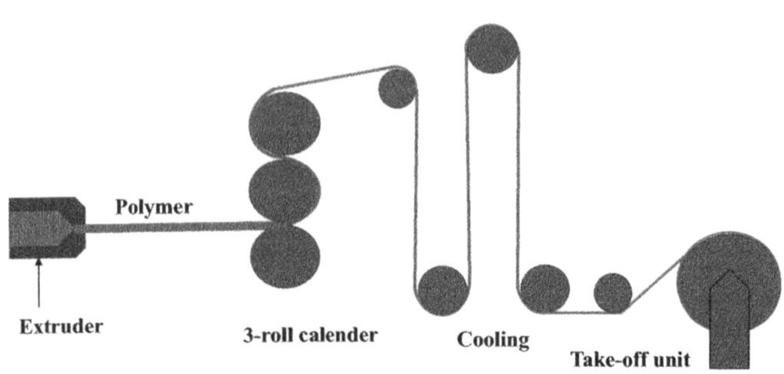

Figure 5.2. *Calender with three-roll system*

Extrusion and calendering are processes used to produce film-type materials, while for rigid forms, the most well-known forming processes used are compression molding, injection molding, blow molding and thermoforming.

In the molding process, polymers are shaped and forced to enter a dedicated mold with a certain shape. The newly shaped forms of a particular material are successively cooled and therefore separated from the connected molds, which are then separated after the process is completed.

Compression molding is a process in which a casting (thermoplastics or thermosets) polymer and additives are placed in a preheated mold cavity (lower movable mold half). Also heated, the upper movable mold half closes the lower cavity by applying pressure so the fluid polymer assumes the shape of a mold, as shown in Figure 5.3. In the end, a sample is ejected by the ejector pin.

Materials commonly processed by compression molding are (Kopeliovich 2018):

- epoxies (EP);
- urea-formaldehyde (UF);
- melamine formaldehyde (MF);
- phenolics (PF).

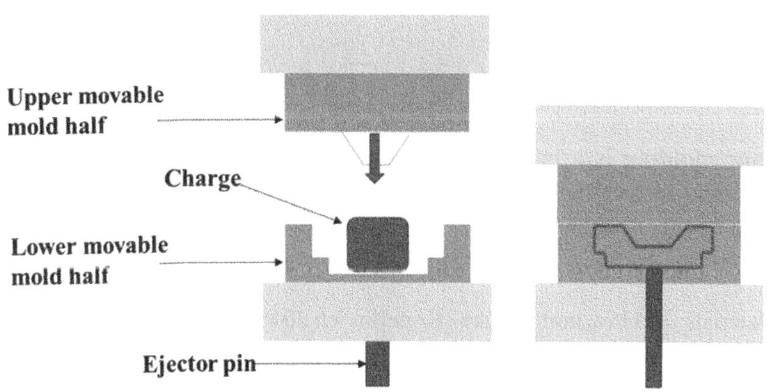

Figure 5.3. *Compression molding*

Injection molding is carried out to create different shapes. During this process, the ram retracts and plastic pellets (thermoplastics and some thermosets) drop from the hopper into the barrel. The ram forces the plastic into the heating chamber (around the spreader) where the plastic melts. As it moves forward, molten plastic is forced under pressure (injected) into the mold (die) cavity where it assumes the shape of the mold (Figure 5.4).

Materials commonly used in injection molding are (Kopeliovich 2018):

- polypropylene (PP);
- polycarbonate (PC);
- acrylonitrile butadiene styrene (ABS);
- nylon 6 (N6) (polyamide).

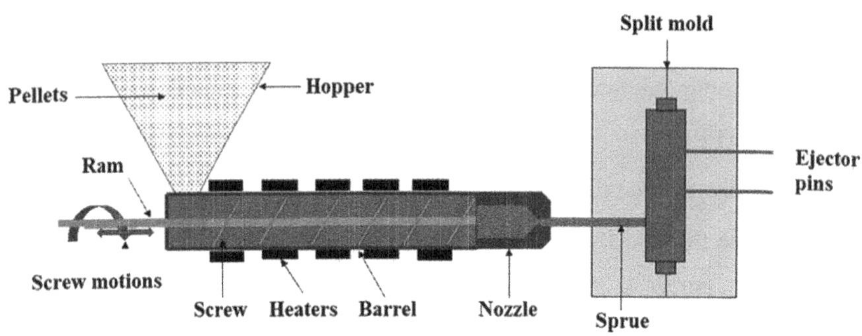

Figure 5.4. *Injection molding*

Blow molding is a process in which air is blown into the closed mold of a certain mold shape, such as a heated hollow thermoplastic tube, as is shown in Figure 5.5. This is accomplished in two steps: (1) fabricate a starting tube of molten polymer (parison) and (2) inflate the tube to a desired final shape.

Materials for blow molding are (Kopeliovich 2018):

- low-density polyethylene (LDPE) and high-density polyethylene (HDPE);
- polypropylene (PP);
- polyvinyl chloride (PVC);
- polyethylene terephthalate (PET).

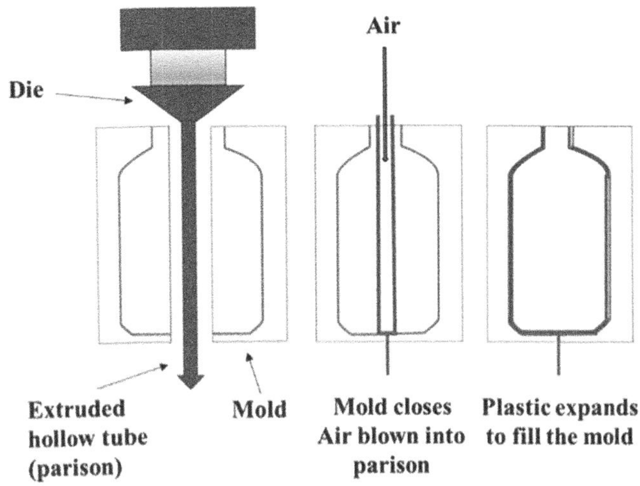

Figure 5.5. *Blow molding*

Thermoforming is a process of shaping a flat thermoplastic sheet or film. Using a mold, the sample is heated and formed into the desired shape. Heating is usually accomplished by radiant electric heaters, located on one or both sides of a starting plastic sheet or film. A vacuum is used in order to assure adequate alignment of a sample to mold (Figure 5.6).

Materials which may be processed by the thermoforming method are (Kopeliovich 2018):

- polypropylene (PP);
- polystyrene (PS);
- polyvinyl chloride (PVC);
- low-density polyethylene (LDPE);
- high-density polyethylene (HDPE).

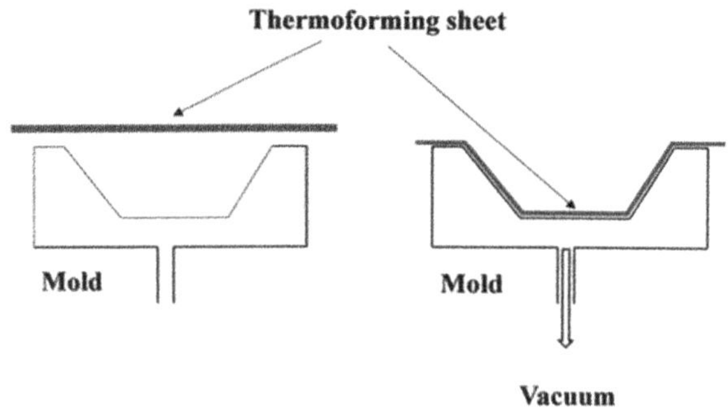

Figure 5.6. *Vacuum thermoforming*

5.3. Plastic films for packaging

5.3.1. *Plastic classes*

Plastic films for packaging need to be flexible and stretchable, lightweight, resistant to breakage, have low-temperature formability yet strong heat sealability, and must be versatile in their barrier properties to oxygen, moisture and light.

By classification, plastic films can be unoriented, cast (uniaxially oriented) or biaxially oriented (Table 5.2), and are defined as sheet materials with thickness values of less than 0.254 mm. Thicker films are referred to as sheets (Ebnesajjad 2013).

Film	Properties	Usage
Unoriented	Casting or blown film processes, easily heat sealed, good heat stability, low flexural moduli, excellent puncture resistance, excellent impact strength, low moisture permeability, poor barriers to gases (oxygen and carbon dioxide)	Polyethylene (PE) Polypropylene (PP)
Cast (uniaxially oriented)	Oriented in the machine direction, with different surface properties on each side, good moisture and barrier properties, scuff resistance, clear and glossy, high tensile strength	Polypropylene (PP)

Biaxially oriented	Stretched in both machine and transverse directions, glossy, transparent, increased toughness, increased stiffness, enhanced clarity, and enhanced barrier properties to water vapor and oxygen	Polypropylene (PP) Heat-shrinkable films in shrink wrap applications Polyethylene terephthalate (PET) Polyamide (PA) Polystyrene (PS) Packaging films, tapes, labels, industrial films

Table 5.2. *Classes of plastic films used for packaging (Ebnesajjad 2013)*

The process of orienting or stretching the film under carefully controlled temperatures can improve certain properties. It causes a change in the orientation of the molecules and results in a much firmer film, with increased strength and barrier properties. The film itself can be stretched in one or two directions with a compaction frame – a device that engages the edges of the weave in clamps that move outward to stretch the sheet in the direction of the cross machine. Stretching in the direction of the machine is done by changing the tension of the film windings. For ambiguous orientation, the film is slightly stretched in the transverse direction (Figure 5.7). Although the strength and barrier properties have been improved by orientation, the materials treated in this way are still not completely impermeable to oxygen, light and moisture.

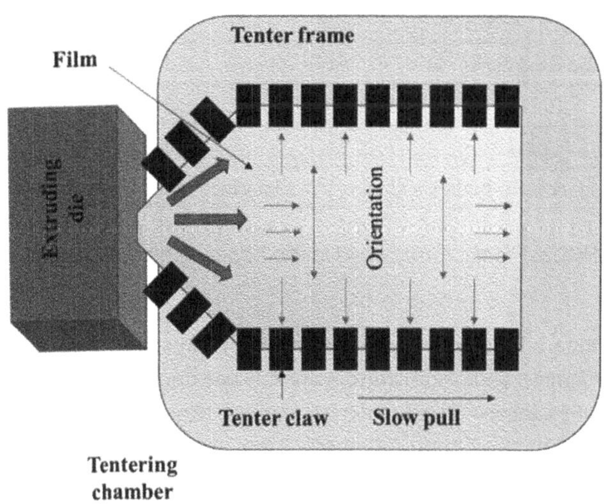

Figure 5.7. *Biaxial orientation of polymeric film*

5.3.2. *Plastic packaging shapes*

Different shapes (such as bags, sachets, pouches and overwraps) can be made from plastic films (Table 5.3). Plastic bags are formed via a number of operations: folding, cutting and sealing. Pouches may be formed either from one or two reels, by folding the film vertically or horizontally using the "Form, Fill, Seal" (FFS) packing machine. With the wrapping, the packages go through a heated tunnel. If a transverse seal is made then there are no end seals. The film shrinks over the ends of the package or product, depending on the width of the film used.

The most common ways of using film wrapping (see also Chapter 11) are for:

– shrink wrap and cling films;

– stretch wrap;

– skin packaging (heating the film and creating a vacuum);

– blister packaging (thermoforming with a vacuum or with pressure).

Plastic	Uses
Rigid containers	Bottles, jars, pots, tubs and trays
Flexible films	Bags, sachets, pouches and heat-sealable, flexible lidding materials, tamper-evident
Plastic combined with paperboard	Liquid (milk, juices, etc.) packaging lamination
Expanded or foamed	Insulation rigidity
Lids and caps	Closures
Bands	External tamper-evidence
Films	Stretch and shrink wraps
Films used as labels for bottles and jars	Flat glued labels or heat-shrinkable sleeves

Table 5.3. *Plastic forms used in food packaging (Coles* et al. *2003)*

Flexible materials include wraps, pouches or bags with a thickness equal to or less than 0.127 mm, while semi-rigid materials are thicker than 0.127 mm. They are usually formed as sheets. Rigid materials include trays, bottles, boxes or crates, and are composed of thick or rigid polymers. They are used for beverages, pasteurized products, pickles, instant coffee, sauces, powdered milk and creams (Ebnesajjad 2013).

Material	Properties	Usage
Polyethylene (PE)	Most common and least expensive, used as a layer for heat sealability	Wide range of applications depending on the particular type
Low-density polyethylene (PE-LD or LDPE)	Heat sealable, inert, odor-free and shrinks when heated. It is a good moisture barrier, but has a relatively high gas permeability, it is sensitive to oils, and has poor odor resistance. It is less expensive than most films	Food wrapping, stretch film, containers (milk, oil, grains, spices, salt), lids for food, squeeze bottles
High-density polyethylene (PE-HD or HDPE)	Stronger, thicker, less flexible and more brittle than low-density polythene (LDPE). It has lower permeability to gases and moisture, and a higher softening temperature (121°C), so can therefore be sterilized with heat	Butter containers, milk and fruit juice bottles, jugs
Polypropylene (PP)	Virtually unbreakable and chemically resistant, with good heat resistant qualities. It has good resistance to oil	Oily and dairy products, bottle caps and closures, preserves the flavors of dried herbs and spices
Polyethylene terephthalate (PET)	Glass-like clarity, excellent gas-barrier properties	Jars (jelly, jam, peanut butter), beverage bottles, microwavable food trays, pharmaceutical products
Polystyrene (PS) and oriented polystyrene (OPS)	Provides a crystal appearance and the stiffness of glass in a plastic container. Good oxygen barrier. **OPS** film: clarity, rigidity, low cost, thinness and high gas permeability	Suitable for packaging fresh produce, shrink wrap and trays (meat), disposable cups and plates, and containers for confectionaries
Polyamide (PA)	High gas and water vapor barrier properties, and greater strength and stiffness	Part of laminated films, used in lidding and pouches (meat and meat products)
Polyvinyl chloride (PVC)	Offers clarity, durability and chemical resistance	Suitable for transporting and dispensing containers (drinking water, cooking oil, blister packs, chemicals and personal care items)
Polyvinylidene chloride (PVDC)	Excellent gas and moisture barrier properties, and heat sealability	Household wraps (Saran wrap) and coated films (food, drugs, cosmetics)

Table 5.4. *Plastic materials properties and applications*

Rigid packaging is defined as being "unable to bend or be forced out of shape" or "simply not flexible" (EPAC 2020). Rigid packaging offers structure, stability and support for products and is known for its strength. Examples of rigid packaging materials include plastic bottles and boxes. The most common plastics dominating the packaging market are listed in Table 5.4.

The global rigid plastic packaging market (Allied Market Research 2019) is projected to reach $340,434.7 million by 2026, growing at a CAGR of 5.6% from 2019 to 2026. Rigid plastic packaging involves the utilization of plastic materials for packaging purposes. Plastic packaging is used in end-user industries such as food and beverages, personal care, household items and healthcare items.

Bottles are made via extrusion blow molding (Figure 5.5). Blow molding is usually used for milk bottles (made from HDPE). First, a preform is stretched and blow molded. This results in biaxial orientation of the polymer molecules, which leads to increased strength, clarity, gloss and gas barrier properties. For example, to make PET bottles for carbonated beverages, injection stretch blow molding is used.

Screw caps and pressure fit closures with accurate profiles are made via injection molding. Packaging on in-line thermoform, fill and seal machines can be combined with thermoforming (Figure 5.6). Thermoforming is used to make trays from extruded foam sheets. Insulated boxes are produced by injection molding, for example, for the distribution of fresh fish. The rotary casting process is used to make plastic containers for bulk material in the food industry for the distribution of ingredients. Plastics such as HDPE and LDPE are used in this process (Coles *et al.* 2003).

Rigid and semi-rigid thermoformed containers are used for packaging different products. Examples include a range of dairy products, single-serving containers, packages for fresh sandwiches, chocolate and confectionery products, etc.

5.4. Properties of plastic packaging

Due to many favorable properties, as well as the low price, plastic packaging is dominating the global market. Depending on plastic usage, different general or specific characteristics are required.

5.4.1. *General properties of plastic*

Optical properties, such as gloss, affect the appearance of the packaging. Each package should be stable and attractive, and highly transparent. This property often defines consumers' preferences in choosing one packed item over another. In general, amorphous polymers are transparent and this transparency is greater when the degree of crystallinity is lower.

Tensile strength and elongation are significant properties for plastic packaging, and are especially important during operations at the fill and seal packaging machines.

Films should be able to withstand the tension of passing through the molding machine without breaking or stretching, while the formed shapes should withstand the force it takes to put the weight of the product in place, without resulting in deformation.

Impact strength is a measure that describes a material's resistance to breakage in the event of a sudden impact that may occur during transport or handling, such as if the product falls.

Tear strength provides information regarding tear resistance during molding on machines and subsequent handling.

Surface friction affects the ease with which packaging materials will go through the whole system of withdrawal to the line of packaging. It also has an impact on the stability of palletized cargo during the transport of primary and secondary packaging. Surface friction can be modified by the addition of additives during the film production phase.

The **density** of the plastic material can change, for example, due to changes in crystallinity, solvent absorption or loss of plasticizer, and varies depending on the temperature.

Viscoelasticity is the main parameter in deciding which plastic to use for a specific product. Plastic materials have a certain degree of permanent deformation which is critical for storage of product-filled containers that are stacked on pallets.

Heat sealability is one of the essential properties in flexible packaging such as in bags, sachets and blister packs, as well as for semi-rigid forms such as those that are thermoformed. The temperature at which the seal softens and begins to flow must be compatible with what can be achieved by a combination of temperature, pressure and time spent on the jaws of the packaging machine, in order to achieve hermetically sealed packages (Emblem and Emblem 2012).

5.4.2. *Barrier properties of plastic*

Barrier properties are determined by the rate of mass transport through the plastic.

Most barrier requirements of packaging materials are related to oxygen, carbon dioxide, moisture, grease/oil, aroma and UV light.

Permeability to gases is expressed in terms of the volume of gas or vapor that can permeate per unit area, per unit time or per unit thickness at a specified temperature and relative humidity as in equation [5.1]:

$$P = D \cdot S = \frac{Q \cdot X}{A \cdot t\,(p_1 - p_2)} \qquad [5.1]$$

where:

P is the permeability coefficient ($cm^3\ m^{-1}\ d^{-1}\ bar^{-1}$);

D is the diffusion coefficient ($cm^2\ s^{-1}$);

S is the solubility coefficient ($cm^3\ cm^{-3}\ Pa^{-1}$, at 273.15 K and $1.013\ 10^5$ Pa);

Q is the quantity (mL or cm^3) of permeant transferred by a unit of area, *A* (m^2), in a time *t* (d^{-1});

X is the thickness of material (m);

$\Delta p = (p_1 - p_2)$ is the partial pressure difference (Pa).

The rate of permeation at which a gas or vapor passes through a polymer consists of three processes: absorption of the permeating species into the polymer, diffusion through the polymer, desorption of the permeating species from the polymer surface, and removal. Test methods for gas permeation measurements can be divided into three categories: (1) total pressure or volume increase methods; (2) isostatic steady-state methods; and (3) quasi-isostatic methods. The most commonly used are isostatic and steady-state methods (Abdellatief and Welt 2012).

Factors that affect the permeability of a polymer are the solubility (*S*) and diffusivity (*D*) of the small molecule in the polymer, such as chain packing and side group complexity, polarity, crystallinity, orientation, fillers, humidity and plasticization. Diffusion is the rate of movement of a molecule through the polymer. The penetrant dissolves in the film matrix at the high concentration side; it diffuses through the material driven by the concentration gradient and evaporates from the other surfaces. The diffusion depends on size, shape and polarity of the penetrant and the polymer-chain segmental motion within the film matrix. The latter is influenced by interchain attractive forces like hydrogen bonding and van der Waals' interactions, the degree of cross-linking and crystallinity.

Diffusion is driven by a concentration gradient between the two sides of the polymer film. It can be described by Fick's law in equation [5.2]:

$$J = -D\frac{\partial c}{\partial x} \qquad [5.2]$$

where:

J is the flux through the film (mol m^{-2} s^{-1});

D is the diffusion coefficient (m^2 s^{-1});

C is the concentration of the permeant (mol m^{-3});

$\frac{\partial c}{\partial x}$ is the concentration gradient of the permeant across a thickness;

X is the thickness of the film (m).

The permeability coefficient is defined as equation [5.3]:

$$P = \frac{Q \cdot X}{A \cdot t\,(p_1 - p_2)} \qquad [5.3]$$

where units of $P = \mathrm{cm^3 \cdot cm \cdot cm^{-2} \cdot s^{-1} \cdot Pa^{-1}}$

or by equation [5.4]:

$$\frac{Q}{t} = \frac{P}{X}A\,(\Delta p) \qquad [5.4]$$

The term P/X is called the permeability or permeance.

Besides the oxygen permeability coefficient, the water vapor permeability coefficient (WVP) is another critical parameter for food packaging materials. In general, water preferentially permeates through the hydrophilic portion of the film. Thus, films prepared from hydrophilic polymers are expected to allow a higher rate of water vapor transmission than those prepared from hydrophobic polymers (Galić *et al.* 2018, Katiyar *et al.* 2015). The water vapor permeability of flexible plastic packaging materials is commonly measured gravimetrically, by the so-called cup method (also known as the desiccant or water method) (ASTM E96-80).

The *WVTR* (water vapor transmission rate) is the steady-state rate at which water vapor permeates through a film at specified conditions of temperature and relative humidity. The *WVTR* is determined using equation [5.5]:

$$WVTR = \frac{\frac{G}{t}}{A} = \frac{G}{t \cdot A} \qquad [5.5]$$

where:

G is the weight change (g);

t is the time (h);

G/t is the slope of the straight line (g h^{-1});

A is the test area (m^2);

WVTR is the rate of water vapor transmission (g h^{-1} m^{-2}).

Permeance (R) is calculated as follows in equation [5.6]:

$$R = \frac{WVTR}{\Delta p} = \frac{WVTR}{p_{sat} \cdot (RH_1 - RH_2)/100} \qquad [5.6]$$

where:

Δp is the vapor pressure difference;

p_{sat} is the saturation vapor pressure at test temperature;

RH_1 is the relative humidity at the source (%) (the test chamber for the desiccant method; the dish for the water method);

RH_2 is the relative humidity at the vapor sink (%).

The *water vapor permeability coefficient* (*WVP*) is then calculated using equation [5.7]:

$$\text{WVP} = \frac{WVTR \cdot X}{\Delta p} \quad [5.7]$$

where:

X is the film thickness (m);

the SI unit used for WVP is $g.m^{-2}. s^{-1}.Pa^{-1}$.

Factors affecting permeability are the chemical structure of the polymer, the physical structure of the polymer, the chemical structure of the permeant, the permeant concentration, temperature and humidity.

The chemical nature of the polymers is the most important parameter determining the ultimate barrier. Polymer chains must "move aside" or "open up" to allow permeation. The weaker the forces holding the chains together, the more rapidly permeation will occur. These chain-to-chain interactions are determined in large part by the chemical nature of the backbone.

Changes in moisture content and weight loss might occur if the product is not packed in a proper packaging material (water vapor permeability barrier); for example, hygroscopic foods like biscuits require protection from what makes them soggy. Loss of water in chilled and frozen food (drying out, dehydration or evaporation) may eventually result in loss of quality and is related to weight loss (Coles *et al.* 2003).

5.4.3. *Food packaging interactions*

Packaging materials may contain a variety of chemical substances that could migrate into the packaged food. They fall into one of two classes:

1) known substances needed to make the packaging (monomers and starting substances, catalysts, solvents and suspension media, additives);

2) known or unknown isomers, impurities, reaction products and breakdown products of these ingredients.

Migration is defined as a diffusion process that is subject to both kinetic and thermodynamic control (described by Fick's law). Diffusion is a function of time, temperature, thickness of the material, amount of migrant in the material, the partition coefficient and the distribution coefficient.

By definition, migration can be global or specific.

Global migration is the summary of all packaging components which can be transferred into the food, under controlled conditions.

Specific migration relates to one individual and identifiable compound only, either with a particular toxicological interest, or as a labeled compound used in experiments designed to elucidate the extent or mechanism of migration.

Migration is measured according to Commission Regulation 2016/1416 (Annex III: Food simulants), amending and correcting Regulation 10/2011 on plastic materials and articles intended to come into contact with food (European Commission 2016). As food itself is a very complex system containing proteins, carbohydrates, fats, micro and macro elements, migration tests are made on so-called food simulants aimed at stimulating food with certain properties (hydrophilic or hydrophobic, acidic, containing alcohol or fats and oils). Currently used food simulants are given in Table 5.5.

Migration is strongly influenced by the type of packed food and the conditions of storage (temperature and relative humidity). Thus, it increases with increased contact time, increased contact/storage temperature, higher levels of additives, etc. in the packaging material and in the intimacy of the contact with aggressive foodstuffs.

Food simulant	Abbreviation
Ethanol 10% (v/v)	Food simulant A
Acetic acid 3% (w/v)	Food simulant B
Ethanol 20% (v/v)	Food simulant C
Ethanol 50% (v/v)	Food simulant D1
Any vegetable oil containing 1% unsaponifiable matter	Food simulant D2
Poly(2,6-diphenyl-p-phenylene oxide), particle size 60–80 mesh, pore size 200 nm	Food simulant E

Table 5.5. *List of food simulants (Regulation (EU) No. 2016/1416)*

According to Commission Regulation 10/2011 (European Commission 2011), the overall migration of the components into a food may not exceed the overall migration limit (OML) of 60 mg/kg food, or 10 mg/dm^2 of the contact material.

Flavor scalping can be defined as the transport of food constituents into packaging. If a chemical compound present in the food has a high affinity for the packaging material, it will tend to be absorbed into or adsorbed onto the packaging until equilibrium concentrations have been established in the food and packaging.

Scalping does not result in a direct risk for the safety of the food, or in the introduction of unpleasant odors or flavors. The loss of volatile compounds that contribute to its characteristic taste affects the perceptible quality of the product itself.

With a certain combination of packaging and flake-sensitive foods, the introduction of an effective barrier layer can reduce the problem. Such food packaging systems must be carefully designed and tested for a high level of sensitivity, to ensure that permeation through the barrier is kept to a minimum (Coles *et al.* 2003). Factors such as RH, temperature, duration of storage and the presence of other food components can also affect the solubility of aroma compounds in packaged foods. Sorption of organic molecules can also change the mechanical properties of polymer films such as tensile strengths, modulus of elasticity and elongation at break (Farhoodi *et al.* 2009).

Light interaction with packaged food (exposure of packaged foods to both ultraviolet radiation and visible light) causes an oxidative deterioration of lipids, vitamins, proteins and colorants in foods, leading to the formation of off-flavors, nutrient losses, and discoloration. Factors influencing these deteriorative effects include the intensity of light, the duration of light exposure and light transmittance of the packaging material. Polymer materials offer some protection against lipid oxidation by filtering out ultraviolet radiation to varying degrees. Pigmenting, labeling or the direct printing of plastic containers can reduce this problem for sensitive food products. Usually, the modification of plastic materials may be accomplished by the addition of dyes, or the application of coatings, which absorb light at specific wavelengths (Robertson 2013). Barrier characteristics to light, water vapor and oxygen of some polymeric films are presented in Table 5.6.

Packaging material (0.025 mm thickness)	Barrier		
	Water vapor (g/m^2 24 h)	Oxygen (cm^3/m^2 24 h)	Light (%)
Polyvinyl chloride (PVC)	1.5–5	8–25	90
Polyvinylidene chloride (PVDC)	0.5–1	2–4	90
Polypropylene (PP)	5–12	2000–4500	80
High-density polyethylene (HDPE)	7–10	1600–2000	–
Low-density polyethylene (LDPE)	10–20	6500–8500	65
Linear low-density polyethylene (LLDPE)	15.5–18.5	200	–
Ethylene-vinyl acetate (EVAC)	40–60	12500	55–75
Ethylene-vinyl alcohol (EVOH)	1000	0.5	90
Polyamide (PA)	300–400	50–75	88
Polyethylene terephthalate (PET)	15–20	100–150	88
Polystyrene (PS)	70–150	4500–6000	92

Table 5.6. *Barrier characteristics of some polymers (Ščetar* et al. *2010)*

5.5. Plastic packaging innovation

While the most common packaging material – PE, is losing market share of late, PP and PET continue to increase with their applications. PVC currently has only a minimal share as a polymer for rigid plastic packaging, and PS has also lost its share compared to PP. Other polymers, including bioplastics and barriers such as ethylene-vinyl alcohol (EVOH), are also increasing market share. Conventional types of packaging are being replaced by innovative and flexible options designed to meet new consumer needs.

Manufacturers have already reached a stage where PET bottles cannot be made with much thinner walls. Therefore, the next step in this process is to replace plastic bottles with lightweight, flexible bags from mono materials whose main goal is to reduce packaging waste (Plastemart 2016).

New functions as an added value of packaging materials have been introduced as active packaging technologies, including oxygen removal and intelligent functions, control of package headspace (see Chapter 8), edibility and biodegradability (see Chapter 7).

5.6. Plastic waste management

A simplified form of the Möbius strip (Figure 5.8) is used for marking material in order to ease material identification and recyclability of used packaging. In addition to indicating that the packaging is recyclable, this symbol identifies the main material that the packaging is made of by a number in the middle of the strip. The numbering and abbreviation system was established by the Commission Decision 97/129/EC (European Commission 1997).

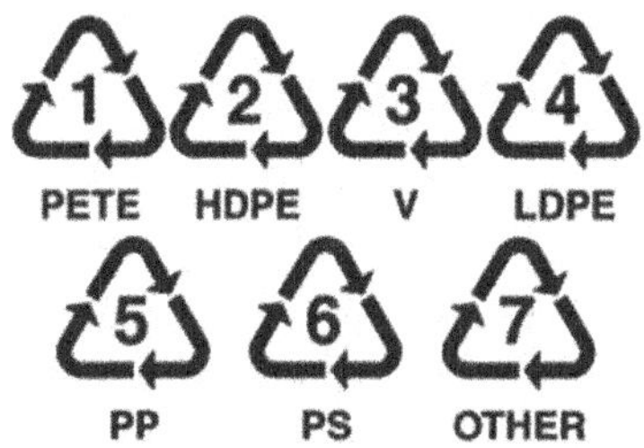

Figure 5.8. *Recycling packaging symbols with Möbius strip (Fidel Fillaud 2015)*

On products and packaging, the conditions for affixing the Möbius strip are defined by the international standard ISO 14021:2016 (2016). With the increasing trends in recycling materials and their possible use in food contact, there are recommendations that, for these materials, the percentage (X%) of recycled material used in packaging material production must be indicated in the middle of the Möbius strip.

Types of packaging placed on the EU market and all packaging waste are covered in Directive 94/62/EC. The same directive regulates heavy metals and chemicals present in packaging. For example, the concentration levels of lead, cadmium, mercury and hexavalent chromium are regulated as follows:

- 600 mg/kg by weight two years after the date referred to in the directive;
- 250 mg/kg by weight three years after the date referred to in the directive;
- 100 mg/kg by weight five years after the date referred to in the directive.

In general, the sum of the concentration levels of heavy metals present in packaging or packaging components shall not exceed 100 mg/kg by weight (Cheng 2019).

Most plastics may cause environmental pollution as they are obtained from fossil fuels and are not biodegradable. Thus, they remain in the environment as pollutants (Alavi 2015).

Generally, three possibilities for recycling exist: mechanical, chemical and energy recovery.

Relatively speaking, the cheapest and simplest recycling method is the mechanical recycling of plastic. More expensive is chemical recycling (feedstock recycling) in a process called depolymerization; monomers as a base unit may then be used in manufacturing new polymer material.

Renewable energy can be obtained from plastics in different ways: by burning waste (combustion) in incinerators, where the heat of plastic waste burned at high temperature is used to produce electricity or steam; also, by the production of alternative fuels from plastic waste (pyrolysis and gasification), where the obtained fuel is used in various production processes and in power plants themselves (Kopeliovich 2018).

In 2019, the EU Commission proposed an EU-wide target of 55% on recycling plastic packaging (by 2025). The indicator is defined as the share of recycled packaging waste in all generated packaging waste. Packaging waste covers wasted material that was used for the containment, protection, handling, delivery and presentation of goods, from raw materials to processed goods, from the producer to the user or the consumer, excluding production residues (Eurostat 2020).

Nowadays, due to growing environmental concerns, the scientific community has focused on researching bio-based sustainable plastic packaging materials. The most promising bio-polymers for future packaging applications are cellulose, chitosan, polylactic acid (PLA) and polyhydroxyl alkanoates (PHA) (Alavi 2015; Siracusa 2012). More information on these materials is provided in Chapter 7.

5.7. Legislation for food contact materials

Legislation for food contact materials (FCMs) determines the safety and suitability of consumer goods (food packaging materials, children's toys, etc.) for human use.

Development and harmonization of experimental approaches to produce reliable data for risk assessment, as well as the monitoring of target contaminants in packaging materials, are given at European level. According to Regulation EC No. 1935/2004 (European Parliament, Council of the European Union 2004), FCMs should be manufactured in such a way that they do not transfer constituents to food in quantities that would: endanger human health, bring about an unacceptable change in the composition of food or bring about a deterioration in the organoleptic characteristics of food. Regulation (EC No. 1935/2004) also prescribes an international symbol for food-safe material: it features a wine glass and a fork, as shown in Figure 5.9.

Figure 5.9. *Food contact materials symbol (Regulation EC No. 1935/2004 (European Parliament, Council of the European Union 2004))*

5.8. References

Abdellatief, A. and Welt, B.A. (2012). Comparison of new dynamic accumulation method for measuring oxygen transmission rate of packaging against the steady-state method described by ASTM D3985. *Packag. Technol. Sci.*, 26, 281–288.

Alavi, S., Thomas, S., Sandeep, K.P., Kalarikkal, N., Varghese, J., and Yaragalla, S. (2015). *Polymers for Packaging Applications*. Apple Academic Press, Florida.

Allied Market Research (2019). Rigid plastic packaging market by material (Polyethylene Terephthalate (PET), Polypropylene (PP), High-Density Polyethylene (HDPE), and others) and end-user industry (food & beverages, personal care, household, healthcare, and others): Global opportunity analysis and industry forecast, 2019–2026 [Online]. Available at: https://www.alliedmarketresearch.com/rigid-plastic-packaging-market [Accessed 15 June 2020].

ASTM E96-80 (1980). ASTM Standard Test Method E96 – 80, Water Vapor Transmission of Materials.

Cheng, V. (2019). Packaging materials regulations in the European Union [Online]. Available at: https://www.compliancegate.com/european-union-packaging-materials-regulations/ [Accessed 17 July 2020].

Coles, R., McDowell, D., and Kirwan, M.J. (2003). *Food Packaging Technology*. Blackwell, Florida.

Ebnesajjad, S. (2013). *Plastic Films in Food Packaging, Materials, Technology, and Applications*. Elsevier, New York.

Emblem, A. and Emblem, H. (2012). *Packaging Technology – Fundamentals, Materials and Processes*. Elsevier, New York.

EPAC (2020). Rigid packaging vs. flexible packaging: Is one really better? [Online]. Available at: https://epacflexibles.co.uk/rigid-packaging-vs-flexible-packaging-is-one-really-better/ [Accessed 22 July 2020].

European Commission (1994). Packaging and packaging waste. European Parliament and Council Directive 94/62/EC [Online]. Available: https://eur-lex.europa.eu/legal-content/en/TXT/?uri=CELEX:31994L0062.

European Commission (1997). Establishing the identification system for packaging materials pursuant to European Parliament and Council Directive 94/62/EC on packaging and packaging waste. Commission Decision 97/129/EC [Online]. Available: https://eur-lex.europa.eu/legal-content/EN/TXT/?uri=celex%3A31997D0129.

European Commission (2011). Plastic materials and articles intended to come into contact with food. Commission Regulation 10/2011 [Online]. Available: https://eur-lex.europa.eu/legal-content/EN/ALL/?uri=CELEX%3A32011R0010.

European Commission (2016). Amending and correcting Regulation (EU) No 10/2011 on plastic materials and articles intended to come into contact with food. Commission Regulation 2016/1416 [Online]. Available: https://eur-lex.europa.eu/legal-content/EN/TXT/?uri=CELEX%3A32016R1416.

European Parliament, Council of the European Union (2004). Regulation (EC) No 1935/2004 of the European Parliament and of the Council of 27 October 2004 on materials and articles intended to come into contact with food and repealing Directives 80/590/EEC and 89/109/EEC.

Eurostat (2020). Recycling rate of packaging waste by type of packaging [Online]. Available at: https://ec.europa.eu/eurostat/cache/metadata/en/cei_wm020_esmsip2.htm [Accessed 9 July 2020].

Farhoodi, M., Emam-Djomeh, Z., Oromiehie, A., and Ehsani, M.R. (2009). Influence of menthol absorption on the mechanical properties of PET bottles. *Iranian Polymer Journal*, 18, 227–232.

Fidel Fillaud (2015). Plastics: How to identify recyclable packaging? [Online]. Available at: https://www.fidel-fillaud.com/en/news/tip-of-the-month/plastics-how-to-identify-recyclable-packaging-xpa1445.html [Accessed 2 July 2020].

Galić, K., Kurek, M., and Ščetar, M. (2018). *Barrier Properties of Plastic Polymers. Reference Module in Food Science*. Elsevier, New York.

Grand View Research (2020). Plastic market size, share & trends analysis report by product (PE, PP, PU, PVC, PET, Polystyrene, ABS, PBT, PPO, Epoxy Polymers, LCP, PC, Polyamide), by application, by region, and segment forecasts, 2020–2027 [Online]. Available at: https://www.grandviewresearch.com/industry-analysis/global-plastics-market [Accessed 3 July 2020].

Hahladakis, J.N., Velis, C.A., Weber, R., Iacovidou, E., and Purnell, P. (2018). An overview of chemical additives present in plastics: Migration, release, fate and environmental impact during their use, disposal and recycling. *J. Hazard. Mater.*, 344, 179–199.

ISO 14021:2016 (2016). Environmental labels and declarations – Self-declared environmental claims (Type II environmental labelling).

Kopeliovich, D. (2018a). Compression molding [Online]. Available at: https://www.substech.com/dokuwiki/doku.php?id=compression_molding_of_polymers) [Accessed 13 July 2020].

Kopeliovich, D. (2018b). Injection molding [Online]. Available at: https://www.substech.com/dokuwiki/doku.php?id=injection_molding_of_polymers) [Accessed 13 July 2020].

Kopeliovich, D. (2018c). Plastics recycling [Online]. Available at: https://www.substech.com/dokuwiki/doku.php?id=plastics_recycling [Accessed 13 July 2020].

Kopeliovich, D. (2018d). Thermoforming [Online]. Available at: https://www.substech.com/dokuwiki/doku.php?id=thermoforming) [Accessed 13 July 2020].

Kopeliovich, D. (2018e). Blow molding [Online]. Available at: https://www.substech.com/dokuwiki/doku.php?id=blow_molding) [Accessed 13 July 2020].

Parisi, S. (2013). *Food Industry and Packaging Materials – Performance-oriented Guidelines for Users*. Smithers Rapra, Ohio.

Plastemart (2016). Trends shift from rigid packaging to flexible packaging [Online]. Available at: http://www.plastemart.com/plastic-technical-articles/trends-shift-from-rigid-packaging-to-flexible-packaging/2229 [Accessed 14 July 2020].

Robertson, G.L. (2013). *Food Packaging Principles and Practice.* CRC Press, Boca Raton.

Ščetar, M., Kurek, M., and Galić, K. (2010). Trends in meat and meat products packaging – A review. *Croat. J. Food Sci. Technol.*, 2(1), 32–48.

Siracusa, V. (2012). Food packaging permeability behaviour: A report. *International Journal of Polymer Science*, 1–11.

Vujković, I., Galić, K., and Vereš, M. (2007). *Ambalaža za pakiranje namirnica*, Tectus, Zagreb.

6

Multilayer Packaging Materials

Mario Ščetar

Faculty of Food Technology and Biotechnology, University of Zagreb, Croatia

6.1. Introduction

Laminate is a product made by bonding together two or more layers or materials in order to create a new material with desirable properties.

Today's trend and desirable target for the packaging industry is to use less material (surface area and thickness) and to create less waste for disposal, while still extending the shelf life of the product.

Demand for laminated packaging foils is frequently increasing with the adoption of high-quality packaging solutions by the food, beverage and pharmaceutical industries. These industries use laminated packaging foils in almost all applications of primary packaging. Thus, manufacturers are focused on developing high-quality and high-grade films that will be easy to produce (and cheaper), using simplified production processes. Strict regulations on the use of plastic material for laminated packaging foils can also be expected to prevent the growth of the global market (environmental pollution) (Fact.MR n.d.).

For a color version of all figures in this book, see www.iste.co.uk/debeaufort/packaging.zip.

Packaging Materials and Processing for Food, Pharmaceuticals and Cosmetics, coordinated by Frédéric DEBEAUFORT, Kata GALIĆ, Mia KUREK, Nasreddine BENBETTAIEB and Mario ŠČETAR.

6.2. Multilayer materials and their production

Multilayer materials (laminates), as the term implies, have a multilayered structure and can be composed of plastic film only or composed of different materials (plastic, paper and/or thin metal foils). Commercial laminates can have up to 9 (or even 12) different layers. Lamination is the process of bringing two or more webs – paper, film, foil or other flexible material that is unwound from a roll and passed through a machine – together and bonding them either with an adhesive or by heating, as shown in Figure 6.1. Structural barriers for food packaging often contain components used for either bonding or sealing, each with specific properties. The most common include (Ebnesajjad 2013):

– poly(vinylidene chloride) (PVDC), which acts as a barrier, sealant and bonding agent;

– ethylene vinyl alcohol (EVOH) serving as a bonding layer;

– polyamide (PA) acting as a barrier layer;

– metallized film or additives such as aluminum and silicon oxide, providing an additional barrier.

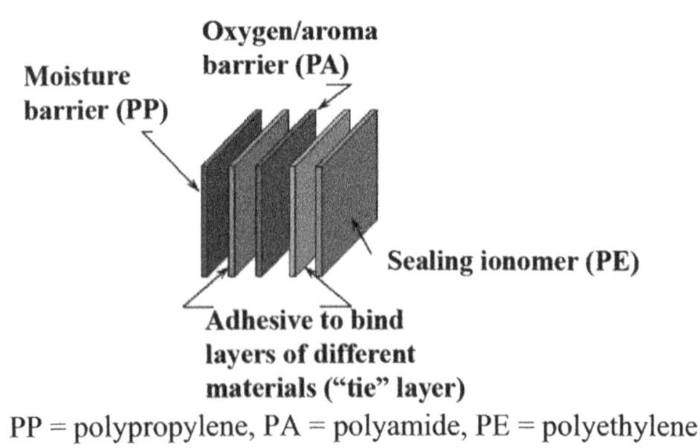

Figure 6.1. *Multilayer material with combined properties from different monofilms*

In the production of food packaging materials, the following lamination techniques are considered to be the most widely used:

– adhesive lamination (dry bonding or wet bonding);

– extrusion and coextrusion coating;

– thermal lamination.

6.2.1. *Adhesive lamination*

Adhesive lamination can be performed in two ways: as dry bonding or as wet bonding.

During the **wet bonding process**, the adhesive is printed onto the substrate and the foil is then laminated onto this substrate, which subsequently passes under a UV curing station. The printed adhesive bonds the foil to the substrate, and the excess is then stripped off (Figure 6.2).

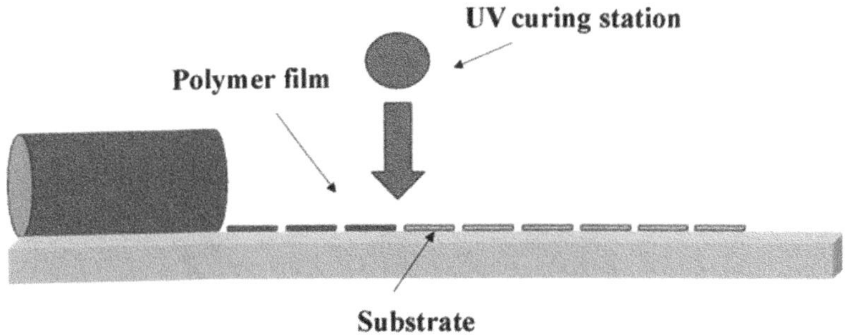

Figure 6.2. *Wet bonding lamination*

During **dry bonding**, the adhesive is printed onto the substrate, which then passes under a UV curing station, making the adhesive tacky. The foil is laminated onto the substrate where it bonds to the adhesive and the excess is then stripped off (Figure 6.3).

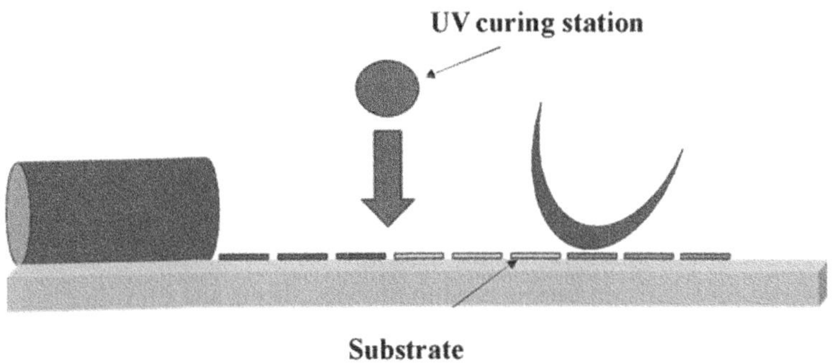

Figure 6.3. *Dry bonding lamination*

6.2.2. *Extrusion and coextrusion lamination*

Extrusion coating and extrusion lamination are processes that involve the use of various materials, in combination, in order to form a single complex structure.

The schema of extrusion coating is given in Figure 6.4. In this process, an extruder forces the melted thermoplastic resin through a horizontal slot-die, onto a moving web of substrate. The resulting product is a permanently coated web structure. Extrusion lamination is a similar process, where resin is extruded between two substrates and acts as a bonding agent. Inside a processing line, the substrates and melt are nipped at a bonding station.

The finished product is wound as a permanently bonded multilayer laminated structure, as shown in Figure 6.4. Most plastic films require a primer coating in the extrusion process before use (BOBST n.d.).

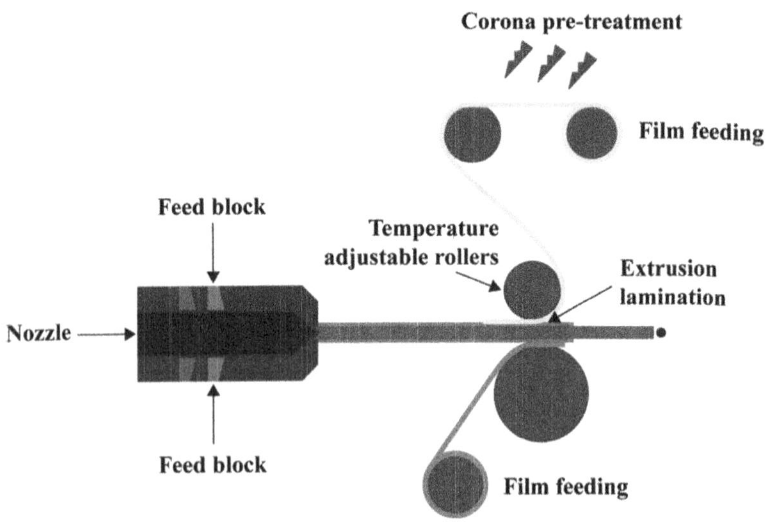

Figure 6.4. *Lamination by extrusion*

Coextrusion lamination is mainly used for the production of tubular and flat films (Figure 6.4). This process is used to produce sheets and films that are suitable for food packaging. The molten polymers from different extruders are molded and combined. The melt flow then expands on the matrix and becomes so thin that it can be evenly shaped as a coextruded film. The problem with coextrusion is the simultaneous processing of polymers of different rheological properties, which directly affect the quality of the product (Wagner 2010).

A three-layer coextrusion (Figure 6.5) can combine many different characteristics of plastic materials in the extrusion process and can significantly reduce production costs. This technology is therefore applied on a large scale nowadays.

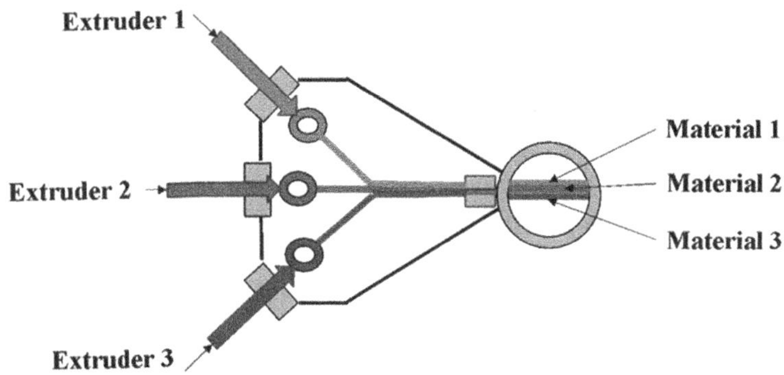

Figure 6.5. *Three-layer coextrusion process*

6.2.3. *Thermal lamination*

The main principle of the thermal lamination process is that the adhesion of different layers of flexible plastic is performed by heat and pressure, without using any chemical adhesion coatings. This lamination method is highly desirable for food packaging where the use of additional chemicals can cause migration and odor problems.

Thermal lamination has many advantages such as: prevention of unwanted taste and odor transmission; the final material is free of diluents and similar chemicals; the process is also free of dangerous vapors that contribute to atmospheric pollution. Furthermore, there is no need for a curing period after production, the process is suitable for hot-fill applications, no stress cracking is observed and good material resistance is obtained. The process enables various possibilities for producing recyclable laminates with excellent gas and moisture barrier properties. Most combinations are the result of a mix of different materials (such as PE/PET and PE/PA) that are used due to their specific properties. It is important to emphasize that the nature of such multilayer materials makes them less suitable for recycling. However, there is the possibility of developing specific laminates consisting of three layers of PE, while retaining the properties required. This also results in a laminate that is easily recyclable and can therefore be re-used (KIVO n.d.).

6.2.4. *Metallized films*

Metallized film can be described as a plastic film that has been coated on one or both sides with a very thin layer of metal (Figure 6.6). The benefits of both metal and plastic films are made use of, which results in much more versatility in application. Direct vacuum metallizing with aluminum on plastic films results in a significant increase in barrier properties. It can also make an opaque barrier to light and offer a highly reflective, glossy appearance.

Metallized (met) films are composed of biaxially oriented polypropylene (BOPP), polyester (PET), polyamide (PA) and cast polypropylene (OPP) and are manufactured by means of high-speed vapor deposition. The vacuum-deposited metal thickness need only be a few micrometers to ensure high barrier properties (Table 6.1).

Material (μm)	Permeability			
	O_2	CO_2	N_2	Water vapor
	$cm^3/m^2 \cdot d \cdot bar$			$g/m^2 \cdot d$
PET/PE (12/75)	**100** (23°C; 0% RH)	**450** (23°C; 0% RH)	**1** (23°C; 0% RH)	**1.2** (23°C; 85% RH)
PET/met/PE (12/75)	**< 1** (23°C; 0% RH)	**2** (23°C; 0% RH)	**< 1** (23°C; 0% RH)	**0.2** (23°C; 85% RH)
PET/Alu/PE (12/9/75)	**< 0.1**			**< 0.1** (23°C; 85% RH)
OPA/PE (15/75)	**35** (23°C; 75% RH)	**200** (23°C; 75% RH)	**8** (23°C; 75% RH)	**2** (23°C; 85% RH)
OPA/met/PE (15/75)	**3** (75% RH)			**0.7** (23°C; 85% RH)
OPP/PE (20/40)	**1,200** (23°C; 0% RH)	**3,500** (23°C; 0% RH)	**250** (23°C; 0% RH)	**0.8** (23°C; 85% RH)
OPP/met/PE (20/40)	**40** (23°C; 0% RH)			**0.3** (23°C; 85% RH)
OPP/Alu/PE (20/9/75)	**< 0.1** (23°C; 0% RH)			**< 0.1** (23°C; 85% RH)

Table 6.1. *Barrier properties comparison for metallized and non-metallized films (Galić* et al. *2009)*

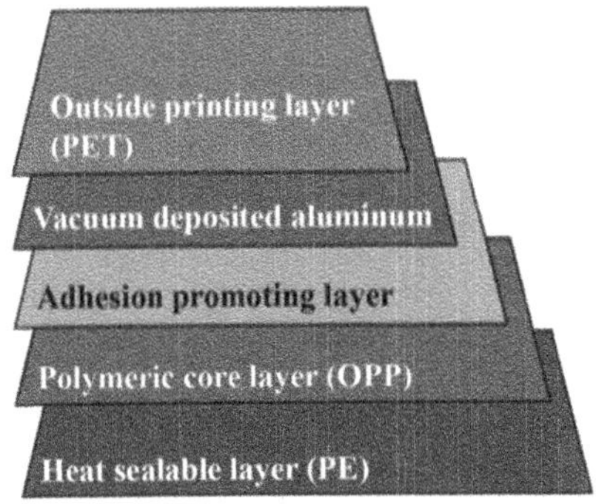

Figure 6.6. *Metallized film laminate with associated layers*

6.3. Properties of laminates

As already stated, laminates are produced in such a way in order that the final product holds desirable properties for a particular application. This is mostly evident in their barrier properties.

6.3.1. *Barrier properties*

When measuring and calculating barrier properties of multilayer materials (laminates), they are considered as a number of membranes in series. The case of three layers in series is shown in Figure 6.7 (Galić *et al.* 2018).

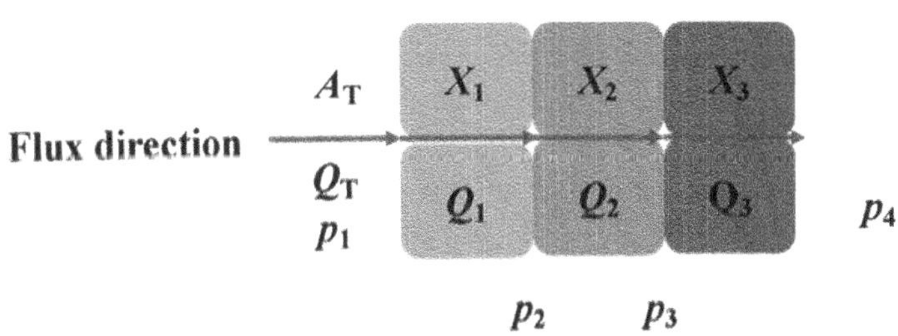

Figure 6.7. *Three layers in series*

The total thickness (X_T) is described as:

$$X_T = X_1 + X_2 + X_3 \quad [6.1]$$

Assuming steady state flux, the rate of permeation (Q) through each layer must be constant, that is

$$Q_T = Q_1 = Q_2 = Q_3 \quad [6.2]$$

Likewise, the areas (A) will also be constant:

$$A_T = A_1 = A_2 = A_3 \quad [6.3]$$

Therefore, by substituting data for multilayered materials in the equation that describes barrier of monolayer plastic material

$$\frac{Q}{t} = \frac{P}{X} A(\Delta p) \quad [6.4]$$

Permeation through the multilayer materials can be written as

$$\frac{Q_T X_T}{t} = p_1 - p_4 \, \Delta p_i \quad [6.5]$$

Since

$$(p_1 - p_4) = (p_1 - p_2) + (p_2 - p_3) + (p_3 - p_4) \quad [6.6]$$

Therefore, equation [6.5] becomes

$$\frac{Q_T X_T}{tA_T P_T} = \frac{Q_T}{tA_T}[\frac{X_1}{P_1} + \frac{X_2}{P_2} + \frac{X_3}{P_3}] \quad [6.7]$$

and [6.8]

$$\frac{X_T}{P_T} = \frac{X_1}{P_1} + \frac{X_2}{P_2} + \frac{X_3}{P_3} \quad [6.8]$$

or [6.9]

$$P_T = \frac{X_T}{\left(\frac{X_1}{P_1}\right) + \left(\frac{X_2}{P_2}\right) + \left(\frac{X_3}{P_3}\right)} \quad [6.9]$$

If the individual thicknesses and permeability coefficients are known for each layer, provided that the P coefficients are independent of pressure, then equation [6.9] can be used to calculate P for any multilayer material. If they are not, then different permeability coefficients will be obtained depending on the positioning of the layers.

As packaging films are almost always used in humid conditions, the oxygen permeability is dependent on the humidity. In such cases, equation [6.9] cannot be used directly since p_2 (P of the center layer) will depend on the average partial pressure at the center. Since the partial pressure of water vapor will not be constant across the multilayer, an expression for transmission rate (TR = Q/At) must be modified to include a term for the partial pressure difference and the thickness:

$$WVTR = \frac{G}{t \leftarrow A} \Delta p \qquad [6.10]$$

Since the area A and time t will be the same for all three layers, the equilibrium WVTR between the outside and center layers can be expressed as:

$$\frac{Q_1}{X_1}(p_1 - p_2) = \frac{Q_2}{X_2}(p_2 - p_3) \qquad [6.11]$$

Similarly, the equilibrium WVTR between the center and the inside layer will be:

$$\frac{Q_2}{X_2}(p_2 - p_3) = \frac{Q_3}{X_3}(p_3 - p_4) \qquad [6.12]$$

The average partial pressure of the center layer (p_c) will be:

$$p_C = \frac{p_2 + p_3}{2} \qquad [6.13]$$

Simultaneous linear solution of equations [6.11] and [6.12] for p_2 and p_3 and substitution in equation [6.13] yields [6.14], that is equal to the average partial pressure of the center layer:

$$P_c = \frac{p_1\left[\frac{X_2}{Q_2} + 2\frac{X_3}{Q_3}\right] + p_4\left[\frac{X_2}{Q_2} + 2\frac{X_1}{Q_1}\right]}{2\left[\frac{X_3}{Q_3} + \frac{X_2}{Q_2}\frac{X_1}{Q_1}\right]} \qquad [6.14]$$

Knowing the p_c, the P_c of the center layer can be determined experimentally at this partial pressure and equation [6.9] can be used to calculate the overall permeability of the multilayer material.

6.3.2. *Interaction between packaging material and food*

Food packaging material is a frequent source of off-flavors found in food products (Reineccius 2013). Off-flavor problems attributed to residues of solvents are often reported after long storage periods. This may be related to the production of the material, to trace residues in recycled materials or to solvents/resins used in adhesives, etc. Even though these substances are mostly released in trace levels and generally do not pose a health risk, their appearance is not accepted by consumers (Robertson 2013). In order to avoid a negative impact on those food products that are frequently complained about by unsatisfied consumers, there are thresholds for such substances. A threshold is defined as the lowest concentration of a compound where an inherent property is detected by a certain proportion of a given group of consumers (usually 50%) (Robertson 2013). Toluene, p-tertiary butyl phenol, ethyl acetate, isopropanol, methyl ethyl ketone, hexanal and hexane are the most commonly reported substances. For example, it is common for extrusion conditions of low-density polyethylene (LDPE) to affect the flavor of the packed product, probably caused by volatile compounds, such as saturated and unsaturated hydrocarbons and aromatic hydrocarbons (Potts *et al.* 1990). Compounds with C3 and C4 alkyl benzenes were reportedly responsible for an "intense plastic off-flavor" described as candle grease, musty, rancid, soapy and pungent (Mathlouthi 1994). "High" levels of toluene from laminated pouches, used for fruity soft drinks, also originate from the adhesive used for lamination of the multilayer structure.

The **migration** issue of polymer materials is described in Chapter 5, and adhesive and ink migration is described in Chapter 10.

6.4. Packaging application of laminates

Laminated films are used in a wide range of flexible packaging applications, such as pharmaceuticals, personal care products, cosmetics, foods (snacks, sugar, seasoning, noodles, beverages and biotic liquid packaging) and many more, as shown in Table 6.2.

Primary packaging sub-type	Film, sheet
Primary packaging type	Flexibles
End use	Snack foods, dry foods, dry powders, herbs, spices, pasta, noodles, seasonings
Market – segment	Cosmetics, personal care, food, beverages (non-alcoholic), pharmaceuticals, beauty, personal care, health

Table 6.2. *Laminates used in packaging applications (Wapo 2020)*

Food Product	Structures	Properties
Dry foods	OPA/PE OPA/OPA/PE	Seal strength, moisture barrier
Biscuits and chocolate	BOPP/BOPP BOPP/CPP PET/PE/BOPP OPP/OPPmet PET/OPPmet	Seal strength, moisture barrier
Candies	PET/BOPP PET/CPP	Seal integrity, high packing speed, moisture barrier
Tea	PETmet	Odor barrier, good seal strength
Coffee (instant)	PET/PE/Alu/PE PET/Alu/PE	High moisture barrier, good seal strength and integrity
Potato chips	PETmet PET/PE PET/PE/PETmet/PE BOPP/PETmet/PE	High seal strength and integrity, gas retention, moisture barrier
Rice and pasta	PE/PE PAP/CPP OPP/CPP OPP/PE	High seal strength and integrity
Frozen food	PET/Alu/PE PET/PE/PE PETmet/PE OPP/PETmet/PE	High seal strength and integrity, gas retention, moisture barrier
Ice cream	OPPmet/Cold seal BOPPmet White BOPP White Two sided acrylic	High seal strength and integrity, gas retention, moisture barrier
Desserts	PAP/PE PAP/PE/PVDC PAP/Alu/PE PETmet/PE PAP/OPPmet/PE	Good seal strength and integrity

PET = poly(ethylene terephthalate), PETmet = metallized poly(ethylene terephthalate), PE = polyethylene, OPP = oriented polypropylene, OPA = oriented polyamide, OPPmet = metallized oriented polypropylene, BOPP = biaxially oriented polypropylene, Alu = aluminum foil, CPP = cast polypropylene, BOPPmet = metallized biaxially oriented polypropylene, PAP = paper, PVDC = poly(vinylidene chloride).

Table 6.3. *Example of laminates used for food packaging (TECHNOCRAFT 2019)*

Extrusion coating and laminating can be configured for a variety of applications including liquid packaging, aseptic drink containers, flexible packaging, toothpaste tubes, medical packaging, industrial wraps, insulation backed materials and plastic-coated tablecloths, as well as sacks for products as diverse as cement, cereal grain, and dry chemicals (BOBST n.d.).

Thermal laminates can be used in common applications including juice bags, Bag-in-Box, drum bags and deep freeze applications. These are often used for filling with liquids such as syrup, juices, wine and oils, as well as for foods that are filled hot and used by the food processing industry (KIVO n.d.).

OPPmet application is a drop-in foil replacement for traditional paper–polyethylene–foil–polyethylene (PAP/PE/PE) packets and stand-up packaging (SUP), used for salty snacks, dried soups, oatmeal, seasonings, hot cocoa, dried cheese, rice, mixes and powdered beverages. It is also used as a substitute for foil packaging structures for nutritional bars.

PETmet films are a foil-replacement solution for coffee, dried fruits, nuts and powdered nutritional or energy drink packaging. It is also a good option for replacing foil in a lidded application for dairy foods, like yogurt and ice cream, although challenges may arise depending on the packaging equipment used (Williamson 2018).

Some of the typical laminate structures for food packaging, with important properties, are listed in Table 6.3.

6.5. Environmental and waste management issues

Although laminates offer a diverse range of applications as well as tailored material properties, there is the huge problem of separating extruded and coextruded layers of polymeric materials. Thus, they pose a problem in terms of their disposal and ability to be recycled. 4% of non-recycled material consists of laminated, flexible plastic packages, concentrated in the sorted and graded films in an amount of 10–15%. These laminated flexible packages are made up of several different polymers and materials, so, by their nature, they cannot be recycled in a circular way into new flexible films.

All of this prevents progress towards a more economical circular economy of plastic packaging. Although laminated, flexible plastic packaging has an important role in our society in protecting the quality of food products and making such food products stable, long enough to allow production and worldwide distribution, manufacturers need to check whether laminates are really needed in their production or whether they cannot be avoided. A separate method of integration into the current collection and recycling system should be then provided.

As a result of environmental issues, laminates should be redesigned in order for them to be identified and sorted much more easily. Therefore, the need for laminates in the supply chain is recognized, and it is accepted that, unfortunately, these packages still cannot be recycled.

In any case, where possible, it would be beneficial to apply materials that are crucial for improving the quality of recycled plastics, making novel and circularly recyclable packaging for the above-mentioned reasons (van Velzen *et al.* 2020). Of course, such changes should also take into account the practical applications of the material for specific purposes, in order to protect quality and ensure safety of the packaged product.

6.6. References

BOBST (n.d.). Extrusion coating & laminating process. BOBST [Online]. Available: https://www.bobst.com/usen/products/extrusion-coating-laminating/process/ [Accessed 25 July 2020].

Ebnesajjad, S. (2013). *Plastic Films in Food Packaging. Materials, Technology, and Applications*. William Andrew, Oxford.

Fact.MR (n.d.). Laminated packaging films market forecast, trend analysis & competition tracking – Global market insights 2018 to 2026 [Online]. Available: https://www.factmr.com/report/2724/laminated-packaging-films-market [Accessed 25 July 2020].

Galić, K., Ćurić, D., and Gabrić, D. (2009). Shelf life of packaged bakery goods – A review. *Crit. Rev. Food Sci. Nutr.*, 49(5), 405–426.

Galić, K., Kurek, M., and Ščetar, M. (2018). Barrier properties of plastic polymers. *Reference Module in Food Science*. Elsevier, 1–21.

KIVO (n.d.). Thermal lamination. KIVO Group [Online]. Available: https://www.kivo.nl/en/processes/thermal-lamination/ [Accessed 24 July 2020].

Mathlouti, M. (1994). *Food Packaging and Preservation*. Springer Science & Business Media, Berlin.

Potts, M.W., Baker, S.L., Hansse, M., and Hughes, M.M. (1990). Relative taste performance of plastics in food packaging. *J. Plast. Film Sheeting*, 6, 31–43.

Reineccius, G. (2013). *Source Book of Flavors*. Springer Science & Business Media, Berlin.

Robertson, G.L. (2013). *Food Packaging Principles and Practice*. CRC Press, Boca Raton.

TECHNOCRAFT (2019). Techno plast & prints. Technocraft Group [Online]. Available: https://www.technokraftgroup.com/cosmo-films.php [Accessed 25 July 2020].

van Velzen, U.T., de Weert, L., and Molenveld, K. (2020). Flexible laminates within the circular economy. Wageningen Food & Biobased Research [Online]. Available: https://edepot.wur.nl/519019 [Accessed 25 July 2020].

Wagner, J.R. (2010). *Multilayer Flexible Packaging: Technology and Applications for Food, Personal Care and Over-the-Counter Pharmaceutical Industries*. Elsevier, Burlington.

Wapo (2020). Laminated packaging film. Wapo [Online]. Available: https://www.webpackaging.com/en/portals/wapo/assets/11063957/laminated-packaging-film/ [Accessed 25 July 2020].

Williamson, T. (2018). Metallized films for foil replacement. Packaging Strategies [Online]. Available: https://www.packagingstrategies.com/articles/90483-metallized-films-for-foil-replacement [Accessed 25 July 2020].

7

Bio-based Materials

Mia Kurek[1] and Nasreddine Benbettaieb[2]

[1] *Faculty of Food Technology and Biotechnology, University of Zagreb, Croatia*

[2] *Institute of Technology, University of Burgundy, Dijon, France*

7.1. Introduction

These days, plastics are materials that are used for food packaging applications, to great extent. Currently, synthetic plastic materials are mostly made from petrochemical sources; a detailed description is given in Chapter 5. The bio-based economy refers to the use of renewable natural resources like crops or trees for fuel and, most recently, for the production of various materials such as packaging foils, in the biomedical field for tissue engineering or even for furniture making. When debating about materials, the largest problem is not their usage or performance, but is hidden in their lifecycle assessment (LCA). When it comes to disposal and waste management of conventional materials, then it becomes a significant concern. Rules about the prevention and reduction of packaging waste are covered under different international standards (EN 13428:2004; ISO 18602:2013). If this is not possible, reuse is proposed as a good alternative (EN 13429:2004; ISO 18603:2013), followed by recycling (EN 13430:2004; ISO 18604:2013), composting (EN 13432:2004; ISO 18606:2013) or incineration with energy recovery (EN 13431:2004; ISO 18605: 2013). Even though in some sectors the reuse of biomass (reproducing or recycling) has been done for several decades – for example, for pulp and paper or textile – in most scenarios, it is just done at the beginning. Plastic waste is one of many types of waste that takes too long to decompose. In answer to the greater issue and with a

---

For a color version of all figures in this book, see www.iste.co.uk/debeaufort/packaging.zip.

Packaging Materials and Processing for Food, Pharmaceuticals and Cosmetics, coordinated by Frédéric DEBEAUFORT, Kata GALIĆ, Mia KUREK, Nasreddine BENBETTAIEB and Mario ŠČETAR.

growing demand for sustainability, there has been a global shift to bio-based chemicals. So, biodegradable, bio-based and compostable materials are capturing market share. Agricultural waste is a large pool of untapped biomass resources. According to European Bioplastics, it is estimated that the global production capacity of biobased plastics will increase by approximately 25% (from 2.05 million tonnes in 2017 to 2.44 in 2022) (European Bioplastics 2017). The global biodegradable material market was valued at $3.02 billion in 2018 and is expected to reach more than $6 billion by the end of 2025, growing at a CAGR of 14% between 2019 and 2025 (Research Reports World 2014). Generally, countries that occupy the largest part of the global market include the UK, China, USA, Brazil and India, with the following most prominent companies: Novamont S.p.A., Plantic, Natureworks, Corbion N.V., Biome Technologies plc, Eastman Chemical Company and Danimer Scientific. The market is segmented depending on the type (polylactic acid (PLA), polybutylene adipate terephthalate (PBAT), polybutylene succinate (PBS), polyhydroxy alkanoates (PHA), starch blends, and others) and application of bio-based plastics for food purposes. These materials have found applications in agriculture, food packaging, shrink films, disposable tableware and 3D printing. The greatest industrial relevance is still on carrier bags. Even though the first knowledge of this kind of material dates back to the 1920s, large-scale production is far from straightforward. Among the proposed materials, PLA and starch blends are the most significant on an industrial level. The global PLA market in 2018 was valued at $673.88 million, with major stakeholder players NatureWorks, Corbion, Pyramid and Hitachi (NKWood Research 2019). Production schemes often include initial demonstrations for proof-of-concept in an academic setting and then the transition to industrial scale. To date, the production costs are significantly higher than that of petroleum-based plastics that can be produced cheaply and at higher volumes. A conservative approach to production methods for bioplastics rarely matches and technical performance is often far away from that of conventional plastics. However, they are quite advantageous selling solutions as more consumers are looking for greener products with a lower environmental footprint, compared to the products used traditionally. It must be pointed out, however, that even though "greener than the conventional", bio-based polymers should not be viewed as a complete solution to plastic waste, as they still require robust waste management and degradation protocols. This means that efforts should be focused on responsible behavior: reducing the consumption of products that generate waste materials that take a long time to completely break down; and on waste management, since these are the two most important features to think about for both petrochemical and bio-based polymers.

7.2. Definitions

Understanding the meaning of the term biodegradability is critical in understanding the issues surrounding biodegradation.

First of all, it must be elucidated that bio-based materials can be (Figure 7.1):

– non-biodegradable (bio-polyethylene (bioPE), bio-polyethylene terephthalate (bioPET), etc.);

– biodegradable (polylactic acid (PLA), polyhydroxyalcanoate (PHA), polyhydroxybutyrate (PHB));

– compostable (chitosan, thermoplastic starch, etc.).

This means that bio-based materials are not always biodegradable or compostable. Just as important, not all biodegradable so-called bioplastics are bio-based. Some, such as PLA or PHA, are biodegradable or compostable under appropriate conditions. BioPE, bioPU and bioPET are neither biodegradable nor compostable. Conversely, some biodegradable or compostable polymers are petrochemical based, such as polycaprolactone (PCL) or PBS. Abbreviations of all materials mentioned in this chapter are given in section 7.5.

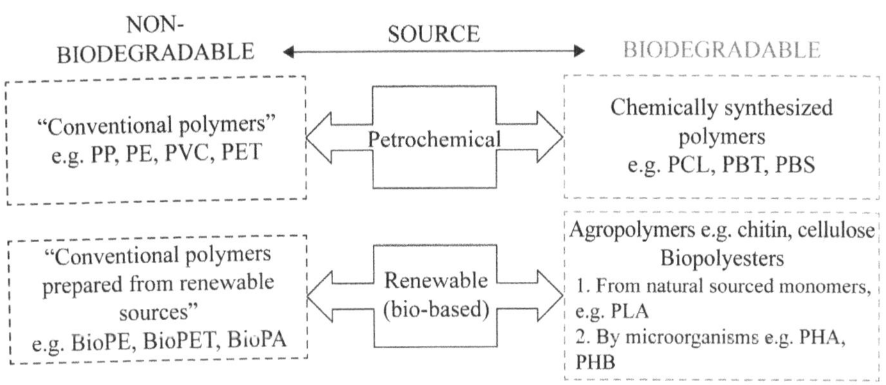

Figure 7.1. *Schematic view of the biodegradability of polymers depending on their source (adapted from European Bioplastics 2017)*

7.2.1. *Bio-based plastics*

According to the EN 16575:2014 standard, bio-based means that a material is made of plastic and is based on renewable and natural raw materials. Bio-based stocks such as corn, potatoes, rice, soy, sugarcane, wheat and vegetable oil can be transformed through chemical or biological processes. Over the last decade, they have found use in various sectors (Table 7.1).

The main materials belonging to this group are produced:

- from starch: PLA – **bio-based** and *biodegradable;*
- through microbial synthesis: PHB – **bio-based** and *biodegradable;*
- produced from sugarcane: bioPE – **bio-based** and *not-biodegradable.*

Bio-based applications	Used materials
Packaging foils and trays, production of paper and paper-based materials	Natural fibers, bio-based plastics and polymers
Engineering, furniture, textile engineering	Natural fibers, bio-based plastics and polymers
Sports accessories, car industry, toy industry	Natural fibers, bio-based plastics and polymers
Personal care and hygiene products, biomedical field, cleaning substrates, edible additives, bio-based fuel	Bio-surfactants, nutraceutical algae and bio-lubricants
Production of bioenergy and biofuels	Cellulose-based, non-edible biomass and agricultural waste

Table 7.1. *Application sectors of bio-based materials (adapted from Overbeek and Hoes 2018)*

Bio-based content is measured following guidelines given in the ASTM D6866-20 standard. Its principle relies on calculating the ratio between: a) fractions of the carbon content newly made up of biological materials or agricultural resources versus b) fossil carbon content. By measuring the presence of carbon-14 in the material, it is possible to differentiate between renewably sourced materials and carbon compounds sourced from long-dead, non-renewable petroleum. This is because carbon-14 is absorbed by plants while they are alive and then they decay over a few thousand years. Bio-based polymers have a lower carbon footprint than petrochemical equivalents or alternatives.

7.2.2. *Biodegradable plastics*

Polymers are considered to be biodegradable if they can degrade into water, carbon dioxide and biomass, through exposure to naturally occurring microorganisms (fungi, bacteria or algae). Methane may also be produced in anaerobic conditions (e.g. buried in landfill sites). Biodegradable plastics are produced using natural plant material (such as corn, fruit peels, starch and plants). They can be broken down by microorganisms, but this does not necessarily mean that they create non-harmful and soil-enriched compostable material. This must be confirmed. A non-negligible aspect is the esthetic appearance of these products. As they are made from renewable sources, they often lack transparency or have a strong odor that is not always well accepted among consumers.

When classifying a plastic as biodegradable, the environmental conditions and time frame must be specified as per accepted industry standards. The issue of being biodegradable or non-biodegradable could harm the perception that consumers have of bio-based plastics and their waste behavior, as they may feel uncertain about how to deal with their end-of-life options. The rate of biodegradation of different materials and its efficiency can vary on an exponential scale. Some examples are given in Table 7.2. Controlled environments are easily managed compared to natural environments that are much more complex. Even though the majority of biodegradable plastics have been designed to degrade in the aerobic environment of an industrial composting plant, they can also be disposed of in landfill due to deficient management practices, misinformation and lack of infrastructure, especially in developing countries (Rujnić-Sokele and Pilipović 2017). It should also be noted that many products are partly bio-based, the so-called "hybrids". This term denotes the partial content of biodegradable polymers and partial non-biodegradables. For example, in the starch/bioPET blend, the bioPET is the non-biodegradable part, while starch is the biodegradable part. However, when mixed (blended), they do not satisfy the biodegradability conditions so can only be considered as a pre-stage of the transition towards bio-based materials and economy.

Biodegradation occurs in three main stages:

1. Abiotic-deterioration and biotic-deterioration

Breakdown of polymers mainly occurs by physical and chemical degradations (combination of mechanical, light, thermal and chemical processes) and rarely by the action of biological reagents, that is, enzymes from microorganisms.

2. Bio-fragmentation

Fragments of polymers are cut into shorter chains that become more sensitive to microbial consumption. It is truly important that the degradation process does not stop at fragmentation, as this results in accumulation of microplastics.

3. Microbial assimilation and mineralization

Fragmented monomers assimilated in microorganisms are generated by cellular biomass, carbon dioxide or methane, depending on the availability of oxygen.

To ensure the integrity of products labeled as "biodegradable", different standards have been established and are given in Table 7.3.

Material	Temperature (°C)	Time (Days)	Biodegradation (%)
Polylactic acid, PLA	35 55 52	75 45 56	21 70 80
Polycaprolactone, PCL	37 55 55	277 150 75	12.5 62 92
Polybutylene succinate, PBS	35 55	139 50	2 0
Poly(3-hydroxybutyrate-co-3-hydroxyvalerate) PHBV	35 35	100 30	80 90
Polyhydroxybutyrate, PHB	55 35	50 225	90 100

Table 7.2. *Variations in biodegradation efficiency in some materials depending on degradation time and temperature (adapted from Quecholac-Piña* et al. *2020)*

It can be taken from these standards that a minimum of 70% (for anaerobic) or 90% (for aerobic) of the material should have biodegraded within 30 days (for anaerobic), or six months (for aerobic), or the duration of the testing procedure.

Country of issue	Degradation conditions	
	Anaerobic	Aerobic
US standards	ASTM D5511-18 ASTM D5526-18	ASTM D6400 ASTM D6868
EU Standards	EN 13432:2000	EN 14046:2004
Australia	AS 4736-2006	AS 4736-2006

Table 7.3. *Standards used for the determination of biodegradation*

Even though proper waste management considers the responsible disposal of any wasted material, it is fairly possible that some originally non-harmful materials may end up in seas, rivers or soil. Biodegradation can occur in different environmental conditions; generally, materials can be degraded in a marine environment, soil or water. As each environment offers different temperatures and microorganisms for biodegradation, its speed may vary from one environment to another. What is important is that after degradation, the resting material does not adversely affect the given environment.

It must be noted that oxo-degradable plastics do not enter into this category and do not meet composability or bioplastics standards (European Bioplastics 2017). These materials contain an additive that promotes their abiotic degradation until they reach a size that can be degraded biotically. They take too long to break down and leave plastic fragments that are not capable of being consumed by microorganisms.

7.2.3. *Compostable materials*

According to EN 1343:2000, compostable polymers can be broken down in an industrial composting plant by at least 95% within 12 weeks. In the US, according to ASTM D6400, compostable polymers should disintegrate for more than 90% of the original dry weight after 84 days. Both specifications require that biodegradable/compostable products completely decompose in a composting setting in a specific time frame, leaving no harmful residues behind.

In calculating the residues, packaging as a whole (with all components, inks and additives) must be taken into account.

Compostable materials are often used for agricultural mulch films, food service and containers.

7.3. Classification of bio-based materials

Bio-based materials are usually classified based on the source of raw materials.

7.3.1. *Polymers directly extracted from biomass*

Polymers directly extracted from biomass are made of molecules that are naturally present in the environment and are often treated as agricultural waste. This eco-friendly packaging option is used in edible food coatings or sachets, which can either be eaten or composted. They are also frequently used in the pharmaceutical industry as supporting matrices for encapsulation of active ingredients, medical and specific drug-releasing systems. For the food industry, they also serve as supporting matrices for the incorporation of functional ingredients in the design of novel functional food. For example, they can be used as coatings/beads for the incorporation of probiotic bacteria in chocolate and sweets. The most relevant actors in the food packaging industry are starch-based materials and regenerated cellulose (cellophane).

Usually, polymers which are directly extracted from biomass are classified either according to their structural material or depending on its source – animal source, marine source or agricultural source. They are composed of hydrocolloids – polysaccharides, proteins or hydrophobic compounds (lipids, waxes and resins) or their mixtures. In the carbohydrate group, there is a hidden subgroup that is very important for packaging design – natural fibers. They can be extracted from various plants such as wood, bamboo, cotton, hemp and wool.

Apart from the great natural character, it is also important that these molecules are extracted from the non-edible parts of food products or from the by-products and waste, thus not competing with food resources. For example, films made from milk proteins, discharged in cheese-making, can serve as wrappers for cheese products, and films which are made from a combination of fibroin protein from silk and chitin from shrimp shells (Shrilk, made by scientists at Wyss Institute, Harvard University in 2011). This is a promising step towards replacing aluminum foil. In addition to conventional shapes and types, such as films and cups, new initiatives include production for various food service utensils and items. One great example is the invention of bottle ring holders that aim to keep together and carry six or eight bottles in a pack made from by-products of the beer industry (wheat and barley).

Due to the different desirable characteristics, biomaterials are intensively investigated for the preparation of edible films and coatings.

The largest market of edible films is the film strip market. Another important usage is the decorative shapes that provide color and flavor in foods. These can also be used to deliver nutrients and other active ingredients. Extensive research studies have been conducted over past decades on the development of edible films and coatings as carriers of various active/functional (antioxidant, antimicrobial, enzymes, etc.) substances and as intelligent food packaging indicators. The most important value of this applications is improving the product shelf life and developing novel functional products with added nutritional and/or health value. However, the environmental aspects of a product as a whole remain quite debatable, since, if applied as edible coatings, these materials become an integral part of a food item. Consequently, most applications require additional packaging materials for retailing purposes.

Various polymers have been used for centuries as waxes for cheese ripening and storage and waxes on fruits, as well as to coat sausages etc. Due to growing interest in packaging-free concepts, novel ideas were principally created in the molecular gastronomy sector. For example, two-layered packaging, made from chitosan, alginate and calcium ions, keeps liquid food together by forming an electrostatic gel. Likewise, a drop of water can be caught in an edible drop so the consumer will just toss it in his/her mouth or it can be filled with a larger volume so that the consumer could drink it by using a straw. Straws can now be made of natural molecules, making them completely eco-friendly and, if needed, edible even. For example, the US brand Loliware makes flavored straws from alginate and agar that, unlike paper straws that get soggy, behave like plastic for 24 hours once they become wet, and can even be eaten. There are also trials for edible wrappers (Evoware, Indonesia) that are essentially made from seaweed and are used for burgers, instant noodle seasoning sachets and coffee pouches. The takeaway message is that the natural concept of bio-based material sources can be designed for use in different areas of the food industry, altogether making a step towards a circular economy.

General properties – Apart from lipids, materials made from natural molecules are made primarily from highly hydrophilic, crystalline, very brittle and rarely sealable materials. This raises the biggest issue regarding bio-based materials – they are difficult to scale up to pilot and industrial level. Natural polymers are unstable biological systems, and reactions with them and with external parameters can be expected, resulting in molecular changes over time (Cuq *et al.* 1997). In order to produce films, the addition of plasticizers, surfactants or other additives, as well as cross-linking agents, is almost always required (Vieira *et al.* 2011). Fast degradability can be considered as a positive characteristic from a biodegradation point of view, but it also means shorter applicability for food storage. Besides fast degradability, their main advantage is that they act as an excellent gas barrier in dry environments (for example for dehydrated food). The main events that occur during storage are: the reorganization of the film matrix (plasticizer migration, protein aggregation, starch crystallization, etc.); chemical and biochemical processes (lipid oxidation, non-enzymatic browning, degradation of active compounds, etc.); or enzymatic (proteolysis) degradation of film components, and microbial growth. Microbial growth remains an important issue because it affects the physicochemical properties (mechanical, solubility, barrier, optical) of the films. It can also reduce the functionality and even impact product safety (Cuq *et al.* 1997).

Hydrocolloid films can be produced by either wet or dry processing (Figure 7.2) (Blanco-Pascual and Gómez-Estaca 2016).

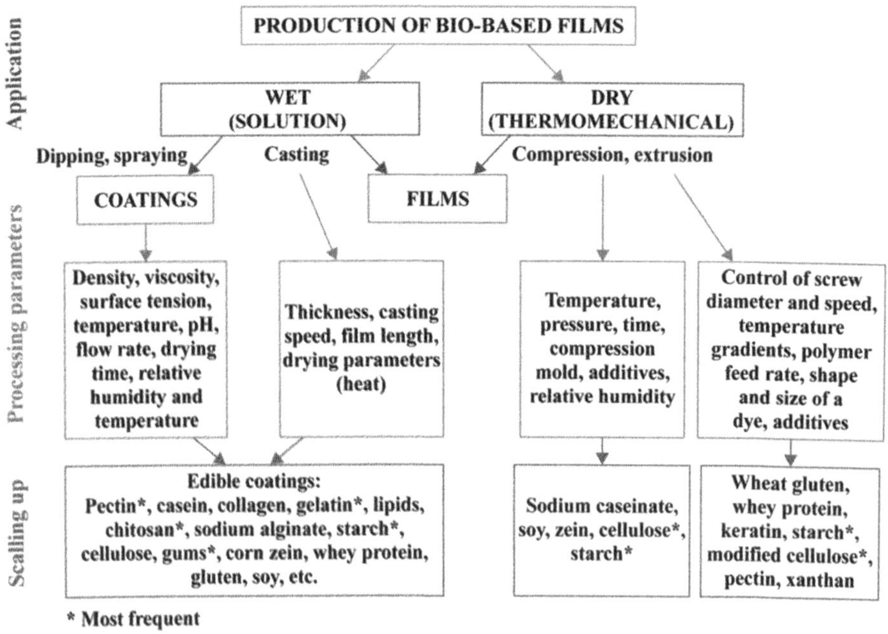

Figure 7.2. *Schematic representation of some of the methods of production of bio-based packaging materials from polymers of natural origin extracted from biomass (adapted from Blanco-Pascual and Gómez-Estaca 2016)*

Wet processing includes solvent casting or direct coating application on the food surface (dipping and spraying), while dry processing is based on the thermoplastic behavior in extrusion and compression molding (Hernandez-Izquierdo and Krochta 2009). Spraying is preferred for larger surface areas, while dipping is more convenient for irregularly shaped food items. As thermoplastic processing is adapted to synthetic materials, the versatile nature of natural macromolecules requires the precise optimization of barrel temperature, high pressure, moisture content (preferably low), plasticizer concentration and residence time (Blanco-Pascual and Gómez-Estaca 2016; Guerrero and de la Caba 2016). Scientists continue to optimize these methods as they have fairly shorter processing times and significantly higher production efficiency.

Over the past few decades, numerous research and scientific studies were carried out on improvements to stability and on the use of naturally issued bio-based polymers. A lot of work has been done on the different molecules, but only a few of them present significant relevance for industrial applications. Thus, only those that already have proof-of-concept, or reach at least Technology Readiness Level 6

(TRL6), will be described in detail within dedicated subsections. Technical challenges still need to be overcome before edible packaging can enjoy more widespread use.

7.3.1.1. *Proteins*

Protein-based films and coatings can be made from plants (wheat, soy, corn zein, silk, etc.) or animal proteins (casein, whey, meat, feather keratin, gelatin, etc.) Generally, **protein films** are characterized by poor water and moisture resistance and brittleness. In order to overcome these issues, they often require one, or a combination, of the following modifications: denaturation and aggregation, plasticization and chemical cross-linking with aldehydes (mainly formaldehyde and glutaraldehyde) and sugars. Water is considered to be the best plasticizer for hydrocolloid-based films, followed by glycerol, sorbitol, oleic and linoleic acids, sugars, etc. They are heteropolymers composed of various amino acids, joined together by peptide bonds which are amide linkages formed by the condensation reaction of amino acids, offering great structural complexity (secondary, tertiary and quaternary). The structure of different molecules is shown in Figure 7.3.

a)

b)

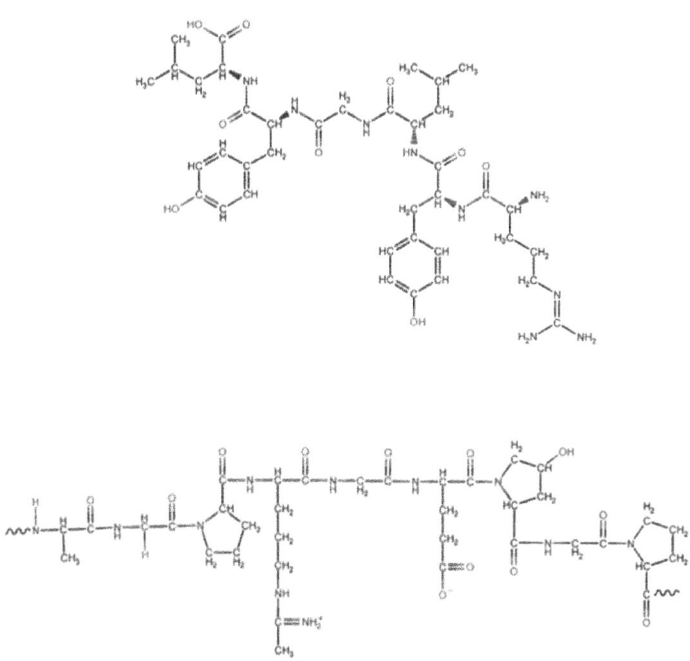

Figure 7.3. *Chemical structure of some protein-based films: a) casein and b) gelatin*

This property provides them with great functional diversity and application diversity in the food and non-food packaging sectors, in reconstructive surgery, tissue engineering, etc. (Guilbert and Cuq 2005). Blending technology uses them as reinforcement agents. They are also well known for improving the barrier properties of blended materials (in combination with carbohydrate polymers) and they facilitate the effective incorporation of bioactive and functional ingredients. In interfacial engineering technology, they serve as natural emulsifiers to create nano-sized emulsions with a lower fat content, or as preventing agents for flocculation of oil droplets under acidic conditions in the stomach (Sekhon 2010). In non-food packaging applications, they could be used to produce various commercial products like shopping bags, mulch films, sanitary products, etc. The most common examples are given in Table 7.4.

Film type	Source	Packaging characteristics
Casein	Milk protein, available as calcium and sodium caseinates and derived as a by-product from the dairy industry	Water-soluble, brittle, UV light-resistant, good barrier against oxygen and water vapor (in low-moisture environments), good barrier against microbiological contamination, possibility of thermomechanical processing
Collagen	Extracted from animal connective tissues	Good barrier against oxygen and solutes, hydrophilic, poor mechanical strengths, biocompatible and non-toxic
Corn zein	Extracted from corn	Glossy appearance, low water solubility, greaseproof, hydrophobic, brittle
Gelatin	Produced from collagen by thermal denaturation, from animal skin and bones (bovine, porcine and fish sources)	Amphiphilic behavior, plasticized films show good mechanical properties, odorless and tasteless, transparent, highly hygroscopic, high content of amino acids glycine, proline and hydroxyproline, poor water vapor barrier
Soy protein	Extracted from soybeans	Poor mechanical strength and high-water sensitivity
Wheat gluten	Extracted from wheat	Poor mechanical strength, biocompatible, soluble in alcohol, brittle, shiny, effective oxygen but poor water vapor barriers, frequently cross-linked with glutaraldehyde or heat cured

Table 7.4. *Some of the most commonly used protein-based hydrocolloids used as edible films, with main characteristics*

7.3.1.1.1. Casein

Casein, Figure 7.3(a), is a water-soluble milk protein, naturally available as calcium and sodium caseinates and derived as a by-product from the dairy industry. These proteins have a very complex structure that provides them with different functionalities (Pérez Gago 2012). Because of high brittleness, films made from casein require the addition of plasticizers, such as glycerol (Cho *et al.* 2014). They can resist UV light due to the presence of amino acids that can form disulfide bonds and result in even better properties than low-density polyethylene (LDPE) or orientated polypropylene (OPP) (Guerrero *et al.* 2011). In a dry environment, due to the presence of hydroxyl and amino groups, they form good barriers against oxygen and to other non-polar molecules (Saez-Orviz *et al.* 2017). They are mainly produced by wet methods, but there is also the possibility of thermomechanical processing. For example, Belyamani *et al.* (2014) produced sodium caseinate films by injection, compression molding and blowing extrusion. In order to overcome mechanical and barrier imperfections, casein films may be chemically modified by alkali-treatments through cross-linking, or they might be mixed with other hydrocolloids such as gelatin, pectin or others (Chambi *et al.* 2006; Wihodo *et al.* 2013). In the non-edible sector, casein is used as a natural glue in the manufacturing of fabrics, adhesives, plastics and protective coatings.

7.3.1.1.2. Whey proteins

Whey comprises 20% of milk protein. It is derived as a by-product during cheese manufacturing. Films can be obtained either from concentrates (WPC) or isolates (WPI). Whey protein and its derivatives have a broad range of characteristics, such as being antioxidant, anti-inflammatory and effective in the management of various human health conditions. It is also used as a food additive and in terms of packaging materials, it is used as an encapsulation agent, in edible coating, in active packaging and for tailored delivery systems (Patel 2015). Whey protein films are transparent, flexible and have good water, gas, aroma and oil barrier properties. Thermal degradation denatures whey protein, resulting in a more cohesive and stronger film than native proteins (Perez Gago *et al.* 1999). It is used as a coating in order to improve the availability of iron in iron-fortified foods (Martin and de Jong 2012). Whey protein microcapsules are used to protect antioxidants, drugs and vitamins such as folic acid (Pérez-Masiá *et al.* 2015). They have also been shown to be good carriers of volatile aroma compounds, which could be released when submitted to a triggering effect at an exact place and time (Wicochea-Rodríguez *et al.* 2019). Layer-by-layer (whey protein isolate/xanthan gum) microencapsulation was shown to enhance the survivability of *Lactobacillus bulgaricus* and *Lactobacillus paracasei* under simulated gastrointestinal juice and thermal conditions (Xiao *et al.* 2020). WP coatings are used as clear and glossy coatings with good mechanical properties on paper and various plastic materials (PP, LDPE, polyvinyl chloride, PVC), and as replacements for conventional petrochemically sourced oxygen barrier materials, such as poly (ethylene-co-vinyl alcohol) (EVOH) and are currently being routed to commercialization. Besides being renewable, they can be removed from a multilayered material (delaminated) by enzymatic removal so they also enhance the recyclability of those materials, contributing to end-of-life product management (Bugnicourt *et al.* 2019). There was significant progress in scaling up within two European projects: WHEYLAYER and WHEYLAYER2. Whey was also shown to serve as molecules for engineering other materials, namely PHB, which is produced via microbial fermentation of surplus whey derived from the cheese industry (EU Wheypack project).

7.3.1.1.3. Gelatin

Gelatin, Figure 7.3(b), is a complex polypeptide, water-soluble protein obtained from bone collagen or connective animal tissue. It is a low-cost material used in the food and food packaging sector, pharmaceutical industry and cosmetics industry. It is well known for its use in capsules, for the delivery of a liquid or powdered medicine. It is also often used for culinary purposes due to its gel-forming properties at approximately 35°C, as a thickener and as a food additive. Two types are mainly used: bovine and fish gelatin. Both types have a large number of various side-suspended groups that can facilitate covalent (with glutaraldehyde or genipin) and enzymatic (transglutaminases) cross-linking and derivatization, which is an important feature in overcoming its moisture sensitivity issue. Gelatin is quite a processable material but its film properties hardly compare to the performance of plastic packaging (Pena *et al.* 2013). It was extensively studied as an edible coating for meat and fish products, and fruits and vegetables (Ramos *et al.* 2016). In order to decrease its brittleness, it is often plasticized by adding water, glycerol, fatty acids and polyethylene glycol. Gelatin films are transparent and typically exhibit good gas barriers. Several attempts of blending it with other polymers have been completed in order to reduce its water sensitivity. So far, it has been blended with PLA in a multilayer sheet manufacture by thermo-molding (Martucci and Ruseckaite 2010) or in a multilayered structure prepared by dip coating (Pena *et al.* 2009). In combination with inorganic fillers like montmorillonite or cellulose nanofibers, it can also be used as good support for the creation of nanobiocomposites (Wastwood *et al.* 2010).

7.3.1.2. *Polysaccharides, gums and fibers*

Polysaccharides are long chains of branched or unbranched carbohydrate molecules that have great film-forming properties (Figure 7.4). All of the common polysaccharides contain glucose, such as the monosaccharide unit. The main characteristic of **polysaccharide-based films** is that they are a good barrier against the transport of oxygen and carbon dioxide and a low barrier against water vapor. In addition, their tensile strength is quite good, approaching that of synthetic polymers. The biggest mechanical concern is the low elongation at break that is far from the desirable values found for synthetic polymers. Some of the most common examples are given in Table 7.5.

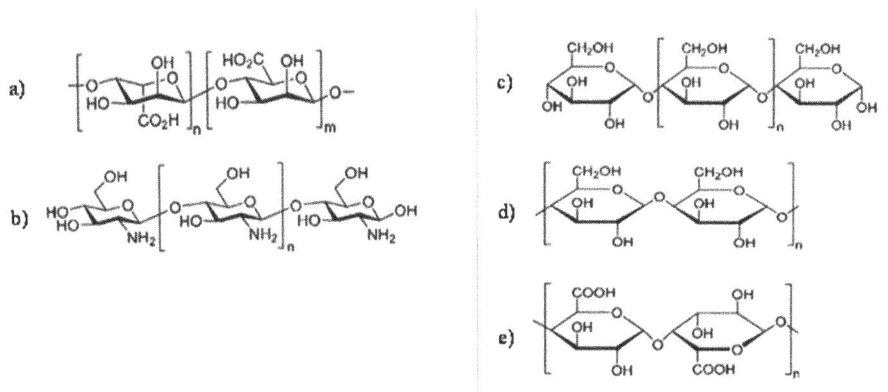

Figure 7.4. *Structural formula of some polysaccharides: a) alginate; b) chitosan; c) starch; d) cellulose; and e) pectin*

Polymer	Source	Packaging characteristics
Chitosan	Extracted from the exoskeleton of crabs, shrimps and shells in the form of chitin	Soluble in weak acids, good barrier against oxygen and carbon dioxide but very hydrophilic, flexible, durable, strong, tough and hard to break
Sodium alginate	Produced by and refined from various genera of brown algae or by some bacteria as exopolysaccharide	Very soluble in water and tends to form lumps with insufficient shear, forms gel and films in the presence of divalent cations (usually Ca^{2+} ions), low tensile strength, clear and transparent
Starch	Isolated from plant roots, stalks, crop seeds and crops such as rice, corn, wheat, tapioca and potato	Tensile properties of starch are suitable for the production of packing materials, and glycerol is added to the starch as a plasticizer

Cellulose	Isolated from plant cells	Great foaming properties, increased solubility in organic solvents with improved sealability, water sensitive but a good gas barrier in low-moisture environments
Pectin	Extracted from plants	Great gelling properties, hydrophilic and has a tendency to form gel in acidic conditions or in the presence of divalent cations, biocompatible, selective gas permeability, poor water vapor barrier and low mechanical properties

Table 7.5. *Some of the most commonly used polysaccharide-based hydrocolloids used as edible films, with main characteristics*

Gums are complex hydrocolloids of different molecular weights that change structural conformation after dissolving in water. They are usually considered to be non-starch, water-soluble polysaccharides with commercial importance (BeMiller 2008). They are made of linked sugar monomers of two or more different sugars. During the solvation and because of the abundance of free -OH groups, complex molecules unwind, imbibe water and cause thickening. They can be solvent cast to make films with various tensile strength, clarity, and solubility. Their main functions include thickening, gelling, film forming and foaming/emulsification. They can be linear or branched, forming from two- up to six-fold helixes; linear are of greater industrial relevance as they have better film-forming properties.

Fibers are used for processing of different packaging but mostly as additives to improve some polymer properties, such as better mechanical resistance. Their main disadvantage is their hydrophilic nature which makes it difficult to bind fibers with hydrophobic polymers, so it is necessary to use chemical or physical treatments to improve the interface between natural fibers and polymers. Materials known for their high fiber content are paper and board. Most fibers come from wood sources; however, there are also different widely-used types that come mainly from agricultural residues, such as straw (of wheat, rye, barley and rice), sugar cane bagasse, cotton, flax, bamboo, corn husks and so on. When using fibers, there are two precursors used: straw and bagasse. Bagasse presents a fibrous residue that remains after crushing the substrate to extract its juices, while straws are considered to be a residue of agricultural production, mainly burned directly in the fields.

7.3.1.2.1. Cellulose and derivatives

Cellulose is the most abundant, cheap (while raw material) natural polymer which is the major component of plant cells. Cellulose fibers have been used as a major component of paper for centuries. It has also been widely used since the 19th century to make cellophane films, viscose fibers and other derivatives. In its raw form, cellulose is highly crystalline, fibrous and insoluble but has poor mechanical properties. Due to its hygroscopic nature, it swells rapidly which limits its film-forming capacity. It is used in food production, food packaging, as a lubricant, for medical purposes, paper and textile sizing, as glue and in cell culture production. To overcome this hurdle, cellulose is often modified.

Two main groups of its derivatives are ethers and esters. Their production requires two steps: a) alkalization and etherification for ethers and b) acidification and esterification for esters. The chemical structures are given in Table 7.6.

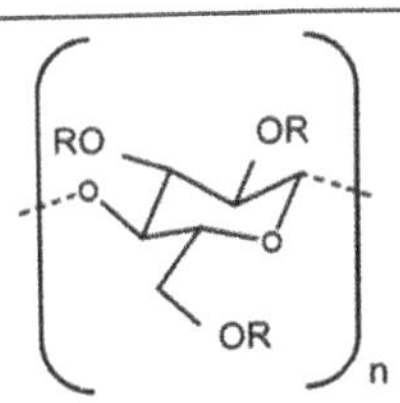	**Cellulose** R=H **CMC** R=H or CH_2CO_2H **HPC** R=H or $CH_2CH(OH)CH_3$ **MC** R=H or CH_3 **HPMC** R=H or CH_3 or $CH_2CH(OH)CH_3$
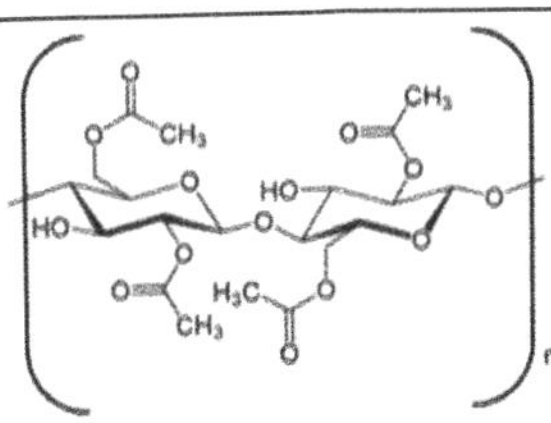	**Cellulose acetate**
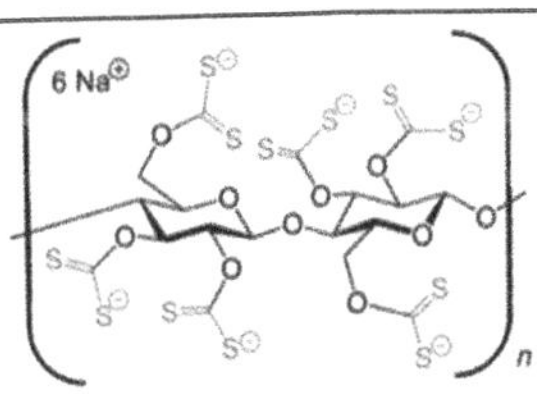	**Cellophane**

Table 7.6. *Structural differences in various cellulose derivatives*

In etherification, different groups are substituted for the -OH groups of sugar, such as methyl, hydroxypropyl or ethyl groups, to make hydroxypropyl cellulose (HPC), hydroxypropyl methyl cellulose (HPMC), carboxymethyl cellulose (CMC) or methyl cellulose (MC). The addition of functional groups results in an increased non-polar character and relatively high resistance to fats and oils. Depending on the degree of substitution, chain length and degree of polymerization, the resulting material has great foaming properties, increased solubility in organic solvents and improved sealability. However, modifications significantly impact its biodegradability; a higher degree of substitution means a larger number of ether linkages and therefore lower biodegradability (Andersson 2008). Apart from HPC, cellulose ethers are non-thermoplastic so they do not provide heat-sealable properties, which is an important feature when they are applied as functional coatings. These materials have been used extensively as food coatings and for the production of water-soluble and edible wrappers (Majid *et al.* 2018). The addition of hydrophobic compounds, such as fatty acids, is a commonly proposed method for increasing the moisture barrier.

The esterification process includes the reaction of an organic or inorganic acid substituent with the three hydroxyls of an anhydro glucose unit (Kuusipalo *et al.* 2008). Two main representatives of this group are cellulose acetate (CA) and cellophane. The main advantage of CA is that it can be processed thermoplastically, which is applicable to injection molding and extrusion, while cellophane cannot. Cellophane is a regenerated cellulose, produced with aggressive chemicals like NaOH, H_2SO_4 or CS_2. It is transparent and is widely used for food packaging, for example, confectionery. It is a pseudoplastic material, so once folded it does not fold back. Because of its highly sensitive nature to moisture, it is often coated with nitrocellulose wax (NC-W) or polyvinylidene chloride (PVDC).

Microcrystalline cellulose, nano-fibrillated cellulose and ligno-nanocellulose are different cellulose derivatives, designed and used as natural performance enhancers for various bio-based packaging materials. In different formulations, they have been shown to improve barrier properties, surface wettability, as well as the mechanical and thermal properties of composite materials.

7.3.1.2.2. Chitosan

Chitosan, Figure 7.4(b), is a natural polymer, produced commercially by deacetylation of chitin, which is the structural element in the exoskeleton of crustaceans (such as crabs, shrimps, oysters and other mollusk shells) and cell walls of fungi. It can either be extracted or enzymatically deacetylated (van den Broek *et al.* 2020), but this process affects its quality. In commercial chitosan, the deacetylation degree varies from 60 to 100% with a molecular weight from 3.8 to 20 kDa. It has a huge number of amino groups (the molecular formula is poly-β-(1 → 4)-2-amino-deoxy-d-glucopyranose) that makes it impossible to solubilize in water, but it is easily soluble in a weak acidic medium (e.g. 1% (v/v) acetic acid). Owing to its structure, it was recognized as a promising polymer that can be spun into fibers, cast into films or precipitated in various microstructures (van den Broek *et al.* 2020). It is also the most widely used polymer for the development of nanocarrier systems (Kashyap *et al.* 2015). The majority of applications are in biomaterials, in pharmaceuticals as a wound-healing agent, in cosmetics, as a flocculent clarifier, as a gas-selective membrane in agriculture and for food packaging and preservation, as an antimicrobial agent or as a carrier of functional compounds in bio-based edible films and coatings. In the food packaging field, its main use is in edible film or coatings. The antimicrobial activity of chitosan is due to the protonation of NH_2 to NH_3^+ which then associates with polyanions to form complexes and bind to anionic sites on microbial cell walls (Goy *et al.* 2009). Chitosan films are flexible, durable, strong, tough and hard to break. They decrease the respiratory rate of food and also inhibit microbial growth. They provide a great gas barrier, particularly under low-moisture conditions. They are either made alone or are mixed with other hydrocolloids.

7.3.1.2.3. Starch

Of the polysaccharide family, starch – Figure 7.4(c) – is the most abundant, renewable and commercially applied biopolymer. It is made by most green plants as an energy source and is recovered from stalks, roots or seeds by wet grinding, sieving and drying. Major plant sources are potato, cassava, corn, tapioca, wheat and rice. Starch is made of linear amylose and branched amylopectin, whose ratio plays an important role in processing (Table 7.7). The amylose content may vary from 20 to 25%, while the amylopectin content varies from 75 to 80% in weight, depending on the type of plant. Amylopectin is a far greater molecule than amylose. It is a crystalline material that undergoes spontaneous destructurization when treated with water and subjected to heat and shear (Versino *et al.* 2016). In other words, it can be thermomechanically processed. It is commonly named thermoplastic starch (TPS) when it is processed at high temperatures in the presence of plasticizers, such as water and glycerol. Its tensile properties are suitable for the production of packing materials when glycerol, sorbitol, urea lipids or other polyols are added as plasticizers (Nafchi *et al.* 2013). The main disadvantages of TPS that limit usage are retrogradation, migration of plasticizers and unsatisfactory mechanical properties, particularly in wet environments and during aging, its fragile nature caused by its relatively low glass transition temperature and the lack of relaxation. Depending on processing temperatures, the type of additives, cross-linking and other reinforcements, different physical, mechanical and thermal properties were addressed (Nafchi *et al.* 2013). For trade applications, TPS is regularly modified in reactive extrusion processes and is mixed with eco-friendly polyesters, polycaprolactone, cellulose acetate, polylactides, etc. (Nafchi *et al.* 2013).

Amylose	Amylopectin
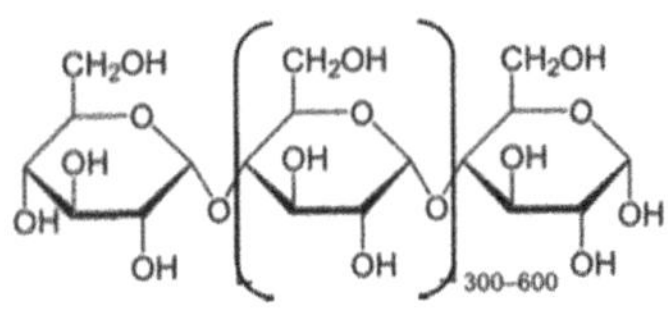	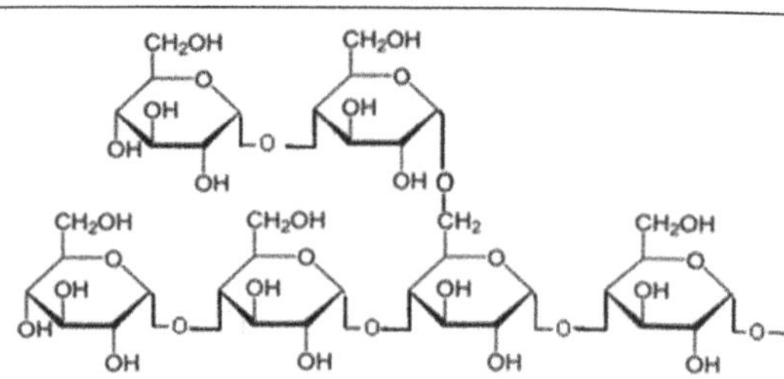

Table 7.7. *Chemical structure of amylose and amylopectin*

7.3.1.3. *Waxes, oils, and lipids*

Waxes, oils and lipids are mostly included in composite systems to improve water resistance, or as plasticizers. Triglycerides, acetylated monoglycerides, fatty alcohols, fatty acid esters of sucrose, fatty acids, oils (vegetable, animal and mineral), beeswax, candelilla wax and carnauba wax are the most commonly used in the manufacture of edible coatings and films.

7.3.2. *Polymers produced by chemical synthesis from biomass monomers*

There are a wide variety of bio-based molecules available for producing polymers with the use of classical chemical polymerization reactions. For a long time, research and development was focused on polyesters. Among them, PLA, bioPE and bioPET have been widely used as packaging.

7.3.2.1. *Polylactic acid*

Polylactic acid, PLA, is a linear aliphatic polyester synthesized from the lactic acid monomer, a basic chemical produced by fermentation from carbohydrate feedstock. During the fermentation process, one molecule of glucose is converted into two molecules of lactic acid. Lactic acid (LA) is optically active and has two enantiomeric forms, L- and D-lactic acids. PLA can be prepared by the direct condensation of lactic acid through azeotropic dehydration, followed by condensation and by the ring-opening polymerization of the cyclic lactide dimer. The direct polycondensation of lactic acid at elevated temperatures, with water removal via vacuum, results in prepolymer PLA chains with low molecular weight (several thousand molecular mass units). These polycondensates do not solidify at room temperature but remain in a highly viscous liquid state. This brittle polymer is, for the most part, unusable unless external coupling agents are employed to increase its chain length. In azeotropic dehydrative condensation, it is possible to produce high molecular weight PLA without chain extenders or adjuvants. However, its major drawbacks are fast degradation and hydrolysis, in addition to toxic catalysts. Finally, in order to produce high molecular materials, the ring-opening polymerization (ROP) of lactides and cyclic dimers of lactic acid have to be used. These intermediates (cyclic dimers of lactic acid) on the other hand, result from the depolymerization of low molecular weight PLA-polycondensates (Kabasci 2020) (Figure 7.6).

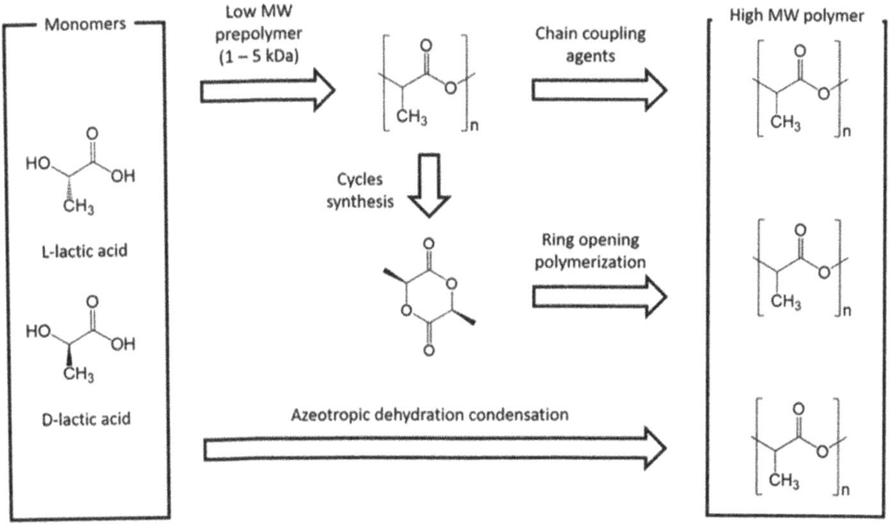

Figure 7.6. *Three processes for synthesis of PLA*

According to commonly-adopted and more specific nomenclature, the polycondensation products of LA are called **polylactic acid**, and the products prepared by the ROP of lactides are called **polylactides**. Regarding the effectiveness of biological production, L-lactic acid has superior productivity compared to D-lactic acid, which presents a small amount. Therefore, poly (L-lactide) (PLLA) is more commonly commercialized (Nakajima *et al.* 2017).

Pure homopolymers, composed of one of the two optical isomers L or D, are optically active and able to crystallize. The properties of polylactides vary, to a large extent, depending on the ratio and distribution of the two isomers and the molecular weight (MW) of the polymer. PLA can either be amorphous or semi-crystalline, depending on the percentage of L or D forms and thermal history. Poly (L-lactide) and poly (D-lactide) are semi-crystalline polymers, while poly (L,D-lactide) is amorphous. For amorphous PLAs, the glass transition temperature (T_g) determines the upper use temperature for most commercial applications. For semi-crystalline PLAs, both the T_g, which is approximately 57°C, and melting temperature (T_m), ranging from 130 to 180°C (depending on the structure), are important. They are key parameters for determining the use of temperature across various applications (Robertson 2016).

Next to cellulose, starch and their derivatives, PLA is the bio-based resin with the largest production quantity in the world. Compared to these natural polymers, PLA is a thermoplastic material with a high modulus (approximately 3.5 GPa) and strength (60–70 MPa), and it can be processed using conventional techniques and equipment. However, its brittleness and low elongation at break point are major drawbacks that limit its application. To overcome these limitations, extensive research and material development efforts have been undertaken concerning the addition of plasticizers, polymer blends and nucleation agents (Kabasci 2020). Plasticizers decrease the T_g of PLA, which results in lower stress at yield and higher elongation at break at room temperature, that is, improved flexibility of films and sheets.

Today, there are several PLA resins available for different applications such as extrusion, injection molding and melt spinning.

Several companies in Belgium, China, Japan, Switzerland and The Netherlands have installed small PLA production capacities. NatureWorks LLC (USA) and Total Corbion (The Netherlands) remain the world's only large-scale commercial producers of PLA. Though PLA prices have fallen significantly in the past years, they are still not competitive with commodity plastics. NatureWorks LLC had assumed it would reach cost parity with PET and PS due to an increased economy by 2015. In terms of its end-of-life fate, PLA can be recycled and degraded. It is even compostable at high temperatures, degrading to lactic acid, which can be metabolized naturally. Degradation of PLA is primarily due to the hydrolysis of ester linkages, which occur more or less randomly along the backbone of the polymer. The hydrolysis of PLA through the backbone ester groups is considered slow, and under conditions of ambient moisture and temperature, it can take several years. However, it is possible to accelerate the process by subjecting PLA to temperatures above 50°C. Even so, PLA is still considered to degrade slowly compared to organic waste in commercial facilities (Lambert and Wagner 2017).

Lifecycle assessment has shown reductions of up to 40% in greenhouse gas emissions and up to 25% in non-renewable-energy use for PLA, compared with those for petrochemical-derived polymers such as polyethylene (PE) or PET. However, the production of PLA might have other environmental impacts, including the use of water and fertilizer, which are more difficult to compare with the impacts of fossil-fuel extraction, purification and storage. Another substantial hurdle is to replace virgin crops (such as maize or sugar cane) with lignocellulosic or waste biomass (Zhu *et al.* 2016).

7.3.2.2. *Bio-polyethylene*

Polyethylene (PE), by annual production volume, is the most important conventional plastic material. It is extensively used in a wide range of applications, especially packaging and construction. Conventional PE is produced by the radical polymerization of the monomer ethylene, which is produced from oil feedstock. Alternatively, bio-based ethylene (Figure 7.7) can be prepared by the catalytic dehydration of bioethanol, which can be made by fermentation of various feedstocks, including sugar cane, sugar beet and wheat grain. Bio-based polyethylene (bioPE) can then be produced using existing polymerization reactors. BioPE is chemically identical to its fossil-based counterpart and can be used in the same applications (foils, films, kitchenware, bags, etc.). In addition, it has a complete infrastructure for processing and recycling, just like PE.

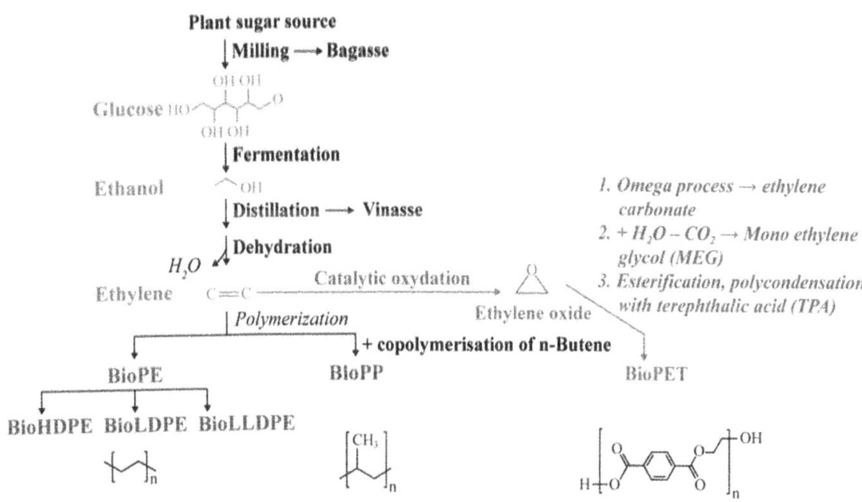

Figure 7.7. *Production processes of bio-based PE, PP and PET*

From BioPE, it is possible to produce all the PE types: high-density polyethylene (HDPE), low-density polyethylene (LDPE) and linear low-density polyethylene (LLDPE).

Like PE, bioPE consists of short repeating monomer units that have strong inter-chain hydrogen bonding, making a highly hydrophobic polymer that limits enzyme susceptibility; this is so that only the polymer surface, with a limited number of free chain ends, is available for microbial action (Lambert and Wagner 2017). Thus, the downside of bioPE is that it is not biodegradable.

The two largest producers of bioethanol from sugar cane in the world are based in Brazil because of favorable geographical conditions, resulting in high yields of sugar cane per hectare. It was there that ethylene and PE production from biomass became economically attractive. The Brazilian petrochemical company, Braskem, and the US company, Dow Chemical, have built their large-scale production facility of bioPE for many applications, including packaging (food and cosmetics), toys and automotive parts (Storz and Vorlop 2013). The environmental footprint for this bio-based material is favorable. One ton of bioPE produced captures 2.5 t of carbon dioxide from the atmosphere (Kabasci 2020). Thus, many companies will probably switch to bioPE to lower their carbon footprint and increase their impact in the circular economy. Hence, the price and availability of the raw materials are the key criteria for siting a bioPE plant.

7.3.2.3. *Bio-poly(ethylene-terephthalate)*

Poly(ethylene-terephthalate) (PET) is a thermoplastic polyester produced by the polycondensation of ethylene glycol (EG) and terephthalic acid (TPA). PET is one of the most employed plastics and is predominantly used in fiber (textiles) and packaging (bottles and food containers) applications. PET production comprises two steps. The first is the esterification of TPA with EG, and the second is polycondensation, in which PET is formed via a transesterification reaction. Bio-based ethylene glycol (bio-EG) production starts from bioethanol, followed by dehydration to ethylene, oxidation to ethylene oxide and finally hydration to EG. Provided bio-based ethylene is available, the production of bio-EG is straightforward and can be done in existing plants. However, the production of bio-based TPA is not as simple. Many processes have been developed and demonstrated on a small scale but none have reached a larger, commercial scale so far (Dodds and Humphreys 2013). The bio-PET used today is partly bio-based, made of bio-based EG and petroleum-derived TPA. The Coca-Cola Company applied this process for the production of their PlantBottle, which was launched in 2009 and was made of partly bio-based PET (bio-EG as 30% of the total PET weight and the remaining 70% is the petroleum-derived TPA). Their express goal is to switch to a fully bio-based bottle by 2025. Meanwhile, other large food producers use the PlantBottle for their products, and their competitor, Pepsi, announced plans to introduce their first fully bio-based bottle; however, they have only provided lab samples of the new bottle so far. In The Netherlands, Avantium is developing a

100% bio-based polyester replacement for PET, poly(ethylene furanoate) (PEF). It can be prepared from EG and 2,5-furandicarboxylic acid (FDCA), a dicarboxylic acid that is structurally comparable with PET and can be produced from renewable resources (Storz and Vorlop 2013). The bioPET has an identical chemical structure to the petrochemical-PET, that is, it can be recycled but is not biodegradable. However, the lifecycle assessment of the present generation of bio-PET shows a reduction of 20–50% in the emission of greenhouse gases in comparison to petrochemical-derived PET and, thus, the advantageous switch to manufacturing this kind of product.

An advantage of developing bio-derived monomers as direct substitutes, like those used in the production of PET and polyethylene, is that the processing and applications of the resulting polymers are identical, thereby simplifying their adoption and accelerating their uptake. This is particularly important for PET, one of the few polymers for which large-scale recycling infrastructures exist, and lifecycle assessment indicates positive contributions to its sustainability if it is recycled (Zhu *et al.* 2016).

7.3.3. *Polymers produced directly by natural or genetically modified organisms*

7.3.3.1. *Polyhydroxyalkanoates*

Polyhydroxyalkanoates, PHAs, are a group of renewably sourced microbial polyesters obtained through the fermentation of sugar and lipids by a very wide range of microorganisms, including the genera *Alcaligenes*, *Bacillus* and *Pseudomonas*. PHAs are accumulated intra-cellularly and serve as carbon and energy reserves. They occur naturally and can be harvested with excellent yields from microorganisms directly, without the need for intermediate monomer isolation. Biosynthesis is achieved by culturing bacteria under growth-limiting conditions and results in the accumulation of considerable quantities of the polymer in the cytoplasm. Under controlled fermentation conditions, some compounds can accumulate up to 90% of their dry mass as polymers (Robertson 2016). PHAs are commercially produced using energy-rich feedstock, transformed into the fatty acids which the bacteria feed on. In addition, hemicellulose hydrolysates, glycerol and methanol are very promising carbon sources for the sustainable production of PHAs. During the industrial production of PHAs, after a few "feast–famine" cycles, the cells are isolated and lysed. The polymer is extracted from the remains of the cells, purified and processed into pellets or powder. In addition to using pure feedstock as a source of energy for PHA production, there are ongoing efforts to use energy-rich wastewater as feedstock and thus as PHAs. Nowadays, it is possible to design bacterial strains to produce PHAs with desired properties. Production of PHAs can be improved by using genetic modification, either by increasing the number of PHA-producing bacteria or by modifying plants to start making PHAs (Nakajima *et al.* 2017).

PHAs are a group of microbial polyesters (Figure 7.8) that are biocompatible, fully biodegradable and optically active. Due to such characteristics, they are suitable for the production of packaging materials (Zhu *et al.* 2016). PHA molecules extracted from bacterial cells have a sufficiently high molecular mass which exhibit characteristics similar to PE and polypropylene (PP). PHAs mainly consist of 3-hydroxyalkanoates as monomeric building blocks. In general, two groups of PHAs can be distinguished: short-chain-length (scl) and medium-chain-length (mcl) PHAs. In the case of scl-PHAs, the monomers consist of three to five carbon atoms, while mcl-PHAs contain 6 to 14 carbon atom monomers. Polyhydroxybutyrate (PHB) is the PHA type most widely synthesized in nature and is classed as a short-chain-length PHA. Using a special growth bacterial medium, 3-hydroxybutyrate (HB) and 3-hydroxyvalerate (HV) can be associated to form a random copolymer poly(3HB-co-3HV) or PHBV. The latter has good chemical and moisture resistance, good O_2 and aroma barrier properties.

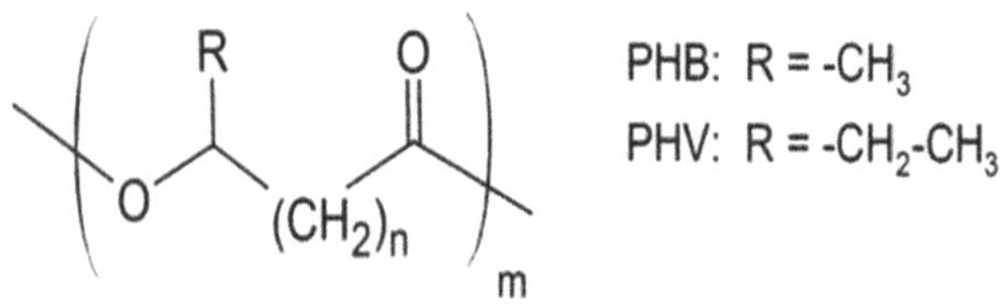

Figure 7.8. *Different types of polyhydroxyalkanoates (PHAs)*

The simplest PHA, P3HB, is very crystalline (up to 60%) and shows thermal and mechanical properties that can be compared to PP and PE. However, the flexibility is two orders of magnitude lower compared to PP (Storz and Vorlop 2013). By changing the ratio of HV to HB, the resulting copolymer has the same properties as PP and LDPE in terms of flexibility, tensile strength and melting point.

PHA-depolymerase detected in several bacteria and fungi in soil can decompose PHAs, and in an aerobic environment, the degradation produces CO_2 and CH_4. PHAs can also be compostable, as seen in composting trials, up to 85% within seven weeks (Robertson 2016). Several commercial PHA grades, optimized for a broad range of applications, including injection molded items, films, foams and fibers, are currently available. However, PHAs are still among the most expensive bio-based plastics. An important cost is the energy required for the intensive extraction and purification steps, which is also adversely contradicts the principles of green chemistry. Although there is a high cost for PHA production, current advances in fermentation, extraction and purification technology, in addition to the development of superior bacterial strains, are likely to reduce the price of PHAs, which remains close to PLA and other aliphatic polyesters. Processing PHAs is challenging compared to conventional petroleum-derived polymers because of their sensitivity to thermal degradation, and slow solidification due to slow crystallization. The degradation temperature of PHAs is approximately 180°C, which is near the optimum processing temperature for polyester (Nakajima *et al.* 2017).

7.3.3.2. *Bacterial cellulose*

Bacterial cellulose (BC) is an extracellular naturally occurring polysaccharide. It is produced in a highly pure form by some bacteria, including the genera of *Gluconacetobacter*, *Acetobacter*, *Agrobacterium*, etc. The yield and properties of BC depend on several factors, including the bacterial strain used, the composition of the culture medium and the operational conditions applied during the cultivation process. The composition of the culture medium determines the material morphology and the physical properties of the resulting material, which affects the range of possible applications. Microorganisms can be genetically modified to produce BC with desired properties. Furthermore, agro-industrial waste can be used as a growth medium so it is economical. To ensure favorable production yields and controlled costs, attention needs to be given to the species and genetic modification of the bacteria used, feedstock type, composition and the type of reactor used for the production process. This kind of polymer is considered to have many potential applications in the food packaging industry but so far remains to be exploited and no BC films have been commercialized for food packaging as yet. But for further application, films of BC can be prepared by air-drying a gel sheet on a flat surface using the heat press method (temperature 150°C, pressure 49 kPa). Recent research studies displayed the potential of BC as a reinforcement in nanocomposites. BC has advantages when compared with conventional cellulose, since it is crystalline in nature, can be obtained in a pure form and does not contain lignin, pectin or hemicellulose. It also possesses specific chemical and physical properties which are better for certain purposes than plant-derived cellulose. The advantages are stable mechanical strength and very low gas permeability in its dry state, which offers applications such as in food packaging, in addition to its non-toxicity, edibility and biodegradability (Shi *et al.* 2014).

7.3.4. *Petrochemical-based polymers*

7.3.4.1. *Polycaprolactone*

Polycaprolactone, PCL, is a semi-crystalline petrochemical polymer created by the ring-opening polymerization of ε-caprolactone, which is produced by the oxidation of cyclohexanone, produced by catalytic oxidation of cyclohexane (Figure 7.9).

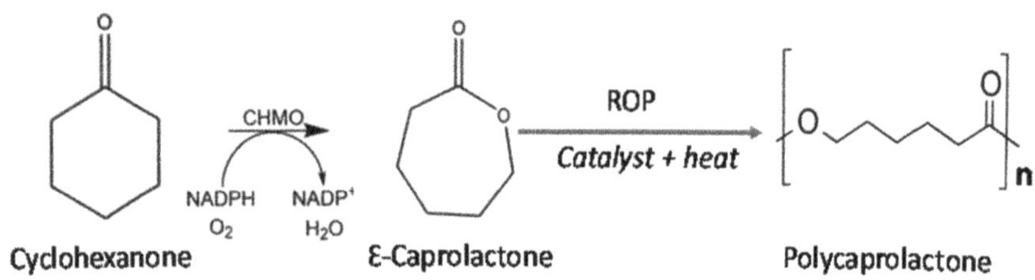

Figure 7.9. *Synthesis of polycaprolactone*

PCL is a thermoplastic polymer with good processing stability, high flexibility, good chemical resistance to water and organic solvents, and a hydrophobic nature; however, there is a problem with its low melting point (59–64°C). The T_g of this polymer is approximately -60°C (Robertson 2016). Moreover, microorganisms (and fungi) easily degrade PCL due to the presence of hydrolytically labile aliphatic ester linkages, and its degradation time is a few months because of its high crystallinity. It is not used for food applications but if mixed with low-cost natural biopolymers like starch, it is possible to obtain a good biodegradable material at a low price, typically used for trash bags. With a PCL content of up to 20%, starch-PCL films have proved the O_2 barrier properties while improving water barrier properties (Robertson 2016). PCL also finds its use as a plasticizer in PVC. PCL exhibits very good processing stability, high flexibility, good resistance to water and organic solvents, and a hydrophobic nature, but its problem is in its low melting point. Blending PCL with PLA increases the brittleness of the obtained film, as well as increasing its barrier properties proportional to the amount of PLA added (Rydz *et al.* 2018).

7.3.4.2. *Polybutylene succinate and copolymers*

Traditionally, polybutylene succinate, PBS, is synthesized via the polycondensation of 1,4-butanediol (1.4 BDO) and succinic acid (Figure 7.10). This is later commercially manufactured through the catalytic hydrogenation of maleic anhydride (which is derived from the oxidation of butane or benzene) to succinic anhydride, followed by hydration to produce succinic acid. Succinic acid can also be prepared through the microbial fermentation of renewable raw materials such as starch and xylose using bacterial strains *Actinobacillus succinogenes* (Rydz *et al.* 2018). After the hydrolytic degradation of PBS, succinic acid, used as a substrate, is one of the compounds involved in the natural cycle of metabolism (Krebs cycle) in the human body. While BDO is produced by a commercial process using petrochemical-based feedstocks, various industrial processes, based on renewable resources, have been developed and are being commercialized. PBS is biodegradable via ester linkages, and it possesses excellent mechanical properties that allow it to be processed through conventional melt techniques. Its applications include mulch film, packaging film and bags.

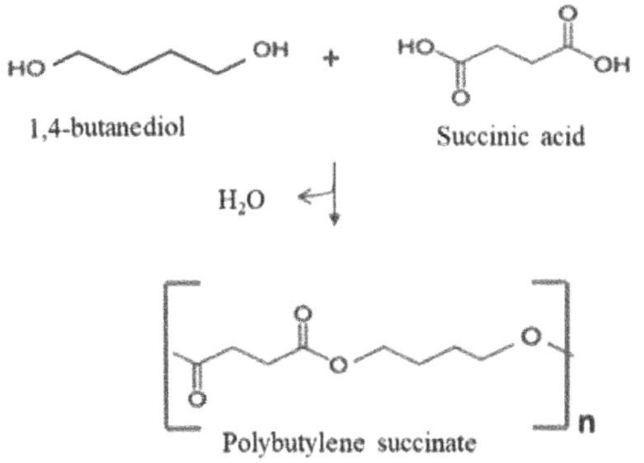

Figure 7.10. *Synthesis of polybutylene succinate (PBS)*

Copolymers are also commercially available, for example, adipic acid poly (butylene succinate-co-adipate) (PBSA) is available under the trade name Bionolle^{TM}. This copolymer is produced through the polycondensation reactions of EG and 1.4 BDO, and aliphatic dicarboxylic acids such as adipic acid (Robertson 2016).

To improve the utility properties of PBS, new aliphatic aromatic polyester (PBST) can be created through chemical polymerization between terephthalic acid, BDO and succinic acids. Aliphatic–aromatic copolyesters and blends combine the good functional properties of aromatic polyesters with the biodegradability of aliphatic polyesters. Additionally, combining the properties of two types of polyesters can match them to specific applications in the packaging industry (Rydz *et al.* 2018).

7.3.4.3. *Poly(butylene adipate-co-terephthalate)*

Poly(butylene adipate-co-terephthalate), PBAT, is a linear statistical aliphatic–aromatic copolyester composed of two co-monomers: a rigid unit consisting of 1,4-butanediol (1,4-BDO) and terephthalic acid monomers, and a flexible unit consisting of (1,4-BDO) and adipic acid monomers (Figure 7.11).

PBAT is available on the market under the trade names Ecoflex (a product of BASF) and Origo-Bi (a product of Novamont, formerly Eastar-Bio/Eastman). PBAT exhibits good mechanical and thermal properties when the content of terephthalic acid is greater than 35%. However, if the content of terephthalic acid increases even more, the biodegradation rate decreases, especially in relation to the number of aromatic units, which makes materials more resistant against microorganisms (Kijchavengkul *et al.* 2010).

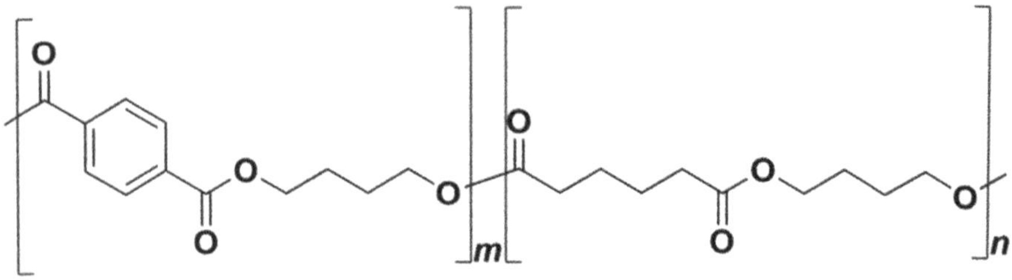

Figure 7.11. *Structure of PBAT*

PBAT is widely used for products that are short-lived, such as compostable, disposable food sacks and bags for fresh meats, vegetables and fruits. For some applications, PBAT is blended with cellulose, starch or biodegradable polyesters such as PHA or PLA. Cellulose is mixed with PBAT to improve hydrophobicity, as well as the mechanical and thermal properties of compounded products. The blends of modified starch or PLA with PBAT are commercially available, respectively, under the trade names Ecobras (Columbia) and Ecovio (BASF, Germany), and they are used for the production of thermoformed and injection molded films (Rydz *et al.* 2018). The packaging from Ecovio is considered to be fully biodegradable (compostable) after use. Finally, in order to increase the benefits of PBAT, petrochemical terephthalic acid sources can be replaced with 2,5-furanodicarboxylic acid from renewable sources.

7.3.4.4. *Poly(glycolic acid)*

Poly(glycolic acid), PGA, is another polymer that can be produced not only from petrochemicals, but also from renewable-derived monomers. It can be prepared from glycolic acid by polycondensation or ROP (Figure 7.12). PGA is a rigid, hydrolyzable, linear, aliphatic polyester with high crystallinity (45–55%) and high T_m (220–225°C), and a T_g range between 35 and 40°C (Robertson 2016). PGA has higher barrier properties against gas (CO_2 and O_2) and excellent mechanical strength. The targeted packaging applications of PGA include its use as a protective layer in multilayer packaging systems, such as PET bottles for carbonated soft drinks and beer (Coles and Kirwan 2011; Robertson 2016). In other packaging applications, PGA multilayer designs have been shown to enhance the gas and moisture barrier of bio-based polymers such as PLA. In addition, according to the Kureha Corporation website, and as reported by Kwiecien *et al.* (2014), the production of PGA for packaging applications, marketed under the trade name Kuredux, has been identified as biodegradable. Its biodegradation occurs in compost over one month and is considered safe for composting because its degradation product is a natural metabolite-glycolic acid.

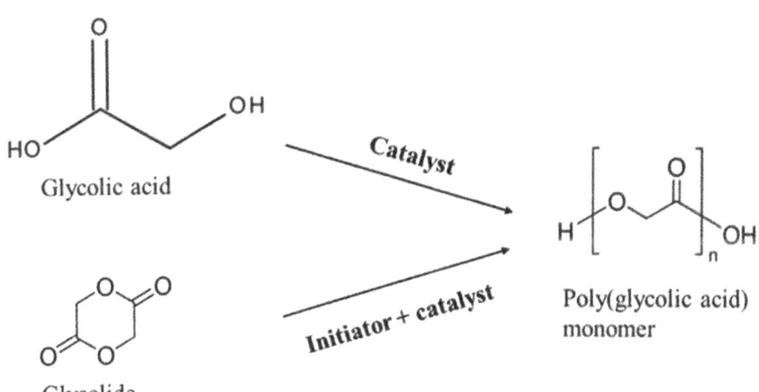

Figure 7.12. *Synthesis of Poly(glycolic acid)*

7.4. Industrial relevance and future trends

The past century was marked by plastic materials because of their widespread ease of use. However, due to their environmental impact, consumer perceptions changed towards plastic packaging, together with changes in the European legislative regarding plastic packaging and waste; there has been a significant increase in scientific and industrial interest in the use of bio-based food packaging materials. The future focuses on compostable, semi-durable and durable bioplastics. It is difficult to estimate how long it will take for bio-based packaging to become a leading material, or indeed if this will ever actually happen; however, its entry into the packaging market is significant. In a nutshell, biopolymer-driven innovations, with the same properties as traditional plastic, will revolutionize the packaging industry. Smart and intelligent sensors, active antioxidant and antimicrobial packaging, with innovative formulations of active ingredients, could allow biodegradable food packaging to prolong the shelf-life of food. This would therefore reduce food waste, helping the world packaging market fight the problems of fossil fuel depletion, health hazards and solid waste management. To put it simply: ideal packaging is yet to come.

7.5. List of abbreviations

Abbreviation	Full name
BC	Bacterial Cellulose
BioPE	Bio-Polyethylene
BioPET	Bio-Poly(Ethylene Terephthalate)
BioPU	Bio-Polyurethane
CA	Cellulose Acetate
CMC	Carboxymethyl Cellulose
EVOH	Ethylene Vinyl Alcohol
HPC	Hydroxypropyl Cellulose
HPMC	Hydroxypropyl Methyl Cellulose
LDPE	Low-Density Polyethylene
MC	Methyl Cellulose
NC-W	Nitrocellulose Wax
OPP	Oriented Polypropylene
P3HB	Poly-3-Hydroxy Butyrate
PBAT	Polybutylene Adipate Terephthalate
PBS	Polybutylene Succinate
PBSA	Adipic Acid Poly(Butylene Succinate-co-Adipate)

PCL	Polycaprolactone
PGA	Poly(Glycolic Acid)
PHA	Polyhydroxy Alkanoate
PHB	Polyhydroxy Butyrate
PHBV	Poly(3-Hydroxybutyrate-co-3-Hydroxyvalerate
PLA	Polylactic Acid
PVC	Polyvinyl Chloride
PVDC	Polyvinylidene Chloride
TPA	Terephthalic Acid
TPS	Thermoplastic Starch
WPC	Whey Protein Concentrate
WPI	Whey Protein Isolate

7.6. References

Andersson, C. (2008). New ways to enhance the functionality of paperboard by surface treatment – A review. *Pack. Technol. Sci.*, 21(6), 339–373.

ASTM D5511-18 (2018). Standard test method for determining anaerobic biodegradation of plastic materials under high-solids anaerobic-digestion conditions.

ASTM D5526-18 (2018). Standard test method for determining anaerobic biodegradation of plastic materials under accelerated landfill conditions.

ASTM D6400 (2019). Standard specification for labeling of plastics designed to be aerobically composted in municipal or industrial facilities.

ASTM D6866-20 (2020). Standard test method for determining the biobased content of solid, liquid, and gaseous samples using radiocarbon analysis.

ASTM D6868 (2019). Standard specification for labeling of end items that incorporate plastics and polymers as coatings or additives with paper and other substrates designed to be aerobically composted in municipal or industrial facilities.

Australian Standard AS 4736-2006 (2006). Biodegradable plastic-biodegradable plastics suitable for composting and other microbial treatment – Australian capital territory.

Belyamani, I., Prochazka, F., Assezat, G., and Debeaufort, F. (2014). Mechanical and barrier properties of extruded film made from sodium and calcium caseinates. *Food Pack. Shelf-Life*, 2(2), 65–72.

BeMiller, J.N. (2008). Gums and related polysaccharides. In *Glycoscience*, Fraser-Reid, B.O., Tatsuta, K., and Thiem, J. (eds). Springer, Berlin, Heidelberg.

Blanco-Pascual, N. and Gómez-Estaca, J. (2016). Production and processing of edible packaging stability and applications. In *Edible Food Packaging Materials and Processing Technologies*, Ribeiro Cerqueira, M.A.P., Pereira, R.N.C., da Silva Ramos, O.L., Teixeira, J.A.C., and Vicente, A.A. (eds). CRC Press, Boca Raton.

van den Broek, L.A.M., Boeriu, C.G., and Stevens, C.V. (2020). *Chitin and Chitosan: Properties and Applications*. John Wiley & Sons, New York.

Bugnicourt, E., Brzoska, N., Kucukpinar, E., Philippe, S., Forlin, E., Bianchin, A., and Schmid, M. (2019). Dispersion and performance of a nanoclay/whey protein isolate coating upon its upscaling as a novel ready-to-use formulation for packaging converters. *Polymer*, 11, 1410.

Chambi, H. and Grosso, C. (2006). Edible films produced with gelatin and casein cross-linked with transglutaminase. *Food Res. Int.*, 39, 458–466.

Cho, S.W., Skrifvars, M., Hemanathan, K., Mahimaisenan, P., and Adekunte, K. (2014). Regenerated cellulose fibre reinforced casein films: Effect of plasticizer and fibres on the film properties. *Macromolec. Res.*, 22, 701–709.

Coles, R. and Kirwan, M.J. (2011). *Food and Beverage Packaging Technology*. Wiley-Blackwell, Hoboken.

Cuq, B., Gontard, N., and Guilbert, S. (1997). Thermoplastic properties of fish myofibrillar proteins: Application to biopackaging fabrication. *Polymer*, 38, 4071–4078.

Dodds, D. and Humphreys, B. (2013). Production of aromatic chemicals from biobased feedstock. In *Catalytic Process Development for Renewable Materials*, Imhof, P. and Waal, J.-K. (eds). Wiley-VCH, Weinheim.

EN 13428:2004 (2004). Packaging – Requirements specific to manufacturing and composition – Prevention by source reduction.

EN 13429:2004 (2004). Packaging – Reuse.

EN 13430:2004 (2004). Packaging – Requirements for packaging recoverable by material recycling

EN 13431:2004 (2004). Packaging – Requirements for packaging recoverable in the form of energy recovery, including specification of minimum inferior calorific value.

EN 13432:2000 (2004). Packaging – Requirements for packaging recoverable through composting and biodegradation. Test scheme and evaluation criteria for the final acceptance of packaging.

EN 14046:2004 (2004). Evaluation of the ultimate aerobic biodegradability and disintegration of packaging materials under controlled composting conditions.

EN 16575:2014 (2014). Bio-based products – Vocabulary.

European Bioplastics (2017). Oxo-degradable plastics [Online]. Available at: https://www.european-bioplastics.org/bioplastics/standards/oxo-degradables/ [Accessed 6 April 2020].

EU WheyPack (2020). EU WheyPack Project: Reduction of CO_2 emissions by the PHB use obtained from whey: Demonstration in dairy products packaging (2014–2017) Available at: www.wheypack.eu [Accessed 1 June 2020].

Gnanasekaran, D. (2019). *Green Biopolymers and their Nanocomposites*. Springer, Basel.

González, A., Strumia, M.C., and Alvarez Igarzabal, C.I. (2011). Cross-linked soy protein as material for biodegradable films: Synthesis, characterization and biodegradation. *J. Food Eng.*, 106, 331–338.

Goy, R.C., De Britto, D., and Asis, O.B.G. (2009). A review of the antimicrobial activity of chitosan. *Polímeros*, 19(3), 241–247.

Guerrero, P. and de la Caba, K. (2016). Protein-based films and coatings. In *Edible Food Packaging Materials and Processing Technologies*, Ribeiro Cerqueira, M.A.P., Pereira, R.N.C., da Silva Ramos, O.L., Teixeira, J.A.C., and Vicente, A.A. (eds). CRC Press, Boca Raton.

Guerrero, P., Stefani, P.M., Ruseckaite, R.A., and de la Caba, K. (2011). Functional properties of films based on soy protein isolate and gelatin processed by compression molding. *J. Food Eng.*, 105, 65–72.

Guilbert, S. and Cuq, N. (2005). Material formed from proteins. In *Handbook of Biodegradable Polymers*, Bastioli, C. (ed.). Rapra Technology Limited, UK.

Hernandez-Izquierdo, V.M. and Krochta, J.M. (2008). Thermoplastic processing of proteins for film formation – A review. *J. Food Sci.*, 73(2), R30–R39.

ISO 18602:2013 (2013). Packaging and the environment – Optimization of the packaging system.

ISO 18603:2013 (2013). Packaging and the environment – Reuse.

ISO 18604:2013 (2013). Packaging and the environment – Material recycling.

ISO 18605:2013 (2013). Packaging and the environment – Energy recovery.

ISO 18606:2013 (2013). Packaging and the environment – Organic recycling.

Kabasci, S. (2020). Biobased plastics. In *Plastic Waste and Recycling*, Letcher, T.M. (ed.). Academic Press, London.

Kashyap, P.L., Xiang, X., and Heiden, P. (2015). Chitosan nanoparticle based delivery systems for sustainable agriculture. *Int. J. Biol. Macromol.*, 77, 36–51.

Kijchavengkul, T., Auras, R., Rubino, M., Selke, S., Ngouajio, M., and Fernandez, R.T. (2010). Biodegradation and hydrolysis rate of aliphatic aromatic polyester. *Polym. Deg. Stab.*, 95(12), 2641–2647.

Kuusipalo, J., Savolainen, A., Laiho, E., and Penttinen, T. (2008). Extrusion coating and products. *Papermaking Sci. Technol.*, 106–166.

Kwiecień, I., Adamus, G., Bartkowiak, A., and Kowalczuk, M. (2014). Synthesis and structural characterization at the molecular level of oligo(3-hydroxybutyrate) conjugates with antimicrobial agents designed for food packaging materials. *Designed Monomers Polym.*, 17(4), 311–321.

Lambert, S. and Wagner, M. (2017). Environmental performance of bio-based and biodegradable plastics: The road ahead. *Chem. Soc. Rev.*, 46(22), 6855–6871.

Li, X., Fang, Y., Phillips, G.O., and Al-Assaf, S. (2013). Improved sugar beet pectin-stabilized emulsions through complexation with sodium caseinate. *J. Agric. Food Chem.*, 61, 1388–1396.

Majid, I., Nayik, G.A., Dar, S.M., and Nanda, V. (2016). Novel food packaging technologies: Innovations and future prospective. *J. Saudi Soc. Agric. Sci.*, 17, 4.

Martin, A.H. and de Jong, G.A.H. (2012). Enhancing the in vitro $Fe^{(2+)}$ bioaccessibility using ascorbate and cold-set whey protein gel particles. *Dairy Sci. Technol.*, 92, 133–149.

Martucci, J.F. and Ruseckaite, R.A. (2010). Three-layer sheets based on gelatin and poly(lactic acid), part 1: Preparation and properties. *J. App. Polym. Sci.*, 118, 3102–3110.

Nafchi, A.M., Moradpour, M., Saeidi, M., and Alias, A.K. (2013). Thermoplastic starches: Properties, challenges, and prospects. *Starch*, 65, 61–72.

Nakajima, H., Dijkstra, P., and Loos, K. (2017). The recent developments in biobased polymers toward general and engineering applications: Polymers that are upgraded from biodegradable polymers, analogous to petroleum-derived polymers, and newly developed. *Polymers*, 9(10).

Nieto, M.B. (2016). Edible film and packaging using gum polysaccharides. In *Edible Food Packaging Materials and Processing Technologies*, Ribeiro Cerqueira, M.A.P., Pereira, R.N.C., da Silva Ramos, O.L., Teixeira, J.A.C., and Vicente, A.A. (eds). CRC Press, Boca Raton.

NKWOOD Research (2019). Global polylactic acid market forecast 2019–2027.

Overbeek, G. and Hoes, A.C. (2018). Biovoices – Connecting biobased forces for a sustainable world, Deliverable 3.1: Synthesis of market perspectives to develop bio-based value chains [Online]. Available at: https://www.biovoices.eu/ [Accessed 20 July 2020].

Patel, S. (2015). Emerging trends in nutraceutical applications of whey protein and its derivatives. *J. Food Sci. Technol.*, 52(11), 6847–6858.

Peña, C., Martin, M.D., Retegi, A., de la Caba, K., Ruseckaite, R., and Mondragon, I. (2009). Gelatin films: Renewable resource for food packaging. *V International Conference on Science and Technology of Composite Materials: COMATCOMP 09*. Donostia-San Sebastián, Spain.

Pérez-Gago, M.B. (2012). Protein-based films and coatings. In *Edible Coatings and Films to Improve Food Quality*, Baldwin, E.A., Hagenmaiier, R., and Bai, J. (eds). Taylor & Francis Group, USA.

Pérez-Gago, M.B., Nadaud, P., and Krochta, J.M. (1999). Water vapor permeability, solubility, and tensile properties of heat-denatured versus native whey protein films. *J. Food Sci.*, 64, 1034–1037.

Pérez-Masiá, R., López-Nicolás, R., Periago, M.J., Ros, G., Lagaron, J.M., and López-Rubio, A. (2015). Encapsulation of folic acid in food hydrocolloids through nanospray drying and electrospraying for nutraceutical applications. *Food Chem.*, 168, 124–133.

Quecholac-Piña, X., Hernández-Berriel, M.C., Mañón-Salas, M.C., Espinosa-Valdemar, R.M., and Vázquez-Morillas, A. (2020). Degradation of plastics under anaerobic conditions: A short review. *Polymer*, 12(1), 109.

Ramos, M., Valdés, A., Beltrán, A., and Garrigós, M.-C. (2016). Gelatin-based films and coatings for food packaging applications. *Coatings*, 6, 41.

Research Reports World (2014). Global biodegradable materials market report, history and forecast 2014–2025, breakdown data by manufacturers, key regions, types and application [Online]. Available at: https://www.researchreportsworld.com/global-biodegradable-materials-market-14391088 [Accessed 23 June 2020].

Robertson, G.L. (2016). *Food Packaging: Principles and Practice*, 3rd edition. CRC Press, Boca Raton.

Rujnić-Sokele, M. and Pilipović, A. (2017). Challenges and opportunities of biodegradable plastics: A mini review. *Waste Manag. Res.*, 35, 132–140.

Rydz, J., Musioł, M., Zawidlak-Węgrzyńska, B., and Sikorska, W. (2018). Present and future of biodegradable polymers for food packaging applications. In *Biopolymers for Food Design*, Grumezescu, A. and Holban, A.M. (eds). Elsevier, London.

Saez-Orviz, S., Laca, A., Rendueles, M., and Díaz, M. (2017). Approaches for casein film uses in foodstuff packaging. *Afinidad – Barcelona*, 74(577), 26–29.

Sekhon, B.S. (2010). Food nanotechnology – An overview. *Nanotechnol. Sci. Applications*, 3, 1–15.

Shi, Z., Zhang, Y., Phillips, G.O., and Yang, G. (2014). Utilization of bacterial cellulose in food. *Food Hydrocolloid.*, 35, 539–545.

Storz, H. and Vorlop, K.-D. (2013). Bio-based plastics: Status, challenges and trends. *Landbauforschung-Ger*, 63, 321–332.

Versino, F., Lopez, O.V., Garcia, M.A., and Zaritzky, N.E. (2016). Starch based films and food coatings: An overview. *Starch, Special Issue: Starch Research in Latin America*, 68(11–12), 1026–1037.

Vieira, M.G.A., da Silva, M.A., dos Santos, L.O., and Beppu, M.M. (2011). Natural-based plasticizers and biopolymer films: A review. *Europ. Polym. J.*, 47, 254–263.

Westwood, M., Gunning, A., and Parker, R. (2010). Temperature-dependent growth of gelatin–poly (galacturonic acid) multilayer films and their responsiveness to temperature, pH, and NaCl. *Macromolecules*, 43, 10582–10593.

Wicochea-Rodríguez, J.D., Chalier, P., Ruiz, T., and Gastaldi, E. (2019). Active food packaging based on biopolymers and aroma compounds: How to design and control the release. *Frontiers Chem.*,7, 398.

Wihodo, M. and Moraru, C.I. (2013). Physical and chemical methods used to enhance the structure and mechanical properties of protein films: A review. *J. Food Eng.*, 114, 292–302.

Xiao, Y., Han, C., Yang, H., Liu, M., Meng, X., and Liu, B. (2020). Layer (whey protein isolate)-by-layer (xanthan gum) microencapsulation enhances survivability of *L. bulgaricus* and *L. paracasei* under simulated gastrointestinal juice and thermal conditions. *Int. J. Biolog. Macromol.*, 148, 238–247.

Zhu, Y., Romain, C., and Williams, C.K. (2016). Sustainable polymers from renewable resources. *Nature*, 540(7633), 354–362.

8

Active and Intelligent Packaging

Nasreddine Benbettaieb

Institute of Technology, University of Burgundy, Dijon, France

8.1. Introduction

Social and economic evolution such as changes in consumer demand, improved standards of living, increase in industrial production, retail practices, new marketing targets and customer lifestyles are the main forces driving the evolution of novel and innovative packaging techniques. The main aims are maintaining and monitoring food safety and quality, and increasing the shelf life of food packaging (Dainelli *et al.* 2008). Therefore, the function of food packaging has evolved from simple preservation methods towards new aspects such as convenience, securing the integrity of the packaging, point of purchase marketing, material reduction, safety and environmental issues (Han 2014). Active and intelligent packaging can, respectively, maintain or monitor food quality and safety, as presented in Figure 8.1.

The design and manufacture of adequate active and intelligent packaging technology is quite challenging from an industrial point of view. In addition, traditional packaging is no longer sufficient, as today's society has become increasingly complex. Like conventional packaging, new packaging must serve a number of important functions, including the containment and protection of food, maintaining its sensory quality, safety and security, and communicating information to consumers. More functions that can secure the integrity of packaging are considered very important, including antimicrobial packaging, freshness indicators,

---

For a color version of all figures in this book, see www.iste.co.uk/debeaufort/packaging.zip.

Packaging Materials and Processing for Food, Pharmaceuticals and Cosmetics, coordinated by Frédéric DEBEAUFORT, Kata GALIĆ, Mia KUREK, Nasreddine BENBETTAIEB and Mario ŠČETAR.

time–temperature integrators and electronic coding systems. Improved security for the user and boosting differentiation and marketing are other advantages of active and intelligent packaging; customer satisfaction and loyalty become important parameters that guide consumers when shopping. Increased competition space on supermarket shelves is often a driving force behind new packaging developments (Ahvenainen 2003). Over the last 10 years, the increasing changes to consumer lifestyle, related to rapid urbanization and the expanded use of personal items with advanced integrated innovation, have been an influencing factor in the market growth of active and intelligent packaging. The evolution of the generations of consumers also influences this development, as they are becoming increasingly well-connected and expect improvements worthy of the evolution of modern technologies, taking into account ecological, practical and esthetic dimensions.

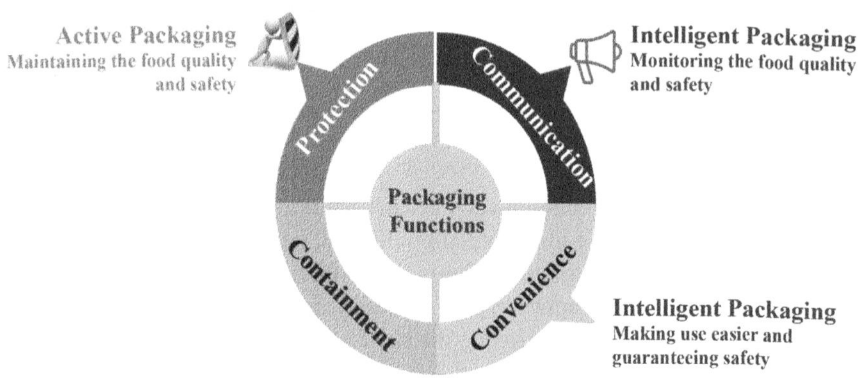

Figure 8.1. *Schematic view of packaging functions (adapted from Yam* et al. *2005)*

As defined in the European Regulation 450/2009 (European Commission 2009) and 1935/2004 (European Commission 2004), active packaging systems are intended to change packaging conditions, in order to extend shelf life and ensure food safety. Active packaging was also designed to deliberately incorporate components into food packaging that would release or absorb substances into or from packaged food, or the environment surrounding the food.

Intelligent packaging (also more loosely described as smart packaging) is packaging that informs the manufacturer, retailer and consumer about the state of some of the properties of the food within the packaging, or within the environment in which it is kept (Han *et al.* 2018). As defined in the European Regulation 450/2009 and 1935/2004 (European Commission 2004, 2009), intelligent packaging systems are intended to monitor the quality of the food product or its surrounding environment and predict or measure the shelf life better than a best before date. They have the ability to communicate the conditions of the packaged food or the environment by providing information about different factors during transportation and storage, without any interaction with the product.

Considering the market share and perspectives of active and intelligent packaging, among innovations, the global market for this type of packaging was estimated at US$5.3 billion in 2017 and should reach US$7.8 billion in 2021, according to an average annual growth rate of +8%. The more mature active packaging market has great opportunities for technological development, while the intelligent packaging market faces strong growth prospects. Technologies that are used for accessing and storing information, verifying the authenticity of products or manufacturing conditions, and delivering messages to consumers and investors are increasingly numerous, thanks to intelligent packaging (ALL4PACK 2016).

The intelligent packaging market has witnessed continuous growth in the past few years and is projected to grow even further during the forecast period 2020–2026. After comprehensive analysis, industry and research have together introduced a new research study: "Global (United States, China, and the European Union) Intelligent Packaging Market Research Report 2020–2026" (Patel 2020). Some of the prominent key players (main manufacturers, distributors and downstream client companies) covered in the intelligent packaging market are Jones Packaging, 3M, DuPont, BASF, Intelligent Packaging, and Amcor and Campden Bri. The report on the intelligent packaging worldwide market highlights key market events like technological innovations, product releases, business plans and development strategies of players, along with industry-oriented growth factors.

8.2. Active packaging

Food packaging is used to protect food from environmental contamination, in order to safeguard the quality of the food, which in turn extends shelf life and reduces food waste. Oxidation, microbial spoilage and metabolism are the main factors for the deterioration of many foods during processing, transport, storage, marketing and at home with end-users. These processes are directly related to the loss of food quality, which can influences consumer buying decisions and impacts consumer health, thus affecting the overall economics of the food supply chain (Zhao *et al.* 2016). Oxidation has many negative effects on the nutritional and organoleptic characteristics of food, such as decreasing nutritional value in terms of the destruction of essential fatty acids, proteins and lipid-soluble vitamins (A, D, E and

K); producing rancidity; and changing the natural colors of foods (degradation of pigments, darkening of fats and oils) (Nerín *et al.* 2008).

Furthermore, the presence of pathogenic microorganisms increases the risk of foodborne diseases in humans and thus presents a problem for public health. To delay oxidation, control foodborne pathogens and respond to growing consumer demand for safe, high-quality food, different strategies have been developed in food packaging technology. Active packaging can effectively improve food quality and safety and extend shelf life by positively affecting the headspace of a packaged product, and/or the product itself, using different absorbing and releasing systems that involve physical, chemical or biological actions.

The first group of active packaging concerns gas absorbing/emitting packaging that use sachets or pads to absorb gases (such as oxygen, water vapor and ethylene) from the package headspace or to emit gases (such as carbon dioxide and ethanol) to the package headspace, so that a favorable internal environment and an extended shelf life can be achieved (Figure 8.2). Controlled release packaging (Figure 8.2) is another group of technologies that use packaging materials as a delivery system to release active compounds (such as antimicrobials, antioxidants, enzymes, flavors, nutraceuticals) to protect against oxidation and microbial spoilage, and therefore improve food quality (Yam *et al.* 2005). Most of the focus within this group has been on antimicrobial and antioxidant packaging that incorporate active compounds into their structure (direct addition, surface immobilization or coating) (Han *et al.* 2018). Antimicrobial polymers (natural or modified) can also be used as active packaging for food and pharmaceutical applications.

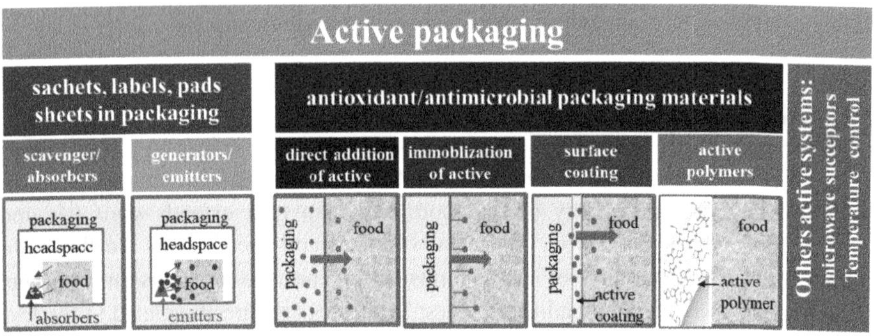

Figure 8.2. *Classification of different types of active packaging systems for food*

Other active packaging systems include microwave susceptors and modifiers, changing gas permeability, removing properties, including catalyzing food component removal (lactose and cholesterol) and temperature control systems that include self-heating and self-cooling.

Excepting the use of natural or modified antimicrobial polymers, active packaging can be divided into two types: scavengers/absorbers or releasers/emitters (Figure 8.3).

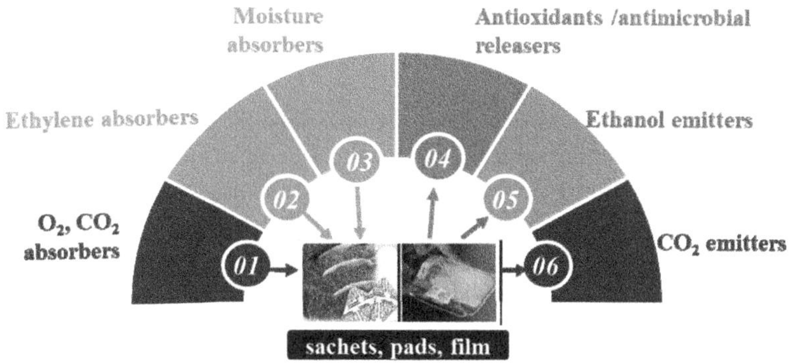

Figure 8.3. *Sachet, pad and film scavenging or emitting systems*

These systems can take the form of sachets and pads, or active materials. Sachets and pads are placed inside the packaging headspace or at the surface of the packaging itself. Active materials, on the other hand, incorporate the scavenging or releasing systems, or the active agent, into the packaging material, for example, extruded or coated on the plastic film.

8.2.1. *Scavenger systems*

8.2.1.1. *Oxygen scavengers/absorbers*

The presence of oxygen in packed food is a critical factor that compromises shelf life. Oxygen present in food packaging accelerates spoilage and chemical deterioration and thus contributes to food wastage. The presence of oxygen, if not desired, may result from inadequate or insufficient evacuation and/or gas flushing during the packaging process, from an existing presence in the food itself or from the packaging headspace (Robertson 2016). Oxygen may also derive from other sources such as oxygen permeation through packaging material, or may be introduced because of small leakages or microperforations in the packaging material, as a result of poor sealing (Kruijf *et al.* 2002).

High levels of oxygen present in food packages may cause food deterioration such as off-flavors (rancidity as a result of lipid oxidation), color changes (discoloration of pigments), nutrient losses (oxidation of vitamin E, β-carotene, ascorbic acid), microbial growth and a decrease in the respiration rate. The removal of the residual oxygen using absorbing systems can minimize the deterioration and quality loss of oxygen-sensitive food products considerably. The combined use of oxygen scavengers with vacuum sealing, or by gas atmosphere modification in the packaging (N_2, CO_2), assures the elimination of oxygen from the package headspace and dissolved in the product (Realini and Marcos 2014).

Structurally, the oxygen scavenging system can take the form of a sachet or pad with an incorporated active ingredient, which is placed inside the packaging headspace. There are different oxygen scavenging technologies commercialized on the market, which are mainly based on iron powder oxidation, oxidation of photosensitive coloring matter, ascorbic acid oxidation, enzymatic oxidation using glucose oxidase, unsaturated fatty acids or immobilized yeast on a solid substrate (Floros *et al.* 1997). The most widely used mechanism today, however – and one of the most effective – is based on the oxidation of iron and ferrous salts using a sachet, pads, strips or labels containing iron powder (Figure 8.4). These systems are incorporated into, or attached to the inside of the packaging, but are separate elements. These are also used in the closures of alcoholic and non-alcoholic beverages to eliminate both the residual oxygen in the headspace and that which has permeated the packaging.

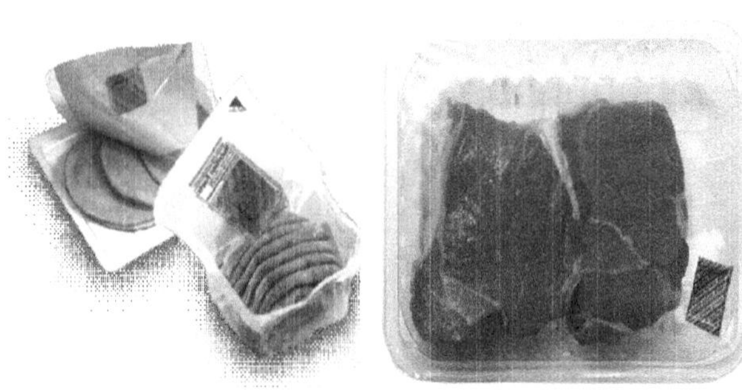

Figure 8.4. *Oxygen scavenger systems based on the oxidation of iron and ferrous salts; sachet containing iron powder (Multisorb Technologies 2020)*

Ageless®, developed by the Mitsubishi Gas Chemical Company in Japan, produced the most commercially available oxygen scavengers based on iron oxidation. This system is used for dry and moist food according to the water activity (aw) level. Furthermore, using oxygen scavenger sachets allows the O_2 concentration in the headspace to be reduced to less than 0.01%, much lower than the typical residual O_2 levels achievable by vacuum or gas flushing (0.3–3%) (Robertson 2016). In addition to the Ageless® system, there are other commercially available oxygen scavenger systems, including TCO® (ATCO/Standa Industrie, France) which is used for wet and dry food, FreshPax® (Multisorb Technologies, USA) which is used for either moist or semi-moist foods (aw > 0.7) and for dehydrated or dry foods (aw < 0.7), FreshMax® and FreshCard® (Multisorb Technologies) which are used for sausages, meat and pizza, Freshilizer® (Toppan Printing) which is used for nuts or chocolate, and O-Buster (Hsaio Sung Non-Oxygen Chemical Company) which is used for dried and oily foods, as well as for fatty foods. These systems also make use of finely divided iron powder with excipients (salts and minerals to provide moisture content) and can reduce residual oxygen levels to approximately 0.01%.

Iron and iron salts (powder) react with water within foods to produce a reaction that moisturizes the iron metal in the product packaging, and irreversibly converts it to a stable oxide (Kerry *et al.* 2006). Oxygen scavenger systems that are based on iron oxidation reactions are explained by the following equation:

$$
\begin{aligned}
Fe &\longrightarrow Fe^{2+} + 2e^- \\
\tfrac{1}{2}O_2 + H_2O + 2e^- &\longrightarrow 2OH^- \\
Fe^{2+} + 2OH^- &\longrightarrow Fe(OH)_2 \\
Fe(OH)_2 + \tfrac{1}{4}O_2 + \tfrac{1}{2}H_2O &\longrightarrow Fe(OH)_3
\end{aligned}
\qquad [8.1]
$$

The selection of oxygen scavengers (shape, size, ability to absorb oxygen, time to reach equilibrium) depends on several interrelated factors, which include: the nature and characteristics of each food (weight, shape, size, liquid, solid, dry, fat, water content, water activity, etc.), the amount of O_2 dissolved in the food, the desired shelf life of the food, the initial level of O_2 in the headspace and the O_2 permeability of packaging materials (Pereira de Abreu *et al.* 2012).

Scavenger systems based on sachets and pads have several drawbacks: they cannot be used in beverages or food containing high levels of aqueous liquids, because they become inactive when wet. They also cannot be used in packaging made from flexible films. In the latter case, the film may cling to the sachet and isolate it from the product, preventing it from functioning (Kerry *et al.* 2006). In order to overcome this problem, the sachet can be glued to the inner wall of the packaging or the active ingredients incorporated into a label, which can be inserted into the inside wall of the packaging (Suppakul *et al.* 2003). New trends in commercial applications involve incorporating oxygen scavengers into the packaging material. This is to avoid negative perceptions in consumers caused by the presence of a non-edible with the food, or to eliminate the risk of accidental rupture of the sachets and unintentional consumption of their contents (Kerry *et al.* 2006). In the case of vacuum packs for cheese and meat, where the presence of O_2 is a primary cause of quality loss, sachets cannot be incorporated into the close-fitting packaging.

Innovation in food packaging technology consists of developing an alternative method whereby O_2 absorbers are incorporated into the plastic components of the packaging film, tray or box. Iron, ascorbic acid and low-molecular-weight ingredients are extruded or coated in the polymer (polyethylene, polypropylene, polyamide, EVA, etc.). Examples of these are manufactured by Shelfplus® O_2 (Albis Plastic, Germany), Oxbar™ (Constar International, USA) and Amosorb 2000 (ColorMatrix, USA).

Cryovac®, the food packaging division of Sealed Air Corporation, developed a UV light-activated oxygen scavenging film, whose structure comprises an oxygen scavenger layer extruded into a multilayer film; this can reduce headspace oxygen levels from 1% to ppm levels in one week, which is comparable to oxygen scavenging sachets (Realini and Marcos 2014). A similar UV light-activated oxygen scavenging polymer, ZERO2™, developed by the CSIRO Division of Food Science and Technology in Australia, in collaboration with VisyPak, forms a layer within a multilayer packaging structure. Applications include the reduction of discoloration of sliced meats.

Bioka (RPC Bebo) has launched a rigid oxygen scavenging system consisting of a high-barrier container made from PP/ethylene vinyl alcohol/PP for the ready-meal market, containing an enzyme-based oxygen scavenger film using a glucose oxidase/catalase reaction (Realini and Marcos 2014). In the same way, OxyVac® (Nutricepts) produced an enzyme-based oxygen scavenger for MAP foods with $aw > 0.65$, as well as cheese products.

Darex® (W.R. Grace, USA) manufactures bottle crowns using ascorbate/metallic salt oxidation. Recently, a rapid oxygen scavenger system has been developed using a catalytic system with palladium. Palladium is coated onto polyethylene terephthalate/SiOx films using magnetron sputtering technology (Yildirim *et al.* 2018). This oxygen scavenger film was able to remove up to 2.5% (in volume) of residual oxygen in food packages, when hydrogen was included in the modified atmosphere of the packaging. This system does not have commercial application as the European Food Safety Authority (EFSA) Panel on Food Contact

Materials stated that palladium should not be in direct contact with food. Rather, it should be incorporated in a passive structure, impermeable to liquids, which prevents migration at detectable levels.

In another approach developed to remove O_2 from beverages, such as beer after it is capped, a multilayer barrier liner is fitted into crown caps and plastic and metal closures. The active ingredients consist of ascorbic acid that is oxidized to dehydroascorbic acid, while sodium sulfite is oxidized to sodium sulfate (Robertson 2016).

8.2.1.2. *Carbon dioxide scavengers*

Carbon dioxide (CO_2) efficiently inhibits the surface growth of a range of aerobic bacteria and fungi by reducing oxygen levels, and/or by direct antimicrobial effects (at CO_2 concentrations between 10% and 80%) in food packaging. CO_2 is often used as a flushing gas in MAP systems to maintain the freshness of some foods, particularly fresh meat, poultry, fish and cheese, in order to prolong their shelf life (Han *et al.* 2018). However, for some CO_2-producing foods, such as fresh horticultural products and fermented foods, increases in CO_2 levels allow their dissolution into foods and may decrease the volume of the package. This decrease in pressure inside the packaging leads to undesirable changes in product quality, such as changes in the flavor and texture of products and/or the development of undesirable anaerobic glycolysis in fruits (Han *et al.* 2018). In addition, reducing the CO_2 concentration in some fermented foods promotes the growth of lactic acid bacteria and improves organoleptic quality. Furthermore, some fruits and vegetables (potato, lettuce, onion, cucumber, apricot, peach, apple and carrot) have negative reactions to CO_2 exposure at levels higher than 5%; these include discoloration, off-flavor development and internal tissue breakdown. In such cases, reducing the level of CO_2 in food packaging can improve food quality. CO_2 absorbers can thus be an effective means of preserving food quality and packaging integrity (Han *et al.* 2018).

Carbon dioxide scavengers, consist of either a chemical absorbent such as silica gel or calcium hydroxide (equation [8.2]), or a physical absorbent such as zeolite or active carbon powder (Robertson 2016):

$$CO_2 + Ca(OH)_2 \longrightarrow CaCO_3 + H_2O \qquad [8.2]$$

These absorbers can be enclosed in a sachet and placed in the food package. ATCO CO_2 – manufactured by Standa Industries in France and by Emco Packaging Systems in the UK – is a sachet capable of absorbing carbon dioxide produced from a mixture of sodium hydroxide, calcium and potassium and is used for CO_2-sensitive products. Ageless Eye (Mitsubishi Gas Chemical, Japan) is another carbon dioxide absorbent recommended for the conservation of roasted coffee, with higher CO_2 absorption capacities available. CO_2 absorber systems can be fabricated as a coating or sheet.

8.2.1.3. *Ethylene absorbers*

Ethylene acts as a phytohormone and is produced during the ripening stage of fruit and vegetables; the effects can either be positive or negative. Positive effects include catalyzing the ripening process (initiating and accelerating ripening). Negative effects include increasing the respiration rate, which accelerates softening and senescence, increasing chlorophyll degradation and promoting a number of post-harvest disorders; these include reducing the shelf life of fresh or minimally processed fruits and vegetables (Robertson 2016).

Ethylene scavengers are used to preserve ethylene-sensitive fruits and vegetables such as apples, kiwis, apricots, mangos, bananas, tomatoes, onions, carrots and asparagus (Pereira de Abreu *et al.* 2012) (Figure 8.5). When ethylene is removed from the immediate environment, the ripening and deterioration processes of plant products are slowed, and storage life is extended.

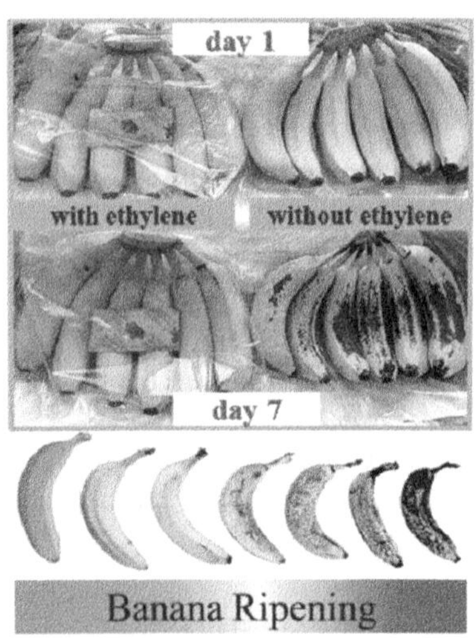

Figure 8.5. *Ethylene absorber sachets to stop ripening of ethylene-sensitive fruits*

Several ethylene scavenging systems have been reported in literature, but the most available and commercialized absorbers are potassium permanganate sachets.

The main mechanism of the action of ethylene scavengers is the use of $KMnO_4$ (in powder form inside a separate sachet) that oxidizes ethylene (C_2H_4) to carbon dioxide and water through the following reaction (Robertson 2016):

$$3C_2H_4 + 12KMnO_4 \longrightarrow 12MnO_2 + 12KOH + 6CO_2 \quad [8.3]$$

Commercial permanganate-based sachets with the ability to absorb ethylene include: Evert-Fresh Green Bags™ (Evert-Fresh, USA), Epax™ (IMPAK, USA), PEAKfresh® (USA), Retarder®, Profresh® and Biofresh® packaging systems. All of these systems are used for fresh produce and flowers (Kerry *et al.* 2006).

Potassium permanganate oxidizes both ethylene acetate and ethanol. $KMnO_4$ changes color from purple to brown after reacting with ethylene, thus, it could also be an indicator of the absorbing capacity of residual ethylene (Pereira de Abreu *et al.* 2012). Potassium permanganate sachets cannot be used in direct contact with food because of their toxicity. To overcome this drawback, another ethylene scavenger system has been developed. Years ago, a few packaging films were commercialized, which used finely dispersed minerals capable of absorbing C_2H_4. These minerals were some kinds of clay (e.g. zeolite, silicates), activated carbon, cristobalite (SiO_2), etc. that were mixed together with a small amount of metal oxide and then dispersed into a plastic film (Robertson 2016). PEAKfresh® and Evert-Fresh commercialized an ethylene absorption system based on zeolites dispersed into polymeric material. PEAKfresh® is a polyethylene bag impregnated with minerals, developed to absorb ethylene and moisture generated during fruit respiration. Evert-Fresh has the ability to absorb ethylene, ammonia and carbon dioxide. PEAKfresh® and Evert-Fresh also provide a controlled humidity atmosphere to absorb moisture generated by the respiration of fruits or vegetables (Kerry, O'Grady, & Hogan, 2006). They act as C_2H_4-absorbing agents, increasing the shelf life of fresh fruits and vegetables. Profresh is an additive made with monolayer or multilayer low-density polyethylene (LDPE), or high-density polyethylene (HDPE) that is able to absorb ethylene, ethanol and ethyl acetate, keeping food fresh for a long time (Kerry *et al.* 2006).

8.2.1.4. *Moisture absorbers*

The presence of high levels of water inside food packaging often promotes the growth of microorganisms, causing softening of dry, crunchy products such as pasta and cookies. It also causes caking and hardening in milk powder or freeze-dried coffee, and wetting of hygroscopic products such as sweets and candies (Kruijf *et al.* 2002). In addition, it is well known that food stability is closely related to the water activity (aw) of a product, which is affected by the relative humidity in the headspace of the packed food. In such cases, high relative humidity in the headspace causes susceptibility to temperature fluctuations during transport or storage. This can encourage the formation of condensation and therefore the fogging of plastic packaging (Kruijf *et al.* 2002).

There are two distinct methods used for regulating the moisture content of packaged foodstuffs. The mechanism of action is based on an absorption process. In this process, excess liquid water can be removed, or the relative humidity in the headspace can be controlled.

Liquid absorbers (pads, sheets): excess water can be controlled by applying drip-absorbent sheets, usually composed of two layers of microporous polymeric materials (e.g. polyethylene, polypropylene), which enclose a layer of superabsorbent hygroscopic polymers, capable of absorbing up to 500 times its own weight of water (Janjarasskul and Suppakul 2018; Kruijf *et al.* 2002). Typical superabsorbent polymers include polyacrylate salts, carboxymethyl cellulose and graft copolymers of starch, which have a very strong affinity for water (Robertson 2016). The sheets can be used as pads (Figure 8.6) under, for example, whole/sliced fresh poultry and meat to absorb drips of water, prevent microbial and mold growth and improve product presentation. Similar sheets are used for the absorption of melting ice water in seafood packaging that has been transported by air.

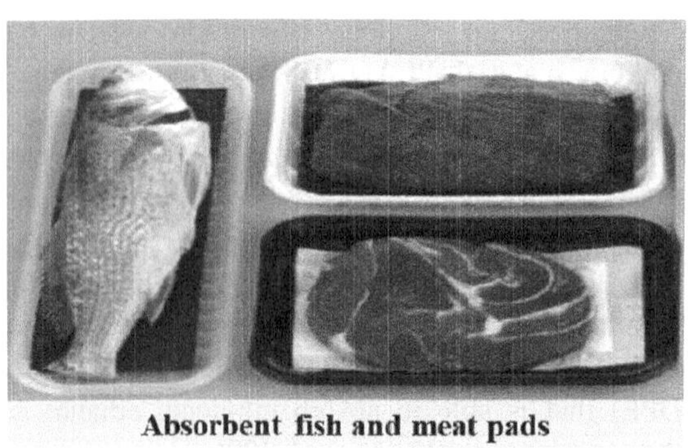

Figure 8.6. *Moisture absorber pads for fish and meat (adapted from Wholesalegroup, 2020)*

Relative humidity regulators (sachets or tags): sachets or tags containing dehydrating agents are used to regulate humidity in a wide range of products, such as cheeses, meats, nuts and spices (Pereira de Abreu *et al.* 2012).

Deciding which moisture absorbers to use will depend on: the size and weight of the food and its initial water activity; water vapor transmission from the package; the temperature and humidity of storage; food sensitivity to moisture; and the length of commercial life.

To prevent any adverse effects during the absorption process that could alter the color of the food surface, which is in contact with absorber, new solutions have been developed. For example, a lack of oxygen on the side of meat that is in contact with the moisture absorber can produce a color change in the product from red to brown. To avoid this, the Sirane Company commercializes moisture absorber pads with a surface permeable to oxygen, which keeps the meat red, while absorbing excess liquid.

Some examples of commercial liquids and moisture absorbers include:

Supa-Loc®, Lite-Loc®Plus consisting of cellulose-based absorbent pads, composed of different layers for meat and fish products (Sealed Air Corporation 2020).

Dri-Fresh® Resolve® is an absorbent pad used to completely eliminate surface discoloration on red meat (Sirane 2020).

Humidipak responds continuously to adapt to changes in humidity and temperature, either by the addition or absorption of moisture in order to maintain a certain level of relative humidity (RH) inside of the food package, for the duration of its shelf life. This system is available in virtually any RH level, from 10% to 95% RH. It is marketed on the basis of weight, volume and water activity of food (Süd-Chemie 2020).

NatraSorb® M is a moisture-absorbing bag made from magnesium and aluminum silicate with strong moisture-absorbing capabilities without swelling or any apparent, substantial deterioration (Multisorb Technologies 2020).

Liquid and moisture absorbers are aimed at products such as meats, fresh fish and seafood, cheeses, spices, baked goods, candies, pizza, frozen and dried food, and some cut or sliced fruits and vegetables.

8.2.1.5. *Flavor/odor absorbers*

Some commercial products have been developed to eliminate undesirable compounds or compounds that release odors. Retarder®, which is marketed as an ethylene absorber, is used to absorb undesirable odors. Minipax manufactured a system that combines clay with activated carbon or silica, in a 50/50 or 60/40 ratio, which has both drying and odor-absorption capabilities. EKA Noble commercializes BHM powder for incorporation into packaging materials, particularly paper and cardboard. It has the ability to absorb the undesirable odors produced from aldehydes (Pereira de Abreu *et al.* 2012).

Volatile amines, formed due to protein breakdown in fish muscle, can also be removed by incorporating acidic compounds, like citric acid, in polymers. ANICO bags (Japan) made from film containing ferrous salt and an organic acid, such as citric acid or ascorbic acid, are capable of oxidizing the amines, thus removing volatile amines (Biji *et al.* 2015).

8.2.2. *Releaser/emitter systems*

8.2.2.1. *Carbon dioxide emitters*

As previously mentioned, relatively high levels of CO_2 have direct antimicrobial effects on many microorganisms. Therefore, a high CO_2 concentration is included in MAP systems for chill-stored non-respiring foods that are susceptible to microbial spoilage (Lee 2016). In the same way, other sachets are available using either ascorbic acid and ferrous carbonate or ascorbic acid with sodium bicarbonate to absorb O_2 and generate an equivalent volume of CO_2. This type of active packaging is used for fresh products where an increased concentration of CO_2, combined with decreased O_2 concentration, reduces the respiration rate and thus increases the product shelf life (Robertson 2016).

Moreover, in the case of MAP systems, when a package is flushed with a mixture of gases, including CO_2 and O_2, the CO_2 dissolves in the product and penetrates into the external environment through the packaging material (the permeability of CO_2 through most plastic films is three to five times higher than that of oxygen). In this case, the concentration of CO_2 is dynamically changed and a loss in CO_2 occurs (Han *et al.* 2018). Therefore, CO_2 emitters are beneficial and necessary for maintaining desired CO_2 concentrations in packaging and ensuring the efficacy of a MAP system (Robertson 2016).

With regard to commercialization, there are different CO_2 emitter sachets and pads already on the market today.

Verifrais™ (SARL Codimer, France), FreshPax (Multisorb Technologies, USA) and Ageless® (Mitsubishi Gas Chemical, Japan) manufacture a sachet containing sodium bicarbonate/ascorbate, which use the dual action of CO_2 generating and O_2 scavenging used for many food products (Figure 8.7). Alongside the sachet system,

CO_2 emitter pads have also been commercialized; the most widely available system is CO2® Fresh-Pads (CO2 Technologies, USA) which are used for meat, poultry and seafood packaging. Drip losses from muscle are absorbed into pads and react with the citric acid and sodium bicarbonate present, resulting in carbon dioxide generation. Active CO_2 pads have also been commercialized in Sweden (Cellcomb) and Germany (McAirlaid's). The principle is based on CO_2-pad-emitter concepts (Figure 8.7) (Yildirim *et al.* 2018). Paper Pak Industries have launched UltraZap®XtendaPak pads, a more evolved version of CO_2 generators. They are designed to act as absorbent pads for fresh meat, poultry and fish, with double the antimicrobial effect due to the incorporation of a CO_2 emitter and an antimicrobial substance. A recent CO_2 emitter pad has been developed by a Norwegian company, Vartdal Plastindustri, for use in fish fillet packaging, as well as for meat packaging (Realini and Marcos 2014).

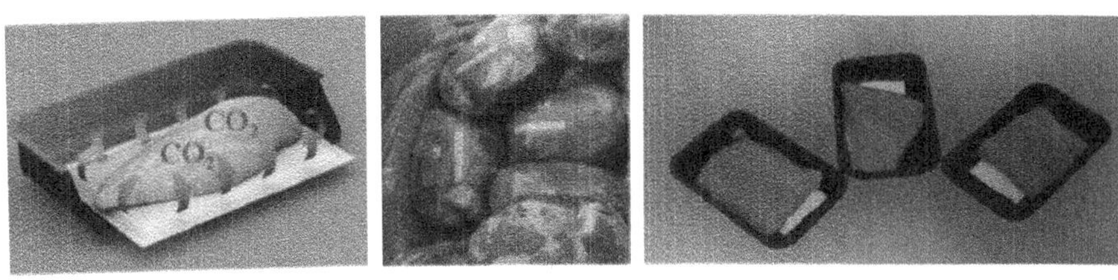

Figure 8.7. *CO_2 emitter sachets and pads*

Although the most widely available carbon dioxide emitters – intended for commercial use – are in the form of sachets and absorbent pads, films that incorporate carbon dioxide emitters have also been developed. The CO_2 emitter market is likely to continue in this direction (Kerry *et al.* 2006). Although research into this concept is still in its early stages, active packaging structures, for controlling the microbial quality of ready-to-eat meat products, have recently been tested by Chen and Brody (2013). Cooked ham samples were packed into three antimicrobial packaging structures, including a nylon/EVOH/polyethylene oxygen barrier bag and an antimicrobial film (3-Phase Activ-Polymer™, CSP Technologies, USA), with the ability to generate CO_2 or allyl isothiocyanate, or for scavenging O_2. Packaging structures with O_2 scavengers or CO_2 generators proved to control bacterial populations, particularly *Listeria*, while the allyl isothiocyanate generator had only limited antimicrobial effects (Realini and Marcos 2014).

8.2.2.2. *Ethanol emitters*

Ethanol has been widely used as an antimicrobial agent to prevent the spoilage of fruit by mold. A novel and innovative method of generating ethanol vapor has been developed in Japan. The system is formed by sachets containing water (10%) and ethanol (55%), which are adsorbed onto SiO_2 powder (35%) and inserted into a paper-ethylene-vinyl acetate (EVA) copolymer sachet (Robertson 2016). The sachet contents absorb moisture from food and releases ethanol vapor. The main drawback of this technology, first, is the high cost of ethanol and, second, the formation of off-flavors and off-odor in the food. To mask the odor of alcohol, some sachets containing traces of vanilla or other flavorings were added.

8.2.2.3. *Ethylene emitters*

In the case of climacteric fruits, harvested at an immature stage, some ethylene releaser systems have been developed to accelerate ripening. Bazzano *et al.* (2016) developed an active polymeric packaging film containing an inclusion complex of α-cyclodextrin. This was done to control the release of ethylene so that fruits that are typically harvested at a slightly immature stage, for example bananas, pears or apples, can reach consumers at the right stage of maturity.

8.2.2.4. *Antioxidant packaging as releasing systems*

After the growth of microorganisms, the oxidation of fats is one of the most important mechanisms leading to food spoilage.

An alternative to oxygen scavenging systems, for the prevention of food oxidation through active packaging, is the incorporation of antioxidants into packaging materials (Realini and Marcos 2014).

Antioxidants can be added directly to food formulations by using different processes such as spraying, immersion or mixing, which may cause changes in food quality parameters (color or taste) and may affect consumer acceptance of the product (Han *et al.* 2018). In addition, today, there is a preference for the exclusion of additives in foods. In this way, active packaging represents an innovative strategy to incorporate antioxidants into a polymer to prevent oxidative processes and extend shelf life (Realini and Marcos 2014). There are several advantages of this technology compared to the direct addition of antioxidants such as: lower amounts of active substances required; localized activity; controlled release of antioxidants; and elimination of processing steps, such as mixing, immersion or spraying (Bolumar*et al.* 2011).

Active agents can be applied to the packaging systems in different forms: coating or immobilization on the packaging material surface, or incorporated into the polymer matrix or multilayer films. They can act either by a direct or indirect contact mechanism with the foodstuff:

– The effect of direct contact with the foodstuff and then controlled diffusion in the food product is called a release mechanism. In this case, the active agent is incorporated into the packaging; it migrates into the food through diffusing and partitioning and then the effect decreases over time.

– The effect of direct contact between the packaging and the food (active agent attached or confined in the packaging materials). In this case, the active agent is chemically or physically immobilized on the packaging, and the desired effect is reached by the contact surface without diffusion.

– The effect occurs by evapo-release in the headspace atmosphere within the packaging (active agents have to be volatile), without direct contact between active packaging and the foodstuff.

In all of the cases mentioned, antioxidant packaging is intended to prevent or slow down the oxidation reactions that affect the quality of food. However, radicals such as hydroxyl and superoxide originate from oxygen and are the main initiators of oxidation. Thus, oxidation can be avoided by eliminating radicals as soon as they are formed (Pereira de Abreu *et al.* 2012). Some antioxidants can react efficiently with the radicals by trapping them, thus avoiding further oxidation. In such cases, neither high-barrier nor vacuum packaging materials are required to avoid oxidation, only the presence of a radical scavenger is needed to protect the food against the oxidation process (Pereira de Abreu *et al.* 2012).

Traditionally, synthetic antioxidants such as butylated hydroxytoluene (BHT), butylated hydroxyanisole (BHA) and propyl gallate (PG) have been used extensively to enhance the shelf life of food products by providing oxidative stability. However, the presence of synthetic antioxidants in food can have potential toxic and carcinogenic effects, and strict statutory controls are required. To avoid potential risks and respond to the consumer need for safe, healthy foods, natural antioxidants can be used as alternatives to synthetic antioxidants, and will likely be used in food packaging to assure antioxidant activity (Benbettaïeb *et al.* 2019; Han *et al.* 2018).

The main natural antioxidants that could be used in active films are organic acids such as ascorbic and citric acids, phenolic compounds such as phenolic acids, coumarins, flavonoids, tannins, essential oil compounds and volatile phenols (thymol, carvacrol, eugenol, cinnamaldehyde, etc.), plant extracts, phospholipids, tocopherols (α, β, γ or δ) and carotenoids (β-carotene, β-cryptoxanthin and lycopene) (Benbettaïeb *et al.* 2019).

8.2.2.5. *Antimicrobial packaging materials as releasing systems*

Microbial growth decreases shelf life and increases the risk of foodborne illness. Traditional methods of preserving foods from the effects of microbial growth include thermal treatment, drying, freezing, refrigeration, irradiation, MAP and the addition of antimicrobial agents or salts (Kerry *et al.* 2006). However, some of these techniques cannot be applied to food products, for example fresh meat and fish, as well as seafood (Quintavalla and Vicini 2002).

Antimicrobial food packaging acts to extend the lag phase and reduce the growth phase of microorganisms, thus extending shelf life and maintaining product quality and safety.

Antimicrobial packaging can take several forms: addition of sachets or pads containing antimicrobial agents; incorporation/dispersion of antimicrobial agents directly into polymers; coating or adsorbing antimicrobial agents onto polymer surfaces; immobilization of antimicrobial agents to polymers by ionic or covalent linkages; and use of polymers that are naturally antimicrobial (Kerry *et al.* 2006; Pereira de Abreu *et al.* 2012; Suppakul *et al.* 2003).

A large number of agents with antimicrobial properties (e.g. ethanol, carbon dioxide, Ag ions, chlorine dioxide, antibiotics, organic acids, essential oils and spices) have been tested for the purpose of inhibiting microorganism growth that can lead to deterioration of foodstuffs (Han *et al.* 2018). However, few of these systems are commercially available.

Active and antimicrobial packaging that involves gas scavenging/releasing systems (sachets and pads) has been reported. Antimicrobial packaging involving the use of active materials and coatings have had relatively few commercial successes. The exception is in Japan, where Ag-substituted zeolite is the most common antimicrobial agent, added as a thin coating onto food contact plastic film at levels of 1–3%. Ag ions inhibit a range of metabolic enzymes and have strong antimicrobial activity (Vermeiren *et al.* 1999). Agion® antimicrobial, MicrobeGuard's antimicrobial paper, FoodTouch® and Ciba® IRGAGUARD® B antimicrobials are based on the incorporation of silver zeolite into the food contact material (PP, HDPE, LDPE, etc.), which then exerts its action onto the surface of the food to inhibit the growth of gram-positive and gram-negative bacteria, molds and yeasts (Pereira de Abreu *et al.* 2012). These are specifically designed to release Ag ions, which are able to disrupt replication of the microorganisms' RNA and therefore prevent microbial growth. In the EU, silver zeolite should not be used in food contact material to extend shelf life but can be incorporated directly into the food matrix. In contrast, this system has a long history of use as antimicrobial packaging for beverages in the USA.

Microban® systems (USA) have become commercially available in the UK recently. These products contain triclosan, an antimicrobial compound allowed for use in cosmetics such as soap, shampoo and toothpaste. The use of triclosan (2,4,4-trichloro-2-hydroxydiphenylether) in food contact applications, however, is not yet regulated at the EU level (Kerry *et al.* 2006).

8.2.3. *Other active packaging systems*

8.2.3.1. *Microwave susceptors*

Packaging materials that absorb microwave energy and convert it to heat are called susceptors (see also Chapter 11). They qualify as active packaging because they enhance performance by achieving localized effects such as the browning and crisping of food (Robertson 2016). Microwave susceptors consist of aluminum or stainless steel deposited on substrates, such as polyester films or paperboard, and serve to dry, crisp and ultimately brown microwavable food (Figure 8.8). Modifiers for microwave heating consist of a series of antenna structures, which alter the way microwaves arrive at the food, thereby resulting in even heating, surface browning and crisping (Gill and Ahvenainen 2003).

Figure 8.8. *Aluminum microwave susceptors (Sirane 2020)*

8.2.3.2. *Temperature control: self-heating and self-cooling packages*

As consumer demand for convenience food is constantly increasing, the use of self-heating active packaging in ready meals is an important future application.

The concept of self-heating packaging is based on an exothermic chemical reaction that generates heat (Robertson 2016). In most cases, the reaction is between CaO (or MgO) and a water-based solution (Figure 8.9).

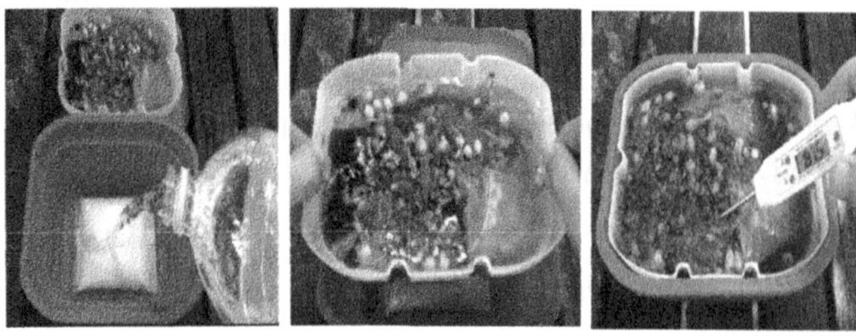

Figure 8.9. *Example of a self-heating package*

Self-heating cans have been commercially available for decades and are particularly popular in Japan for sake, coffee, tea and ready meals.

Another concept for self-heating packaging is based on the electrochemical reaction between Mg and H_2O, as follows (Robertson 2016):

$$Mg + 2H_2O \longrightarrow Mg(OH)_2 + H_2 + heat \quad [8.4]$$

Self-cooling cans have been commercialized in Japan, using an endothermic reaction based on the dissolution of ammonium nitrate and ammonium chloride in water.

8.3. Intelligent packaging: concepts and commercial applications

As previously defined, the aim of intelligent packaging is to monitor the product and transmit information to consumers. This can be information about the state of the package and its contents, time of manufacture or storage conditions. Depending on whether it is simple or reactive intelligent packaging, this can be part of the primary (outside or inside), secondary or tertiary packaging (Müller and Schmid 2019). An intelligent packaging system is first an extension of the communication function of traditional packaging. It is also a system capable of detecting, sensing and recording external or internal changes within the product's environment (Figure 8.10) throughout the entire supply chain (Yam *et al.* 2005).

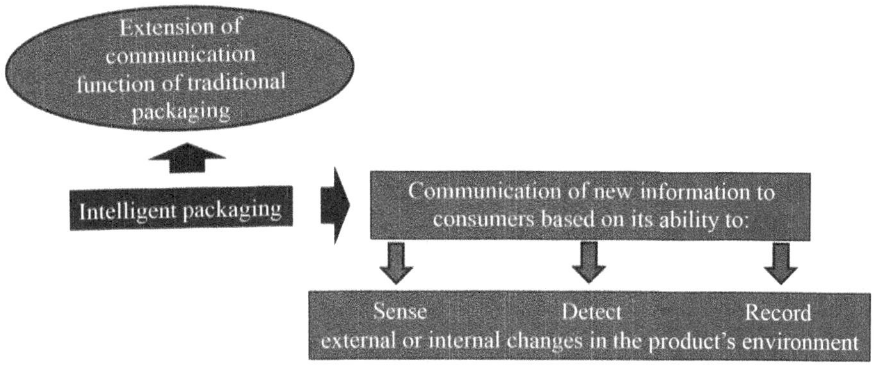

Figure 8.10. *Communication functions in intelligent packaging systems*

It is widely accepted that intelligent packaging systems fall into four categories (Figure 8.11):

Indicators: more convenience and inform consumers about food quality.

Sensors: definite quantification of the analytes in foods.

Data carriers: specifically intended for storage, distribution and traceability.

Providing more convenience, and controlling theft, counterfeiting and tampering: new value-added functions (microwave "doneness" indicators and thermochromic inks for protection against theft, etc.).

A subdivision according to the following types is also possible (Müller and Schmid 2019):

Environmental conditions: this system monitors the conditions that can lead to changes in the quality characteristics of the food. Examples of these types are time–temperature and gas leakage indicators. Depending on the monitoring factor, these systems can be placed outside or inside the packaging.

Quality indicator: this is used for the direct monitoring of the quality attributes of the food itself. Examples are biosensors and freshness sensors/indicators. These are usually located inside the packaging.

Data carriers: these systems are only used to store and transfer data, while indicators and sensors are used to monitor the external environment and display the information afterwards.

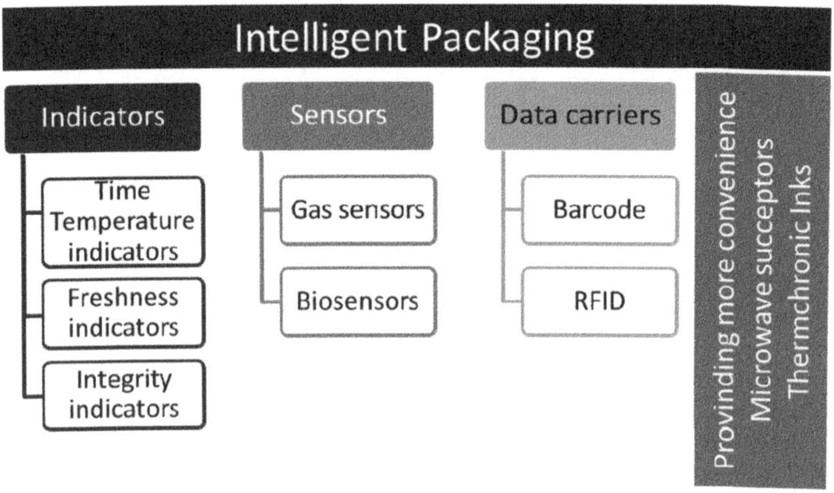

Figure 8.11. *Different categories of intelligent packaging systems*

8.3.1. *Indicators*

Indicators determine the presence or absence of a substance, the extent of a reaction between different substances or the concentration of a particular substance. Despite the large variety of indicators, all of them can effectively be grouped into three categories: time–temperature indicators, quality or freshness indicators, and integrity or gas concentration indicators (Hogan and Kerry 2008). All of them fall within the main category of "product quality and value-improving systems", which are undoubtedly the most widely used devices for food packaging applications (Ghaani *et al.* 2016).

8.3.1.1. *Time–temperature indicators*

Time–temperature indicators (TTIs) are the first generation of indicators intended to monitor any detrimental change in temperature, over time, along the food supply chain (Han *et al.* 2018). TTIs are typically small tags or labels attached onto shipping containers or individual consumer packages. These labels provide visual indications of temperature history during distribution and storage. The labels usually respond by displaying visually distinct changes that are temperature dependent, such as an increase in color intensity and diffusion of a dye along a straight path. Their use is essentially for meat and poultry products, where monitoring of the cold distribution chain, microbial safety and quality is of vital importance (Realini and Marcos 2014).

The basic idea underlying visual indicators is that the quality of food deteriorates more rapidly at higher temperatures because chemical and biochemical reactions, as well as microbial growth, favor higher temperatures. The indicators change color in response to cumulative exposure to temperature. TTIs available on the market are based on physical or chemical responses, which depend on acid–base reactions or polymerization towards time and temperature. In contrast, biochemical and biological responses are based on biological changes such as microorganisms, spores or enzymes, in relation to time and temperature (Kerry *et al.* 2006). Different examples according to the nature of the response are presented below.

Chemical TTIs

OnVu™: indicators contain benzopyridines (pigment that changes color over time in relation to temperature). When activated by UV light during packaging, it becomes dark blue (Realini and Marcos 2014). It then discolors gradually depending on the ambient temperature (Biji *et al.* 2015). The drop in color intensity indicates the gradual degradation of the product during the cold chain (Robertson 2016).

This system can be applied as a label or printed directly onto the package. It monitors product quality.

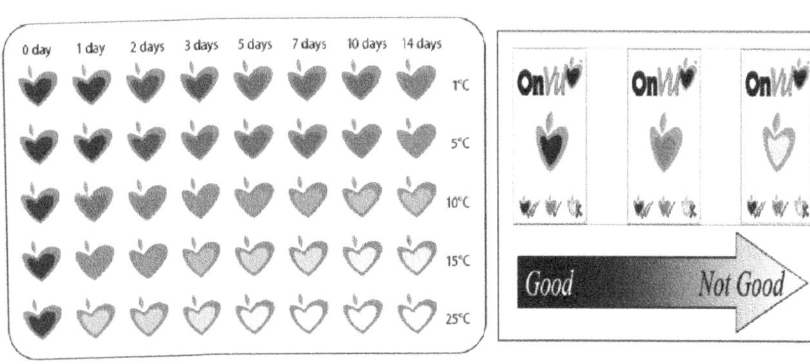

Figure 8.12. *Chemical TTIs (OnVu™ indicators 2020)*

Fresh-Check®: this is another chemical TTI based on a solid-state polymerization reaction, and consists of a small circle of polymer surrounded by a printed reference ring (Robertson 2016). The system is coated onto a pressure-sensitive label and can be adhered to the packaging of perishable food, such as meat and poultry. The indicator contains diacetylene monomers, which is polymerized through a lattice-controlled solid-state reaction, resulting in a highly colored polymer. The response of the TTI is the color change, evaluated as the decrease in reflectance, which reflects the cumulative temperature exposure (Han *et al.* 2018). Consumers do not consume food if the polymer center is darker than the surrounding reference ring (Kruijf *et al.* 2002).

Figure 8.13. *Chemical TTIs: Fresh-Check® indicators (Temptimecorp 2020)*

Physical TTIs

3M™ Monitor Mark™: this is a physical TTI that provides a visual history of time/temperature exposure. In this case, a TTI contains a fatty acid ester, with a selected melting point, mixed with a blue dye (Realini and Marcos 2014). Initially, the indicator track is white, but upon exposure to temperatures exceeding the specified threshold, the substance melts and migrates from a reservoir along a carrier. This causes blue coloring to appear at the left edge of the first viewing window that gradually moves across to the right (Figure 8.14) to the end of the track. The diffusion can be observed through a transparent window. The rate (or the distance) of the blue color movement is dependent on the temperature. The melting temperature of the colored fatty acid ester determines the range of temperatures which the food is to be stored at. They serve as an effective check by giving workers guidance on whether or not the cold chain has been maintained.

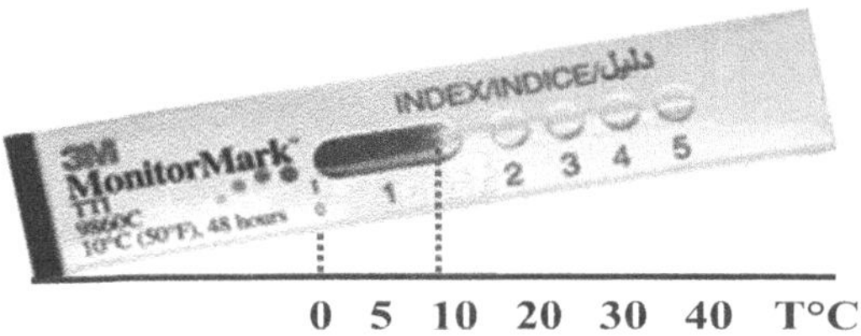

Figure 8.14. *Physical TTIs from 3M Monitor Mark® (Infoagro, 2020)*

Biochemical TTIs

Vitsab CheckPoint® is an enzymatic TTI that is dependent on color changes. It is stimulated via a reduction in pH due to the release of acid after the hydrolysis of lipid (triglycerides) substrates by enzymes (lipase) (Robertson 2016). This indicator is made up of two different compartments with a lipid substrate and an aqueous solution of lipolytic enzymes, in an aqueous medium (Kerry *et al.* 2006). The color changes from dark green to bright yellow, and then to an orange-red color (Figure 8.15) as a result of a pH drop. A visual scale of the color changes allows an assessment of freshness related to pH (Pereira de Abreu *et al.* 2012).

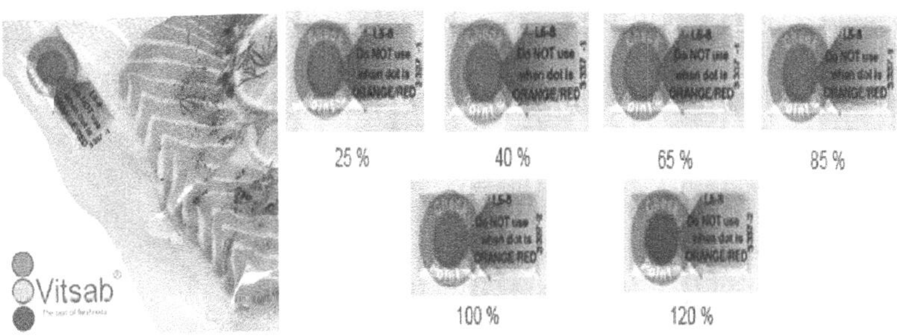

Figure 8.15. *Vitsab CheckPoint® biochemical TTIs (VITSAB, 2020)*

Biological TTIs

TopCryo™ is a biological TTI based on bacterial growth. Using lactic acid bacteria, an indicator dye and a nutritive medium gel are incorporated into a multilayer plastic sachet, glued onto the outer layer of the food package (Robertson 2016). Depending on the time–temperature profile, the bacteria grow and, as a result, the pH of the medium decreases and the indicator turns from green to red (Figure 8.16).

Figure 8.16. *Biological TTIs (TopCryo™ indicators)*

8.3.1.2. *Quality and freshness indicators*

Increasing consumer demand for healthy, fresh foods over the last two decades has led companies and research labs to develop new freshness and quality indicators. These are used to indicate if the quality of the product has become unacceptable during storage, transport, retailing and in consumers' homes (Ghaani *et al.* 2016). Freshness indicators are devices used to provide direct information about quality or the loss of freshness in packaged foods, resulting from physical or chemical changes, or microbial proliferation.

Freshness indicators typically undergo a color change that remains permanent and is easy to read and interpret by consumers. This color change is simply monitored and correlated with food product freshness.

Most of these indicators detect volatile metabolites produced during the aging process of foods such as CO_2, diacetyl, amines, ammonia and H_2S (Robertson 2016). Others respond to organic acids or volatile biogenic amines formed by microbial degradation of protein-containing foods, such as cadaverine from lysine and histamine from histidine (Poyatos-Racionero *et al.* 2018).

Some of the more well-known, commercialized quality and freshness indicators are presented below.

Ripeness indicator ripeSense®

This kind of indicator allows consumers to choose fruit that best appeals to their tastes by detecting aroma compounds or gases involved in the ripening process (e.g. ethylene) released by the fruit. In this case, the sensor changes color when it reacts with aromatic compounds developed by the fruit and indicates the state of the product to the consumer. The indicator is red in color for unripe (crisp) fruit, changing through orange (firm) to yellow, when the fruit is very ripe (juicy) (Robertson 2016). There is a good correlation between the level of aroma produced (ethylene, for example) and the actual softening of fruits. The more ripe a fruit is, the more aroma is produced (Figure 8.17).

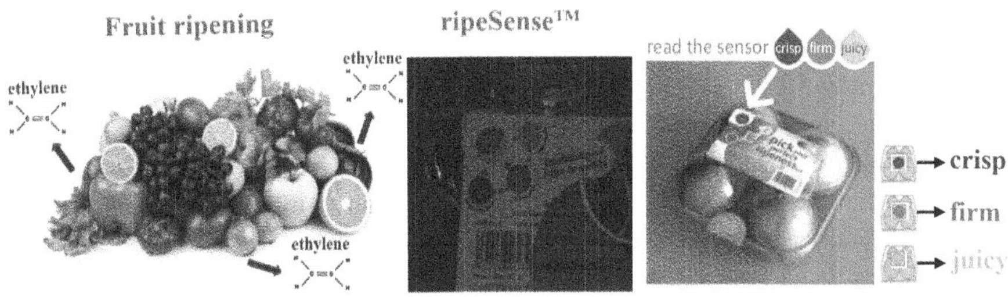

Figure 8.17. *ripeSense™ indicator (ripeSense, 2020)*

SensorQ™ indicator

This indicator is a type of label designed to detect spoilage in uncooked meat and poultry products, after packaging by the meat packer, distributor or grocer. SensorQ™ is a pH-sensing technology based on anthocyanins, which is able to inform on the formation of biogenic amines from microbiological origin in packed meat and poultry (Poyatos-Racionero *et al.* 2018). A "Q" printed on the label acts as a visual tool to determine freshness. When the inside of the Q is orange, the product is fresh. As the meat or poultry spoils, naturally occurring bacteria becomes too high and emits gaseous by products, including molecules called biogenic amines. These molecules change the pH and cause the inside of the Q to turn from orange to tan (Figure 8.18), indicating the product is not fresh, and then to dark green, which denotes "beyond spoiled".

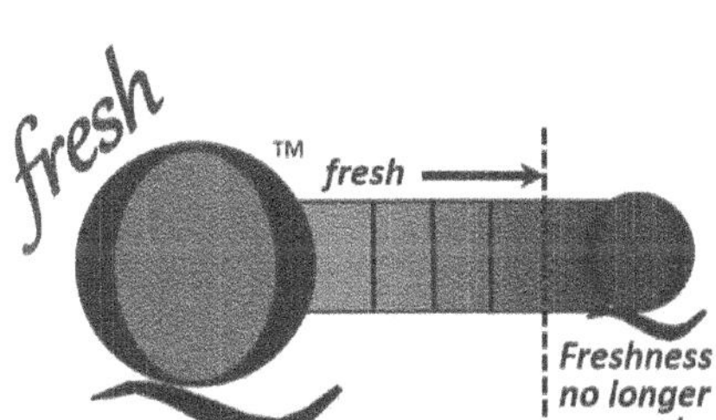

Figure 8.18. *SensorQ™ indicator (Packworld, 2020)*

8.3.1.3. *Integrity or gas concentration indicators*

It is very difficult to maintain the quality of food materials within the packaging system for many reasons, including respiration of fresh fruits and vegetables, changing gas concentrations and gas leakage from inside or outside the packaging materials, or due to gas produced by microbial growth within the package (Sohail *et al.* 2018). In order to solve this problem, manufacturing and research labs developed gas indicator systems. The most widely known gas indicators are used to check oxygen and carbon dioxide concentrations (Ghaani *et al.* 2016) within the packaging material by changing color as a result of specific chemical or enzymatic reactions. The gas composition in the package headspace often changes due to the activity of the food product, the nature of the package or the environmental conditions (Fan *et al.* 2017). For example, respiration of fresh products, gas generation by spoilage microorganisms or gas transmission through the packaging material or package leaks may cause a change in the gas composition inside the package (Yam *et al.* 2005). Gas indicators in the form of a package label or printed on packaging films can monitor changes in gas composition, thereby providing a means of monitoring the quality and safety of food products (Yam *et al.* 2005).

Among the various types of gas indicators, oxygen indicators are most commonly used for MAP packaging applications (Yam *et al.* 2005). These are based on colorimetric redox dye, such as methylene blue, and a strong reducing agent, such as glucose in an alkaline medium (Mills 2005). When the redox dye of the indicator is oxidized by oxygen, a color change can be observed. The most known commercial application of gas indicators is Ageless Eye® tablets (Mitsubishi Gas Chemical, Japan), which is a reversible oxygen indicator used in combination with AGELESS® O_2 absorbers. The color of the tablets initially inserted inside the packages changes from pink at a concentration of $O_2 \leq 0.1\%$, to blue (Figure 8.18) at a concentration $\geq$ 0.5% (Figure 8.19). The color change rate is temperature dependent (Realini and Marcos 2014).

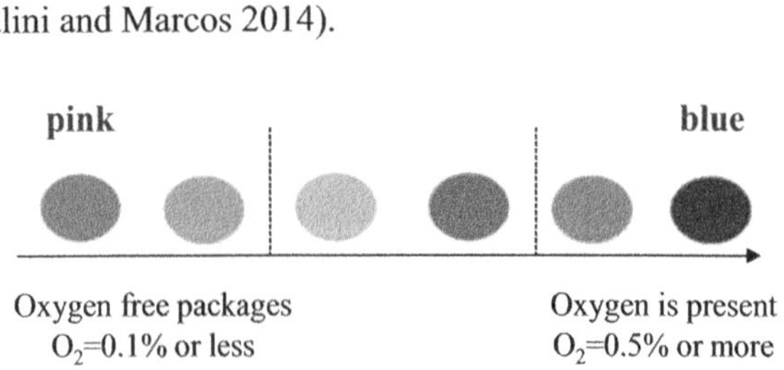

Figure 8.19. *Ageless Eye® O_2 absorbers*

The CO_2 indicator designed by Moonstone Co. consists of a label containing a gas-sensitive dye that can be inserted into a package, producing different colors according to gas concentrations. When carbon dioxide has leaked or diffused out of the MAP, the dye changes from dark blue to a permanent yellow color (Han *et al.* 2005). The function of the dye is to indicate any leakage through the seal, as well as any increase in the level of carbon dioxide, due to microbial growth.

8.3.2. *Sensors*

A sensor is defined as a device used to detect, locate or quantify energy or matter, giving a signal for the detection or measurement of a physical or chemical property to which the device responds (Müller and Schmid 2019). A biosensor is a compact analytical device that detects, records and transmits information relating to biochemical reactions. It consists of two primary components: a bioreceptor that recognizes a target analyte, and a transducer that converts biochemical signals into a quantifiable electrical response. The bioreceptor is an organic or biological material (Fang *et al.* 2017). The transducer can take many forms: electrochemical, optical, thermal, etc., depending on the parameters being measured. The two most well-known sensors are gas sensors and biosensors.

8.3.2.1. *Gas sensors*

The extent of food spoilage can be determined by the concentration of certain gases such as CO_2 or H_2S. The sensors quantitatively monitor the presence of a gas by changing the physical parameters. One example is non-dispersive infrared (NDIR) CO_2 sensors, which are spectroscopic sensors that measure the CO_2 content by gas absorption at a certain wavelength (Müller and Schmid 2019).

8.3.2.2. *Biosensors*

Biosensors are another type of sensor that, unlike chemical sensors, have a receptor made of biological material such as enzymes, antigens, hormones or nucleic acids (Ghaani *et al.* 2016). Biosensors can monitor food freshness in a more specific way than freshness indicators, as they can detect degradation in products and can be tailored to the type of product being packaged (Kuswandi 2017). As taste and quality of fruit is closely linked to the content of sugars, the quality of fruit and vegetable beverages can be determined by a glucose oxidase- or lactate-based biosensor (Pereira de Abreu *et al.* 2012). In the presence of glucose oxidase, glucose is transferred to gluconolactone and H_2O_2, and the transducer then gives information in relation to the concentration of H_2O_2. Several prototypes of commercial biosensors for intelligent packaging are currently being developed.

Food Sentinel System™ (SIRA Technologies) is a biosensor developed to detect food pathogens with a specific pathogen antibody attached to a membrane, which forms part of a barcode. The presence of contaminating bacteria causes the formation of a localized dark bar (Figure 8.20) and the barcode therefore becomes unreadable upon scanning (Fang *et al.* 2017; Realini and Marcos 2014; Yam *et al.* 2005).

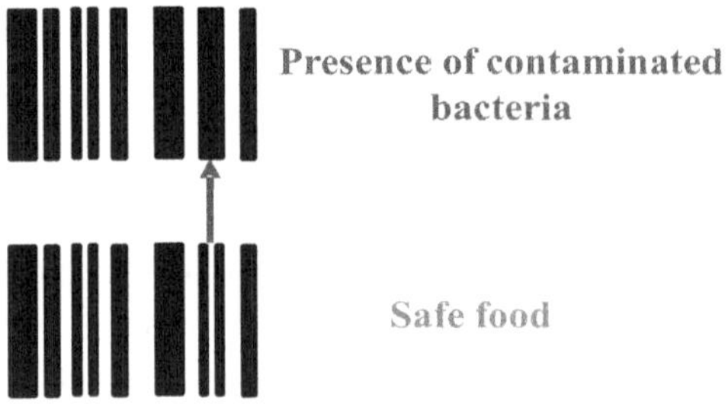

Figure 8.20. *Principle of the Food Sentinel System™ biosensor (SIRA Technologies, 2020)*

Toxin Guard™ technology (Toxin Alert) is another biosensor in which the functional system is based on antibodies. It is integrated into plastic packaging and makes it possible to detect pathogens such as *Salmonella*, *E. coli*, *Listeria* and *Campylobacter*. A positive result is indicated by a visual signal (Fang *et al.* 2017).

8.3.3. *Data carriers*

Data carriers help to make information within the supply chain flow more efficiently. More specifically, data carrier devices do not provide any kind of information on the quality status of food but rather, are intended for automatization, traceability, theft prevention or counterfeit protection (McFarlane and Sheffi 2003). Data carriers are most often placed on tertiary packaging. The most frequently used data carriers are barcode labels and radiofrequency identification (RFID) tags.

8.3.3.1. *Barcodes*

Barcodes are the least expensive and most popular form of data carrier; they have been widely used to facilitate inventory control, stock reordering and checkout. Barcodes have different storage capacities depending on type. One-dimensional (1D) barcodes were developed first. These consist of a pattern of parallel spaces and bars (Figure 8.21). The different arrangement of the bars and spaces results in the coding of data (Ahmed *et al.* 2018). A barcode scanner and an associated system can translate the coded information. Two-dimensional (2D) barcodes offer more memory capacity (e.g. packaging date, batch number, packaging weight, nutritional information or preparation instructions), compared to 1D barcodes, by combining dots and spaces arranged in a matrix (Ghaani *et al.* 2016). The more recent 2D barcodes are quick response (QR) codes that enable an even larger amount of data to be stored using four different encoding modes: numeric, alphanumeric, byte/binary and kanji; the latter referring to logographic Chinese characters (Ghaani *et al.* 2016).

Figure 8.21. *Different examples of barcodes*

8.3.3.2. *Radiofrequency identification tags*

Radiofrequency identification (RFID) tags are the most advanced example of a data carrier device. They can notify consumers, in real time, about which foodstuffs to consume as a priority, according to their freshness in the refrigerator, when connected through a smartphone (Figure 8.22) (Ahmed *et al.* 2018).

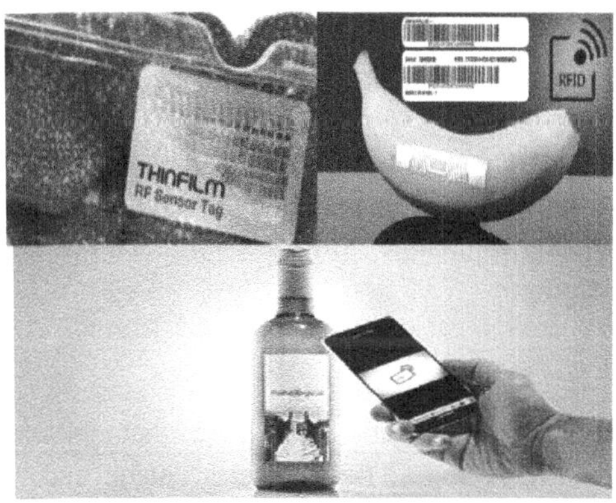

Figure 8.22. *RFID tag application*

RFID tags are electronic information-based systems. They use radiofrequency electromagnetic fields to transfer data from a tag attached to an object, in order to trace and identify the object automatically. RFID is an improvement on the previously mentioned barcodes (Mohebi and Marquez 2015). It offers further advantages for the entire food supply chain. These include traceability, inventory management and promotion of quality and safety, cold chain monitoring, livestock management and shelf life prediction. An RFID system includes three main elements: a tag formed by a microchip connected to a tiny antenna; a reader; and middleware that bridges the RFID hardware and enterprise applications. In a typical RFID system (Figure 8.23), a reader emits radio waves to capture data from an RFID tag (1), an antenna receives the signal and transfers it to the microchip (2), and the microchip returns the information to the antenna, which transfers it to the reader (3). Finally, the data is passed onto a host computer (which may be connected to a local network or to the Internet) for analysis and decision-making (Ghaani *et al.* 2016).

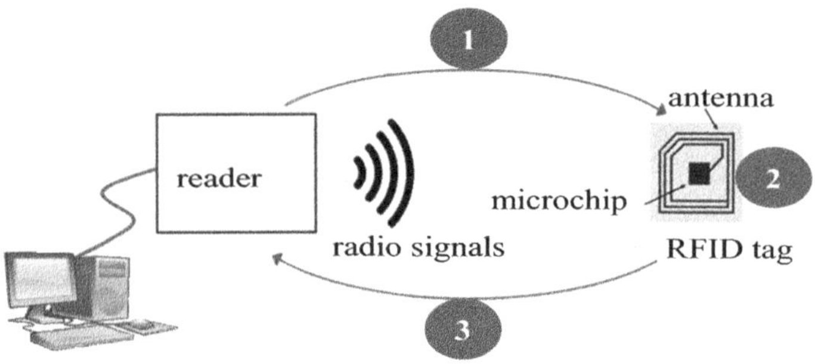

Figure 8.23. *The working principle of a radiofrequency identification (RFID) tag*

8.3.4. *Providing more convenience*

The packaging industry has always attempted to improve convenience for the consumer because convenience is a value-added function that consumers are likely to pay extra for, as lifestyles change. More convenience therefore becomes another value added for packaging.

8.3.4.1. *Microwave doneness*

The first example of intelligent packaging that increases convenience is microwave "doneness" indicators. These devices detect, and visually indicate, the state of readiness of food heated in a microwave oven, informing the consumer when the food is ready (Robertson 2016).

8.3.4.2. *Thermochromic inks*

The second and most well-known type of packaging that increases convenience is thermochromic inks. Thermosensitive inks are printed onto the package (e.g. shrink sleeves of beverage cans), which can be heated or cooled, before consumption, to indicate the ideal drinking temperature for the product (Robertson 2016). The color of the ink changes according to the temperature. One example of thermochromic inks is a smart lid on a coffee cup, infused with a color changing additive that allows it to go from a dark brown to a bright red when exposed to higher temperatures (Figure 8.24). If the red color is very intense, it indicates to consumers that the coffee in the cup is not yet at a comfortable drinking temperature.

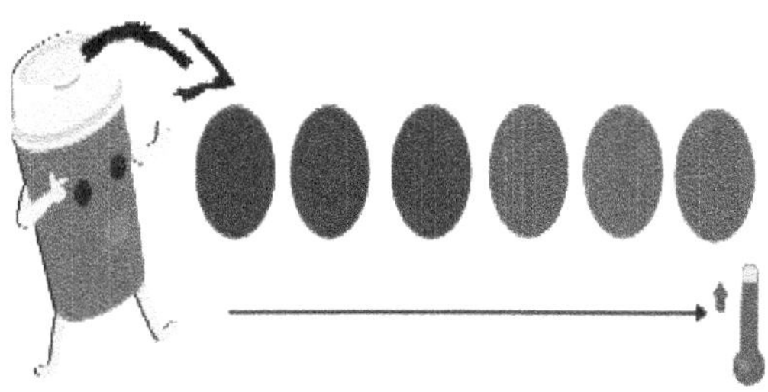

Figure.8.24. *Thermochromic smart lid inks*

Another example comprises a cold-activated label in which thermal ink changes according to the temperature. Figure 8.25 shows an example of a beer bottle with mountains on the label that turn from white to blue when the beer is appropriately chilled and ready for consumption.

Figure 8.25. *Thermochromic smart label inks*

Based on the temperature, the color pattern of these inks is either reversible or irreversible. An irreversible ink remains constant on exposure to a certain temperature and does not change once it becomes a specific color. Reversible ink has the ability to change color once it has reached a certain temperature; as the temperature falls below the threshold, the ink reverts to its original color (Sohail *et al.* 2018).

8.3.5. *Controlling theft, counterfeiting and tampering*

Theft, counterfeiting and tampering is a worldwide concern both for the food industry and consumers. To reduce incidences of theft and counterfeiting among food and pharmaceutical products, special inks, dyes, laser labels, barcodes and RFID systems are used. Special intelligent closures (see Chapter 9) are being developed in order to identify food tampering. These plastic- and aluminum-based closures are broken during any attempt of counterfeiting or opening before reaching the end consumer (Sohail *et al.* 2018). However, this technology has not found widespread commercial use in the food industry because of the comparatively low unit value of packed food as well as the cost of this technology.

Intelligent tamper-evident technologies are being developed on the basis of labels or seals that remain transparent, as long as the package is not opened or tampered with. After opening or tampering, they change permanently to colored labels or seals. Aluminum and plastic closures can help consumers to identify whether or not the package has been opened. The tamper-proof band will split when the package is opened, providing tamper evidence to the consumer. Alcoa CSI of Germany has invented a special cap with "skirts" that curl under the bottom edge of the closure. The segments of the tamper-evident ring ("skirt") rupture when the container is opened (Han *et al.* 2005) (see also Chapter 9). Recently, Alfian *et al.* (2017) developed a food traceability system based on RFID technology, which was used to track and trace food product location, and a wireless sensor network to collect temperature and humidity during storage and transportation. The system was used for a supply chain in Korea; it produced positive results for both industry and consumer in detecting real-time location of the product and complete temperature and humidity history. This traceability system helps to optimize food distribution while also increasing customer satisfaction as it can monitor product freshness.

Electronic article surveillance (EAS) is a form of electronic tagging developed and used as an efficient technology against theft, generally for high-priced goods (Han *et al.* 2005). But theft and counterfeiting are usually not too common in the food industry; they do pose a huge economic burden in other industries (pharmaceutical, cosmetic, perfumes, clothes, etc.).

8.4. Consumer safety and related regulations

Current packaging legislation requires the migration level of substances from food contact materials to be low. The adequacy of active and intelligent packaging technology is related to the migration level and risk. General requirements stated in the EU Regulation 1935/2004 (European Commission 2004) and Directive 89/107/EEC (European Commission 1989) for the safe use of active and intelligent packaging have recently been integrated into the EU Regulation 450/2009 (European Commission 2009). The new regulation also establishes specific requirements for the marketing of active and intelligent materials and items intended to come into contact with food. Active packaging systems that intentionally release substances into the package must comply with (direct) food additives legislation. The released substance must be in the positive list of additives and the use of the substance must fulfill a technological need. It follows that, substances deliberately incorporated into active materials to be released into the food, or the environment surrounding the food, do not need to be listed in the Community list of EU Regulation 450/2009 (European Commission 2009). They shall be used in full compliance with the relevant community and national provisions applicable to food, and shall comply with the provisions of the EU Regulation 1935/2004 (European Commission 2004) and its implementation measures (Heckman 2005).

Intelligent packaging systems can be divided into two groups: the first involves external indicators fixed onto the outer surface of a package, such as time–temperature indicators. The second group consists of internal indicators intended to be positioned in the headspace of a package, such as gas concentration indicators (Han *et al.* 2005). Migration does not occur in the first group of indicators, since there is no direct contact between the indicator and the food product. Indicators in the second group are not intended to come into direct contact with the packaged foods (placed in the free headspace of a package or fixed to the inner surface of the lid) and therefore also have no migration risk (Han *et al.* 2005). As a result, the EU Directive 89/107/EEC (European Commission 1989) and the EU Regulation 450/2009 (European Commission 2009) do not apply to such systems (Restuccia *et al.* 2010). However, in the case of sachets that are in direct contact with the food product – particularly with moist or fatty foods – migration can occur as sachet materials are generally very porous. In the latter case, the EU Directive 89/107/EEC (European Commission 1989) and the EU Regulation 450/2009 (European Commission 2009) must be applied.

8.5. References

Ahmed, I., Lin, H., Zou, L., Li, Z., Brody, A.L., Qazi, I.M., Lv, L., Pavase, T.R., Khan, M.U., Khan, S., and Sun, L. (2018). An overview of smart packaging technologies for monitoring safety and quality of meat and meat products. *Pack. Technol. Sci.*, 31(7), 449–471.

Ahvenainen, R. (2003). Active and intelligent packaging: An introduction. In *Novel Food Packaging Techniques*, Ahvenainen, R. (ed.). Woodhead Publishing Ltd. Elsevier, Cambridge, UK.

Alfian, G., Rhee, J., Ahn, H., Lee, J., Farooq, U., Ijaz, M.F., and Syaekhoni, M.A. (2017). Integration of RFID, wireless sensor networks, and data mining in an e-pedigree food traceability system. *J. Food Eng.*, 212, 65–75.

ALL4PACK (2016). L'emballage dans le monde. "*Packag'in*" *Actualites-packaging* [Online]. Available: www.all4pack.fr [Accessed 7 February 2020].

Bazzano, M., Barolo, C., Buscaino, R., D'Agostino, G., Ferri, A., Sangermano, M., and Pisano, R. (2016). Controlled atmosphere in food packaging using ethylene–α-cyclodextrin inclusion complexes dispersed in photocured acrylic films. *Ind. Eng. Chem. Res.*, 55(3), 579–585.

Benbettaïeb, N., Debeaufort, F., and Karbowiak, T. (2019). Bioactive edible films for food applications: Mechanisms of antimicrobial and antioxidant activity. *Crit. Rev. Food Sci. Nut.*, 59(21), 3431–3455.

Biji, K.B., Ravishankar, C.N., Mohan, C.O., and Srinivasa Gopal, T.K. (2015). Smart packaging systems for food applications: A review. *J. Food Sci. Technol.*, 52(10), 6125–6135.

Bolumar, T., Andersen, M.L., and Orlien, V. (2011). Antioxidant active packaging for chicken meat processed by high pressure treatment. *Food Chem.*, 129(4), 1406–1412.

Chen, J. and Brody, A.L. (2013). Use of active packaging structures to control the microbial quality of a ready-to-eat meat product. *Food Control*, 30(1), 306–310.

Dainelli, D., Gontard, N., Spyropoulos, D., Zondervan-van den Beuken, E., and Tobback, P. (2008). Active and intelligent food packaging: Legal aspects and safety concerns. *Trends Food Sci. Technol.*, 19, S103–S112.

European Commission (1989). Food additives authorised for use in foodstuffs intended for humain consumption. Commission Regulation 89/107/EEC [Online]. Available: https://eur-lex.europa.eu/legal-content/en/ALL/?uri=CELEX%3A31989L0107.

European Commission (2004). Materials and articles intended to come into contact with food and repealing Directives 80/590/EC and 89/109/EC. Commission Regulation 1935/2004 [Online]. Available: https://ec.europa.eu/food/safety/chemical_safety/food_contact_materials/ legislation_en [Accessed 24 October 2017].

European Commission (2009). Active and intelligent materials and articles intended to come into contact with food. Commission Regulation 450/2009 [Online]. Available: https://eur-lex.europa.eu/legal-content/EN/ALL/?uri=CELEX%3A32009R0450.

Fang, Z., Zhao, Y., Warner, R.D., and Johnson, S.K. (2017). Active and intelligent packaging in meat industry. *Trends Food Sci. Technol.*, 61, 60–71.

Floros, J.D., Dock, L.L., and Han, J.H. (1997). Active packaging technologies and applications. *Food Cosmetics Drug Pack.*, 20, 10–17.

Ghaani, M., Cozzolino, C.A., Castelli, G., and Farris, S. (2016). An overview of the intelligent packaging technologies in the food sector. *Trends Food Sci. Technol.*, 51, 1–11.

Gill, C. and Ahvenainen, R. (2003). Active packaging in practice: Meat. *Novel Food Pack. Tech.*, 365–383.

Han, J.H. (2014). A review of food packaging technologies and innovations. In *Innovations in Food Packaging*, 2nd edition, Han, J.H. (ed.). Academic Press, San Diego.

Han, J.H., Ho, C.H.L., and Rodrigues, E.T. (2005). 9 – Intelligent packaging. In *Innovations in Food Packaging*, Han, J.H. (ed.). Academic Press, London.

Heckman, J.H. (2005). Food packaging regulation in the United States and the European Union. *Regulatory Toxicol. Pharmacol.*, 42(1), 96–122.

Hogan, S.A. and Kerry, J.P. (2008). Smart packaging of meat and poultry products. In *Smart Packaging Technologies for Fast Moving Consumer Goods*, Kerry, J.P. and Butler, P. (eds). John Wiley and Sons, London.

Infoagro (2020). TTIs monitors [Online]. Available: https://www.infoagro.com. [Accessed 18 August 2020].

Janjarasskul, T. and Suppakul, P. (2018). Active and intelligent packaging: The indication of quality and safety. *Crit. Rev. Food Sci. Nut.*, 58(5), 808–831.

Jia-Wei, H., Ruiz-Garcia, L., Jian-Ping, Q., and Xin Ting, Y. (2018). Food packaging: A comprehensive review and future trends. *Compr. Rev. Food Sci. Food Safety*, 17(4), 860–877.

Kerry, J.P., O'Grady, M.N., and Hogan, S.A. (2006). Past, current and potential utilisation of active and intelligent packaging systems for meat and muscle-based products: A review. *Meat Sci.*, 74(1), 113–130.

Kruijf, N.D., Beest, M.V., Rijk, R., Sipiläinen-Malm, T., Losada, P.P., and Meulenaer, B.D. (2002). Active and intelligent packaging: Applications and regulatory aspects. *Food Additives & Contaminants*, 19(S1), 144–162.

Kuswandi, B. (2017). Freshness sensors for food packaging. *Ref. Module Food Sci.*

Lee, D.S. (2016). Carbon dioxide absorbers for food packaging applications. *Trends Food Sc. Technol.*, 57, 146–155.

McFarlane, D. and Sheffi, Y. (2003). The impact of automatic identification on supply chain operations. University of Cambridge, Department of Engineering. *International Journal of Logistics Management*, 14(1), 1–17 [Online]. Available: https://doi.org/10.1108/09574090310806503.

Mills, A. (2005). Oxygen indicators and intelligent inks for packaging food. *Chem. Soc. Rev.*, 34(12), 1003–1011.

Mohebi, E. and Marquez, L. (2015). Intelligent packaging in meat industry: An overview of existing solutions. *J. Food Sci. Technol.*, 52(7), 3947–3964.

Müller, P. and Schmid, M. (2019). Intelligent packaging in the food sector: A brief overview. *Foods*, 8(1), 16.

Multisorb Technologies Inc. (2020). Oxygen absorbers [Online]. Available: https://www.multisorb.com/oxygen-absorbers/freshmax/ [Accessed 19 August 2020].

Nerín, C., Tovar, L., and Salafranca, J. (2008). Behaviour of a new antioxidant active film versus oxidizable model compounds. *J. Food Eng.*, 84(2), 313–320.

Onvu (2020). TTis indicator Onvu [Online]. Available: http://www.onvu.com [Accessed 18 August 2020].

Packworld (2020). Packaging World [Online]. Available: http://www.PackWorld.com.

Patel, C. (2020). Global (United States, China, and European Union) Intelligent Packaging Market Research Report 2020-2026 [Online]. Available: www.bccourier.com/global-intelligent-packaging-market-top-growing-companies-analysis-during-2020-2026-jones-packaging-3m-du-pont-basf-intelligent-packaging [Accessed 17 February 2020].

Pereira de Abreu, D., Cruz, J.M., and Paseiro Losada, P. (2012). Active and intelligent packaging for the food industry. *Food Rev. Int.*, 28(2), 146–187.

Poyatos-Racionero, E., Ros-Lis, J.V., Vivancos, J.-L., and Martínez-Máñez, R. (2018). Recent advances on intelligent packaging as tools to reduce food waste. *J. Cleaner Prod.*, 172, 3398–3409.

Quintavalla, S. and Vicini, L. (2002). Antimicrobial food packaging in meat industry. *Meat Sci.*, 62(3), 373–380.

Realini, C.E. and Marcos, B. (2014). Active and intelligent packaging systems for a modern society. *Meat Sci.*, 98(3), 404–419.

Restuccia, D., Spizzirri, U.G., Parisi, O.I., Cirillo, G., Curcio, M., Iemma, F., Puoci, F., Vinci, G., and Picci, N. (2010). New EU regulation aspects and global market of active and intelligent packaging for food industry applications. *Food Control*, 21(11), 1425–1435.

Ripense (2020). Ripesense indicator [Online]. Available: http://Ripensense.com. [Accessed 16 August 2020].

Robertson, G.L. (2016). *Food Packaging: Principles and Practice*, 3rd edition. CRC Press, Taylor and Francis Group, Boca Raton.

Sealed Air Corporation (2020). Sealed Air Corporation [Online]. Available: http://www.sealedair.com/index.htm.

SIRA Technologies (2020). SIRA technologies food sentinel system thermal barcode for packaging [Online]. Available: https://www.adazonusa.com/blog/barcode-industry/sira-technologies-food-sentinel-system [Accessed 18 August 2020].

Sirane (2020). Sirane [Online]. Available: http://www.sirane.co.uk/.

Sohail, M., Sun, D.-W., and Zhu, Z. (2018). Recent developments in intelligent packaging for enhancing food quality and safety. *Crit. Rev. Food Sci. Nut.*, 58(15), 2650–2662.

Süd-Chemie (2020). Humidipak [Online]. Available: http://www.sud-chemie.com/scmcms/web/pagees-2991.htm [Accessed 18 August 2020].

Suppakul, P., Miltz, J., Sonneveld, K., and Bigger, S.W. (2003). Active packaging technologies with an emphasis on antimicrobial packaging and its applications. *J. Food Sci.*, 68(2), 408–420.

Temptimecorp (2020). FreshCheck indicator [Online] Available: http://www.temptimecorp.com [Accessed 18 August 2020].

Vermeiren, L., Devlieghere, F., van Beest, M., de Kruijf, N., and Debevere, J. (1999). Developments in the active packaging of foods. *Trends Food Sci. Technol.*, 10(3), 77-86.

VITSAB (2020). Welcome to Vitsab International AB [Online]. Available: http://Vitsab.com.

Wholesalegroup (2020). Absorber pads for meat [Online]. Available: https://www.wholesalegroup.com.au/Meat-Soaker-Pads.html [Accessed 18 August 2020].

Yam, K.L., Takhistov, P.T., and Miltz, J. (2005). Intelligent packaging: Concepts and applications. *J. Food Sci.*, 70(1), R1–R10.

Yildirim, S., Röcker, B., Pettersen, M.K., Nilsen-Nygaard, J., Ayhan, Z., Rutkaite, R., Radusin, T., Suminska, P., Marcos, B., and Coma, V. (2018). Active packaging applications for food. *Compr. Rev. Food Sci. Food Safety*, 17(1), 165–199.

Zhao, C.-J., Han, J.-W., Yang, X.-T., Qian, J.-P., and Fan, B.-L. (2016). A review of computational fluid dynamics for forced-air cooling process. *Appl. Energy*, 168, 314–331.

9

Packaging Caps and Closures

Kata Galić

Faculty of Food Technology and Biotechnology, University of Zagreb, Croatia

9.1. Introduction

Closures used with food containers can be defined as a method for closing a pack so that the product is properly contained and protected. There is a huge variety of closure types. Furthermore, material used for closures can range from metal to glass, plastics, cork, rubber and others. Metal caps and closures are used for containers made from glass and plastic, as well as from metal itself.

Closures are required to provide different functions (hermeticity, easy opening and closing, and tamper evidence), without affecting, or being affected by, the packaged content (Theobald 2006b; Emblem 2012; Robertson 2013).

At the beginning of this century, the global caps and closures market value was approximately US$15 billion, with an expected increase in market size by about 4% per annum. In 2005, the caps and closures market was split as follows: 5% for cork, 48% for metals and 47% for plastics (Theobald 2006a). With respect to light-metal packaging in the food sector, approximately 45 billion beverage cans, 20 billion food cans and 100 billion closures (20 billion closures for jars and 80 billion crowns for bottles) are used in Europe per year. Of the 80 billion crowns used per year, beer bottles account for 80% (64 billion), while the other 20% (16 billion) is used for carbonated/still drinks, juices and water (Oldring *et al.* 2014).

For a color version of all figures in this book, see www.iste.co.uk/debeaufort/packaging.zip.

Packaging Materials and Processing for Food, Pharmaceuticals and Cosmetics, coordinated by Frédéric DEBEAUFORT, Kata GALIĆ, Mia KUREK, Nasreddine BENBETTAIEB and Mario ŠČETAR.

The natural corks market represents 62% of the total value (367.2 million euros), followed by champagne stoppers with 20% (117.1 million euros) and the remaining 18% accounts for other types of stoppers (APCOR 2015).

9.2. Closure types

In general, closures can be classified under four main headings: (1) closures to retain internal pressure; (2) closures to contain and protect contents; (3) closures to maintain a vacuum inside the container; and (4) closures to secure contents inside the container (Robertson 2013). With regard to function, another group covers specialized closures.

9.2.1. *Closures to retain internal pressure*

This type of closure is required to contain pressure from about 200 to over 800 kPa, as is typical for carbonated drinks and beer, for hot- or cold-filled, pasteurized and pressurized products, and for sterilized vacuum products.

The crown cork is a type of pressure-retaining closure. These are made from tinplate or tin-free steel (TFS), with adequate coatings on both sides. They have a sealing pad inside the closure, which could be the cork (hence the name "crown cork") or a soft plastic material. The closure is preformed and then placed over the neck of the bottle. Crown cork closures are widely used for beer bottles (Figure 9.1).

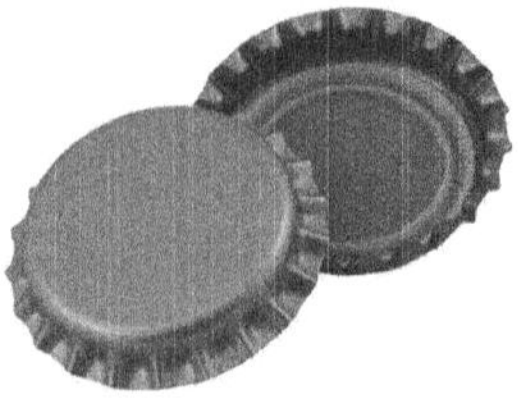

Figure 9.1. *Crown cork closure*

The roll-on (spin-on) tamper-evident closure is made of aluminum with a partially perforated ring at its lower edge. It is placed on the filled container and the threads are rolled into the outside surface of the metal closure, conforming to the threads on the container. This type of closure is used with rigid containers such as glass bottles (for wines and spirits). When the closure is open (unscrewed), the perforated ring is detached, thus leaving evidence of opening (Figure 9.2).

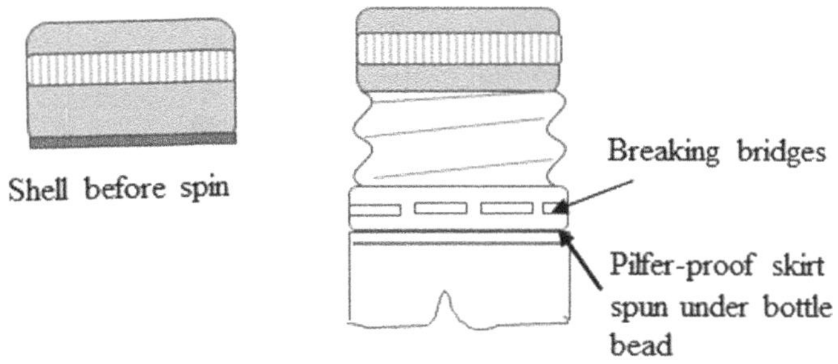

Figure 9.2. *Roll-on tamper-evident cup applied to the finish of a glass container (Paine 2012)*

9.2.2. *Closures to contain and protect contents*

This type of closure is designed to contain and protect the contents with no internal pressure (e.g. a wine bottle). The most common closure for wine is a traditional bark cork (obtained from the cork oak and often called a natural cork stopper), which provides a friction-hold seal (Figure 9.3).

Some wines sealed with natural cork show signs of "cork taint", an objectionable characteristic caused by TCA (2,4,6-trichloroanisole) and other organohalogens, which can result in undesirable musty aromas and flavors. In order to overcome the problem of cork taint in wine, different solutions have been proposed, such as the bleaching of corks, applying a polymeric-based membrane to each end of the cork, and using a synthetic cork to regulate oxygen permeation through the cork and retain cork moisture (Karbowiak *et al.* 2010; Poças *et al.* 2010).

Technical corks (Figure 9.3) are made using a combination of agglomerated natural cork granules, a binding agent and other optional elements. These corks, due to their efficiency in preserving sulfur dioxide concentrations within the bottle, are usually used for wines that will be consumed within two to three years.

Synthetic closures (Figure 9.3) imitate the external appearance and function of natural corks. As they are made of plastic (extruded or injection-molded), they do not risk causing TCA contamination. Some of the problems associated with synthetic closures are related to difficulty in removing them from the wine bottle (or difficulties resealing it), as well as higher oxygen permeation compared to natural cork, although the latter has been improved.

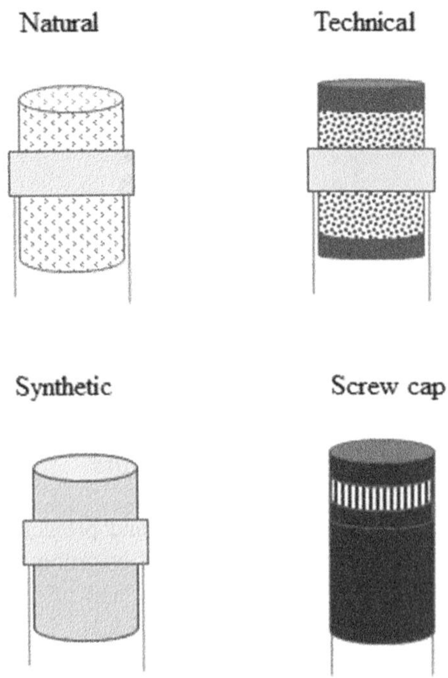

Figure 9.3. *Wine closures*

The specialty polymer cork, Nomacorc, is derived from sugarcane that helps in the efforts to replace the production of synthetic closures with plant-based polymers.

Screw caps (Figure 9.3), also known as "Stelvin caps", ROPP (roll-on pilfer-proof) caps or ROTE (roll-on tamper-evident) caps, are sealed onto the neck of a wine bottle. The Stelvin capsule is a long printable circle and a liner designed for wine bottles. The liner consists of three or four components (Figure 9.4), depending on the expected shelf life of the wine. For wines with a shelf life of up to five years, the liner consists of expanded low-density polyethylene (LDPE) foam, covered on both sides by a co-extruded LDPE-polyvinylidene chloride (PVDC) copolymer (Saran)-LDPE film. For wines with a shelf life of more than five years, expanded LDPE is laminated to white kraft paper that is close to a tin foil layer, to provide a gas barrier, and a PVDC copolymer, which isolates the tin foil from the wine and provides an additional O_2 barrier (Robertson 2013).

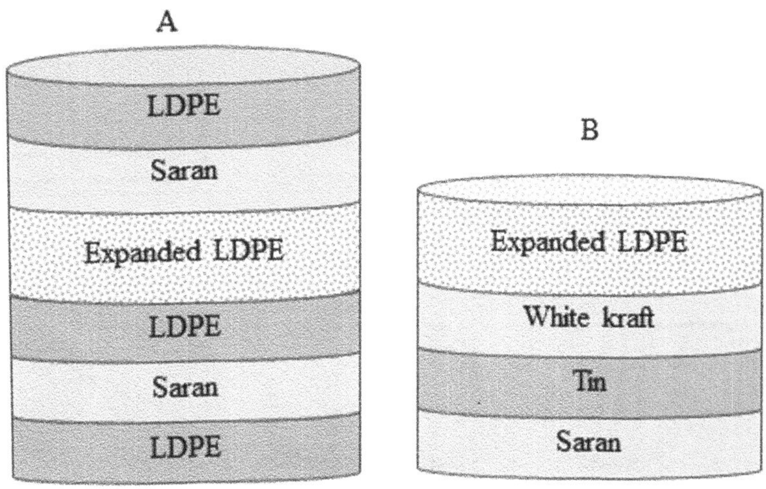

Figure 9.4. *Composition of liners in Stelvin® capsules for shelf life: (A) up to five years and (B) greater than five years (Robertson 2013)*

As the cost of tin has increased significantly, there is limited use of the premium brands. Cheaper solutions are found in composite materials, such as polyethylene (PE) capsules, which have three layers, with aluminum on the outsides and PE in the middle. Other materials include heat-shrink plastic (polyvinyl chloride – PVC), aluminum or wax (Figure 9.5).

Figure 9.5. *PVC bottle capsule*

9.2.3. *Closures to maintain a vacuum inside the container*

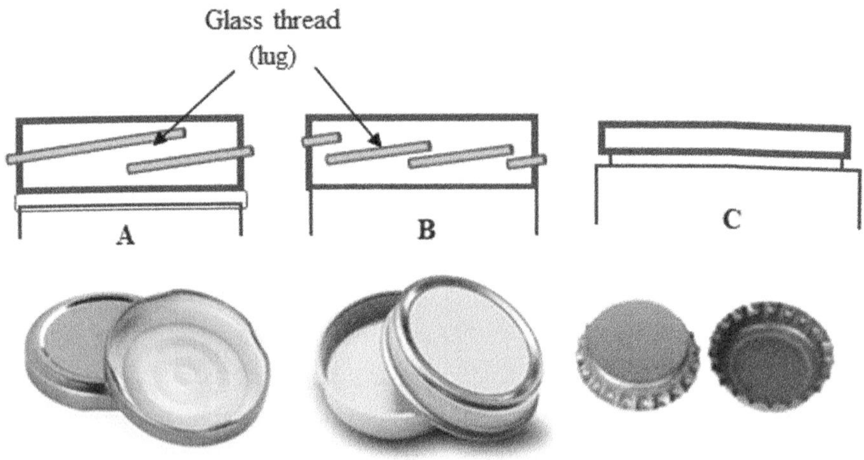

Figure 9.6. *Examples of glass finishes for vacuum closure with (A) lug cap (regular twist-off) closure, (B) press-twist closure and (C) pry-off crown*

These closures are used for heat-processed food and are designed in a way that keeps a vacuum inside a glass container. Three types (Figure 9.6) of vacuum closure are used for processed food packaging in glass jars (Emblem 2012; Robertson 2013):

– **Lug-type** or twist-off cap: this can be removed without a special tool and forms a good reseal for storage. The lug cap consists of a steel shell and has multiple (three to six) metal lugs (threads), depending on its diameter, and a flowed-in plastisol gasket (liner). Plastisol is a suspension of PVC or other polymer particles in a liquid plasticizer. It flows as a liquid and can be poured into a heated mold. The top of the glass finish makes contact with the gasket on the inside of the cap. The caps are usually heated with steam to soften the compound and facilitate sealing.

– **Press-twist** (PT) cap: this cap has a steel shell and no lugs although the finish is threaded. The gasket is molded plastisol and covers the outer edge of the panel and the curl of the cap. It is mostly used on baby foods. Other possible applications include all hot-fill, aseptic, cold-fill, pasteurized and sterilized food. The PT closure is held on mainly by a vacuum with some assistance from the thread impressions in the gasket wall.

– **Pry-off** (side seal) cap: this cap is mainly used on retorted products. It consists of a cut rubber gasket held in place by being crimped under the curl. The pry-off cap is held in place by a vacuum with slight assistance from the friction of the rubber gasket against the side of the container finish.

9.2.4. *Closures to secure contents inside a container*

Some closures are designed to secure the product inside the container. They are designed as screw caps with a minimum thread arrangement of three quarters of a turn. These closures are made either from metal or plastic and can have either a plastic-coated paperboard pad inside or may be of the linerless plastic style (Robertson 2013).

9.3. Specialized type of closures

Specialized types of closures have different functions, which include: (1) dispensing and metering closures; (2) tamper-evident closures; (3) child-resistant closures; and (4) easy-open closures (Emblem 2012).

9.3.1. *Dispensing and metering closures*

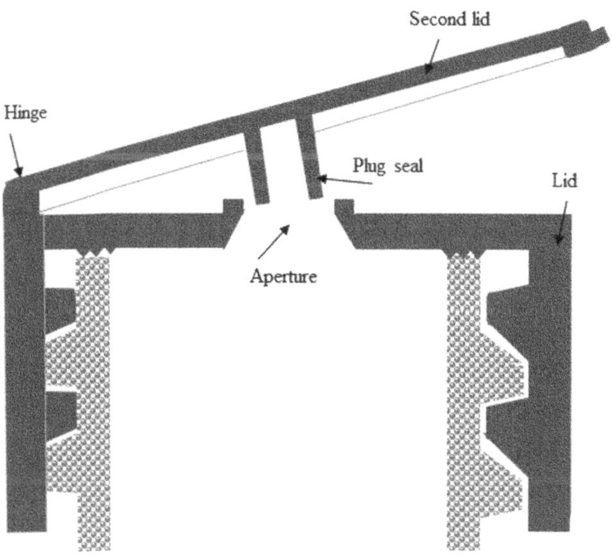

Figure 9.7. *Flip-top closure*

Flip-top closures are used in applications to control the dispensing of the product in the container. This type of closure is usually made of polypropylene (PP), which ensures repeated use without breaking. These closures can be either screw-threaded or push-on. The top of the inner part has an aperture of appropriate size for the packaged product, over which a hinged lid is well fitted (Figure 9.7). A plug on the second lid hermetically seals the aperture when the lid is closed. In the closed position, the lid remains in place in the same manner as push-on lids. By flicking it upwards, with one hand, it is in the open position.

9.3.2. *Tamper-evident closures*

Tamper-evident closures indicate to consumers whether the product has been opened, but breaking the seal in order to access the product must still be convenient.

Some examples of tamper-evident closures are:

– roll-on pilfer-proof metal closures and screw-threaded plastic closures – when opening this closure, a ring is broken (Figure 9.2);

– plastic closures with a tear-off component that is removed for opening;

– metal closures with pop-up buttons;

– shrink seals around the container/closure body (Figure 9.8);

– adhesive labels and tapes.

Figure 9.8. *Shrink seals*

9.3.3. *Child-resistant closures*

Packaging that is difficult for children to open (Figure 9.9) was introduced at the beginning of the 1970s and is subject to international standards in order to meet special requirements. Such packaging includes pharmaceuticals, household cleaning products and any other product that could be harmful to young children (Theobald 2006c).

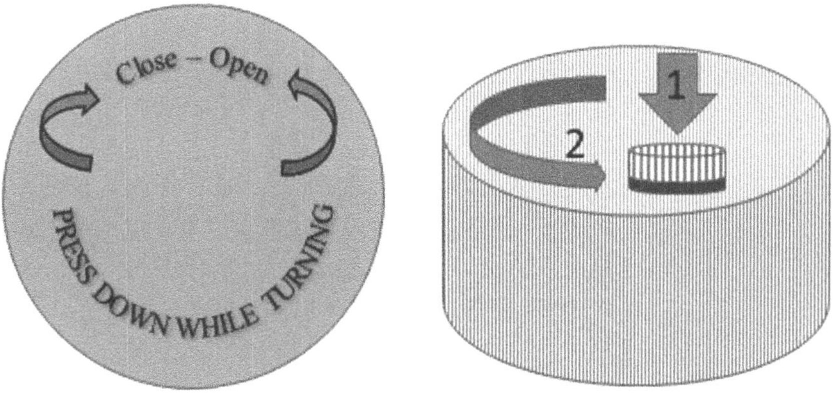

Figure 9.9. *Child-resistant closure*

9.3.4. *Easy-open closures*

Metal cans for beverages have an easy-open lid (EOL) consisting of a scored portion in the end panel and a levering tab (formed separately) that is riveted into a bubble-like structure, fabricated during pressing. EOL for drink cans are usually made from aluminum, and the top is usually referred to as a stay-on-tab (SOT), enabling the opening tab and pierce-open end section to be retained on the can (Figure 9.10).

For food cans (Figure 9.11), the top may either be plain (requiring an opening tool), full-aperture easy-open, or have a peelable membrane design (Theobald 2006a; Emblem 2012; Robertson 2013).

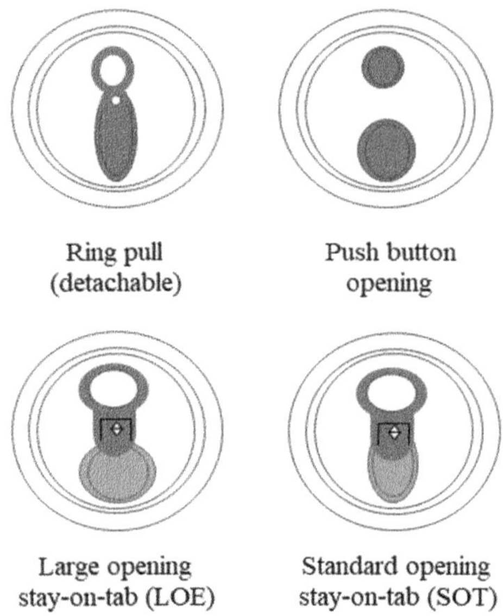

Figure 9.10. *Easy-open ends for drinks*

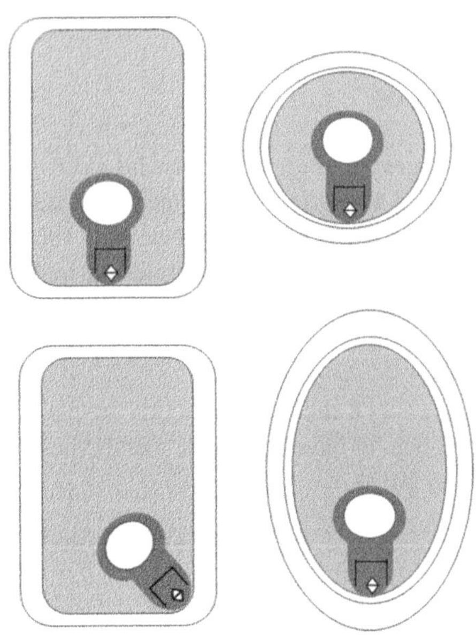

Figure 9.11. *Easy-open ends for food*

Peelable (easy-open) sealing materials are used as the lid on some products such as yogurt (Figure 9.12). Typical lidding materials include aluminum foil, paper, metallized polyethylene terephthalate (PETmet) or PP, often combined. A polymer layer on the inside is essential to enable the required heat seal. The seal must be sufficiently strong to protect the product throughout its shelf life but not so strong in order to still be easily opened by consumers. These lids are removed by peeling them back.

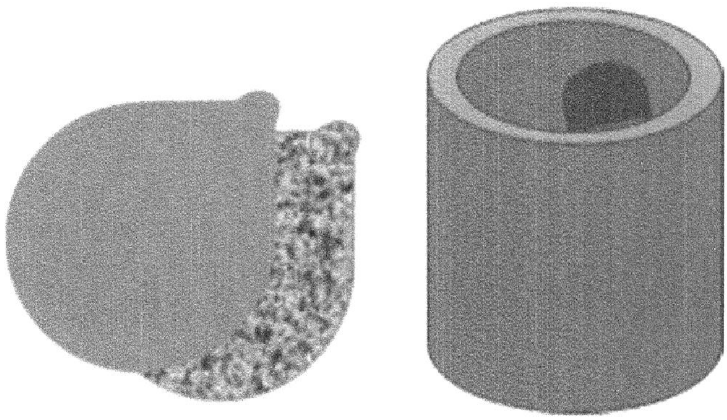

Figure 9.12. *Peelable seal lids*

9.4. References

APCOR (2015). *Cork Yearbook 2015*. Portuguese Cork Association, Santa Maria de Lamas, Portugal.

Emblem, A. (2012). Packaging closures. In *Packaging Technology – Fundamentals, Materials and Processes*, Emblem, A. and Emblem, H. (eds). Woodhead Publishing Limited, Cambridge.

Karbowiak, T., Gougeon, R.D., Alinc, J.-B., Brachais, L., Debeaufort, F., Voilley, A., and Chassagne, D. (2010). Wine oxidation and the role of cork. *Crit. Rev. Food Sci. Nutr.*, 50, 20–52.

Oldring, P.K.T., Savrij Droste, F., Whitaker, R., Smith, D., O'Mahony, C., and Castle, L. (2014). FACET: Light metal packaging methodology. *Jct. Coatings Tech.*, 30–40.

Paine, F.A. (2012). *The Packaging User's Handbook*. Springer, Boston, MA.233.

Poças, M.F., Ferreira, B., Pereira, J., and Hogg, T. (2010). Measurement of oxygen transmission rate through foamed materials for bottle closures. *Packag. Technol. Sci.*, 23, 27–33.

Robertson, G.L. (2013). *Food Packaging: Principles and Practice*, 3rd edition. CRC Press, Boca Raton, FL.

Theobald, N. (2006a). Introduction. In *Packaging Closures and Sealing Systems*, Theobald, N. and Winder, B. (eds). Blackwell Publishing Ltd, Hoboken, NJ.

Theobald, N. (2006b). Closures for glass containers. In *Packaging Closures and Sealing Systems*, Theobald, N. and Winder, B. (eds). Blackwell Publishing Ltd, Hoboken, NJ.

Theobald, N. (2006c). Child resistance, tamper evidence and openability. In *Packaging Closures and Sealing Systems*, Theobald, N. and Winder, B. (eds). Blackwell Publishing Ltd, Hoboken, NJ.

10

Auxiliary Materials

Mia Kurek and Mario Ščetar

Faculty of Food Technology and Biotechnology, University of Zagreb, Croatia

10.1. Introduction

Auxiliary compounds enhance the properties of the initial material. The design and processing of food packaging materials is becoming more and more challenging. The choice of auxiliary ingredients or additives makes significant changes, as well as the choice of the material. In food packaging, all auxiliary ingredients must have received clearance by the appropriate food regulatory authority. Printing ink for food contact materials, whose main characteristics are low migration potential and unwanted transfer to food, is applicable in the labeling or branding of packaging items. Great attention is paid to the regulative restrictions in food contact materials and in boosting the demand for low impact/migration auxiliary materials. The types and properties of barrier coatings, inks and adhesives used for food packaging will be covered in this chapter. According to a report given by Sinha (2017) in Allied Market Research, Europe is the main global packaging market on coatings. In this report, it is stated that the coating market's value is expected to reach US $3,865 million by 2023. Coatings are mostly used in the food and beverage industry, in personal care and health care as they are functional (protection) and have esthetic appeal. Fortune Business Insights (Market Research Report 2020) reports that the food and beverage coating industry is expecting growth for the can coating market.

For a color version of all figures in this book, see www.iste.co.uk/debeaufort/packaging.zip.

Packaging Materials and Processing for Food, Pharmaceuticals and Cosmetics, coordinated by Frédéric DEBEAUFORT, Kata GALIĆ, Mia KUREK, Nasreddine BENBETTAIEB and Mario ŠČETAR.

10.2. Coatings

The coating process can be described as the application of a very thin layer (from several nanometers to a few micrometers) of a specified functional material to a substrate. It can be applied as a liquid or a powder and upon solidification it must firmly adhere to the surface. Even though it is applied in a very thin layer, the coating needs to contribute to packaging value and improve the properties as well as all other layers. In flexible packaging materials, coatings may either be external or sandwiched between two substrates. They are recognized as powerful tools to improve many properties of packaging materials (Farris and Piergiovanni 2012). From the food packaging aspect, as well as taking into consideration the safety of consumers, the coating of packaging materials could improve the resistance of materials to degradation, and could impact and increase the packed food's shelf life and resistance to an aggressive environment. Today, water-based coatings are used as an environmental factor. Coatings may provide various benefits and, depending on the substrate, they could be applied to metals, plastics and other materials as described in the following subsections. The functional material may be metallic or non-metallic; organic or inorganic; solid, liquid or gas.

10.2.1. *Coatings on metals*

Coatings applied on metal are principally aimed at separating metal from a corrosive environment. On metal cans, they could be applied on the inner and exterior parts of a can, depending on the packed item and the envisaged storage conditions. As an exception, tin cans without internal coatings are used for light colored, acidic juices and fruits (e.g. pineapple, pears, peaches, etc.). Under these conditions, tin is more easily oxidized than the food, thus preventing the darkening and flavor changes caused by the oxidation of the food. However, for the majority of food, metal can coatings are applied on both sides of planar metal sheets or coils before cans are formed, or, sometimes, on preformed objects. In the case of three-piece cans, the side seam stripe (coating on the connective part) is applied after the can is fabricated to protect the welded seam from corrosion. The majority of coatings get their final properties after the application process (that is mainly wet) followed by curing. Curing means that further chemical reactions are established. It can be attained either by stoving or baking. In reality, resin(s) react with one or more cross-linking agents (or resins), which join individual resin molecules together to form a three-dimensional cross-linked network (Oldring and Nehring 2007).

Generally, food and beverages that are packed in cans are aimed to be stored for several years, and during this period, the taste and nutritional value of packed items should be preserved. As a consequence of this long time period, there is a high risk of interaction between the can material and the food item. In order to protect consumer's safety and the nutritional value of the packed item, this interaction should be minimized. One method of doing this is to apply different types of coatings in order to form a physical barrier between these two systems. Ordinary metal materials would normally react with food components, especially food rich in corrosion enhancers (phenols, tannins, pro-oxidants, acids, etc.), resulting in a corrosion process that both damages the material and contaminates the food. As such, coatings do not only aim to minimize migration, but also costs, because if an adequate coating is used, then a thinner layer of tin (considered to be an expensive material) can be used. There are important features to look out for when identifying the product needs and application trends to choose appropriate can coating that will make durable packaging. In order to fulfill the technical and legal requirements, can coatings should withstand the production and sterilization processes and severe mechanical deformations. They should be applicable for various food types with different pH and aggressiveness towards the corrosion process, and should prevent chemical migration from coatings into food at a lower level than one that could pose health risks. Key benefits of a good coating also include good adherence, even after the potential deformation of material due to handling and transport; they should be stable, even in contact with aggressive foodstuffs, over several years. They should also offer a range of products from high gloss to matte appearance, preferably solvent-free and with low volatile organic compounds (VOC). Thus, depending on the material and the application, it is sometimes necessary to use several layers of coatings (generally with thickness from 1 to 10 μm), for example, to overcome low adhesion to the metal, secure high protection, etc. (Oldring and Nehring 2007). The amount of applied coating is measured in weight per area (e.g. g/m^2, normally used for sheets for food packaging) or weight per can (e.g. mg/can used for preformed objects such as beverage cans). For food and beverages, typically 5–15 g/m^2 and 110–180 mg/330 ml cans are used, respectively.

The corrosiveness of food is maybe the most important factor to think about when choosing the coating type. Food can be classified in order: strong detinners (tin dissolving compounds from tinplate, for example, green beans, spinach, asparagus, tomato products)→ non-corrosive (dehydrated soups, frozen foods, nuts, etc.) → mildly corrosive (peas, corn, meat and fish) → moderately corrosive (apricots, figs, grapefruits, apples, citrus fruit) → highly corrosive (apple and grape juice, berries, cherries, prunes, pickles, etc.). Commercially, various coatings are available, varying in their chemical compositions, production processes, costs and technical properties. The major constituents in a can coating include resin(s), cross-linking agents (almost always present), additives and, if needed, solvents (Oldring and Nehring 2007). The types and characteristics of commonly used coatings are given in Table 10.1.

Coating type	Composition/ synthesis	Properties (advantages (A), disadvantages (DA))	Application/products
Vinyl	Vinyl chloride and vinyl acetate Need plasticizers and stabilizers	A: Excellent flexibility, stable in acidic and alkaline DA: Bad adhesion on metals, not good for high temperatures	Applied as second coating layer ("top coat", for example, soft drinks) Good for covers
Phenolic	Composed of phenols and aldehydes	A: Highly corrosion-resistant, good sulfide staining protection DA: Very poor flexibility, bad adherence, may change the odor and flavor of foods	Drums and pails where flexibility is not crucial
Acrylic	Synthesized from ethylacrylate; usually pigmented white (with TiO_2)	A: Corrosion and sulfide stain resistance, very good flexibility, attractive clean appearance in opened cans DA: Brittleness, may change organoleptic food properties	Commonly used as external coatings Vegetables, soups, prepared foods containing sulfide stainers
Epoxy	Mostly synthesized from bisphenol A and epichlorohydrin Subgroups: epoxy-phenolic, epoxy-anhydride, epoxy-amino	A: Universal applicability, high stability, good processability, low cost DA: Toxicological concern, poor sulfide resistance	Meat, fish, vegetables, soups, beer and beverages (first coat)
Polyester	Polyester resins (IPA and TPA) May contain lower M_w epoxy resin	A: Very good flexibility, easy to handle during production, good adhesion DA: Not stable in low pH, poor corrosion resistance	Currently used as first-generation alternatives to epoxy coatings
Organosol	PVC dispersed in varnish, stabilized with low M_w epoxy resin or novolac epoxy resin	A: Very good flexibility, very good pack resistance	Drawn cans Easy-open ends Top coat for epoxy-phenolic basecoat
Oleoresins	Mixture of natural oils and resin extracted from plants with synthetic modification	A: Flexible, easily applied DA: Bad adherence, not sulfide-resistant, long curing, may change organoleptic properties	Very limited use: acid fruits Topcoat over epoxy-phenolic

Polyolefins	Based on dispersions of polyolefin	A: Corrosion protection, adhesion, flexibility, no impact on flavor	Various foodstuffs

Table 10.1. *Types, characteristics and application of different can coatings (IPA – isophthalic acid, TPA – terephthalic acid, M_w – molecular weight) (Oldring and Nehring 2007)*

10.2.2. *Coatings on plastics*

According to the packaging optimization concept, the use of multifunctional thin layers applied on plastic materials would minimize the amount of packaging materials used. Novel coated multilayer is aimed at either keeping or, in most cases, even improving the functional properties of the final package. For plastic packaging materials, coatings are applied with different aims. They are principally used to improve gas, water vapor, grease and flavor resistance or to reduce reflection in an optical application in food storage. A method for applying plastic coating is extrusion. Coatings for improving the properties of plastic containers and films are given in Table 10.2.

Coating type	Properties	Application
Acrylic	Glass clear, hard, heat-sealable and very glossy	OPP (coating thickness of about 1 g/m^2)
PVDC	Good heat-sealing polymer or a high-barrier polymer	Plastic films (coating thickness of about 3 g/m^2)
PVOH	Receives printing inks with low absorption or retention of solvents	BOPP, part of a laminate (coating thickness of about 1 g/m^2 or higher)
LTSCs	Seals at lower temperatures	OPP
Metallizing with aluminum	Higher barrier properties	PET, PA and OPP
SiO_x	Excellent barrier properties	PET film, plastic bottles
DLC	Improvements in barrier properties	PET bottles
Extrusion coating with PE	Heat-seal	PET and PA

PVDC – polyvinylidene chloride, PVOH – poly(vinyl alcohol), LTSCs – low-temperature sealing coatings, DLC – diamond-like coating

Table 10.2. *Types, characteristics and application of coatings for plastic containers and films (Coles* et al. *2003)*

Since the recommendation in the European Union is to reduce food packaging-associated waste and to ensure that all plastic packaging is reusable or recyclable by 2030, there are numerous newly designed barrier coatings that can contribute to making packaging more environmentally compatible. Of the latest technology formulation improvements are lacquers, which have inorganically or organically functionalized biopolymers, mainly from lignocellulosic biomass, specially designed to be applicable to all kinds of plastic packaging materials, with excellent barriers against water vapor, oxygen or flavors; they are applied automatically, with improved end-of-life (recyclable) (HyperBioCoat, BBI Project).

10.2.3. *Coatings on paper and glass*

Paper-based materials offer plenty of good properties, such as physical and mechanical strength, flexibility of processing, recyclability, biodegradability and composability. However, their poor properties are characterized by bad barrier performance because of the high porosity of their fiber network. To overcome this hurdle, surface treatments are commonly used. Coatings on paper are aimed at improving grease resistance and hydrophobicity, the barrier to oxygen, water vapor and aroma compounds, plasticization, pigmentation and functional properties (antioxidant, antimicrobial).

Resistance to grease is relevant to greasy and fatty food packaging, guaranteeing the integrity of the structure of the material. It is also important to avoid the penetration of oils and grease through paper-based products, because of the lipophilic nature, fatty substances could damage the print, and generate the separation of laminated materials. The most common method to fabricate oil-resistant paper is beating and refining the pulp in a way that induces structural changes, facilitates the formation of inter-fiber connections, increases pulp density and higher mechanical resistance. The most common and industrially used method is coating with compounds with low surface energies, such as perfluorocarbons (PFCs), polysiloxanes or higher alkanes that have great water and fat repellence. This type of coating is commonly used for butter. Paper can also be used in multilayered packaging, known as lining, often in combination with plastic materials (LDPE, EVOH, etc.) with a significantly higher barrier than paper substrate, which is especially important for liquid food. This process can be achieved either by extrusion or heat lamination. Food containers made from lined paperboard are often used as kitchenware for single-use applications, for example, cups or take-away food containers. Polyvinylidene chloride (PVDC) coatings are commonly used on paper for fatty products, dairy products, cured meats, smoked fish and dried products. These days, it is fashionable to coat paper with natural oil-repellent bio-based coatings as a replacement for synthetic treatments, providing interesting functionalities while maintaining the environmentally friendly characteristic of the final material. A wide variety of biomass products, for example, polysaccharides, proteins and lipids and bacterial cellulose are already scaled-up to an industrial level.

A barrier to oxygen, water vapor and aroma compounds is essential to extend product shelf life. Paraffin wax or polyethylene waxes are commonly applied in order to improve the water vapor barrier (Robertson 2013).

Paper and paperboard can be coated to enhance printability, visual properties (brightness, opacity, gloss) or functionality (ink holdout, the ability of it to be glued) of the paper sheet. Pigments improve the printing and optical properties of the sheet, and make up 70–90% of dry matter in paper coatings (Robertson 2013). The type of pigments used as coatings are similar to those used for paper, with kaolin clay and calcium carbonate as the most significant ones. Binders adhere the pigment particles to each other and to the sheet, and additives either assist in the coating process or enhance sheet properties. There are also plastic pigments based on polystyrene (PS) that can improve the gloss when used as coatings in combination with mineral pigments (Robertson 2013).

Methods that are used for paper coating include surface sizing, solution coating, compression molding and curtain coating. Chemical (layer-by-layer, grafting approach) and physicochemical modification (plasma etching, laser, etc.) have recently been proposed as novel techniques, but their large-scale production is limited due to high costs, non-suitability for food contact and the use of non-environmentally friendly solvents. The selection of the method depends on the properties of the coating and the type of cellulosic material used for paper-making.

Even though glass is considered as a tough and all-purpose material, it is often coated, in order to improve its durability and esthetic value. Coatings are used to reduce breakage or scratching that can occur during filling and transportation. These are known as cold-end coatings and they provide a shiny look and smooth touch. In order to preserve the life span and appearance of returnable glass bottles for drinks, there are coatings applied which help to delay the formation of scuff rings and to mask those that have already been formed.

10.2.4. *Surface treatments and coating application*

In the packaging industry, adhesion and wettability play key roles in successful coating applications. Typical surface treatments include alternating the wettability of the substrate, improving its bendability or the elimination of accumulated static charge. Achieving adequate adhesion is crucial, since the type and duration of treatment will impact how well and for how long a bond will hold. In order to obtain good coating adherence and thus efficacy, treatments on surfaces may involve preparation (cleaning), pre-treatment (surface activation) and post-treatment (application of adhesion promoters or primers to improve adhesion to the adhesive and/or protect the surface). They may only use one of these steps, or all of them.

Lately, in addition to technological improvements in material production, coating technology has been proposed as an additional strategy for the rational use of materials. To date, new production and degradation pathways, new environmentally friendly (biodegradable) raw materials and non-conventional deposition techniques are being developed. Coatings may be deposited in gas phase, in liquid phase or in a molten/semi-molten state (Table 10.3) (Holmberg and Matthews 2009).

Coating method	Advantages (A) and disadvantages (DA)
Gaseous coating	
Chemical vapor deposition (CVD)	Process involves the reaction of a volatile precursor which is injected into a chamber (typically under vacuum) A: Applied to various materials, precision surfaces can be coated including seal areas, withstands low/high temperatures and high stress DA: Difficult to mask surface, so coating either all or nothing, size limited to reaction chamber capacity (O'Brien 2001)
Physical vapor deposition (PVD)	Material goes from a condensed phase to a vapor phase and then back to a thin film condensed phase; sputtering and condensation A: Improved hardness and wear resistance, reduced friction, improved oxidation resistance DA: High cost and slow production rate (Rane *et al.* 2018)
Ion beam-assisted deposition (IBAD)	Combines sputtering or electron beam evaporation with ion implantation A: Final coating with a highly intermixed interface and with a less built-in strain compared to other PVD techniques DA: High cost, plasma activation processes are not operational for reactive deposition

Solution coating	
Chemical solution deposition	Any technique where a chemical precursor solution is used to create a film; precursor chemicals are either dissolved or suspended in a solvent that, when deposited and heated, will result in the formation of the material of interest (Dorrey 2012) A: Flexible and cost-effective, does not need vacuum processing and is relatively easy to scale up to meet the demands of industrial production processes (Elshof 2015)
Electrochemical deposition	Takes place in an electrochemical cell where coating layers are deposited on selected conducting substrates A: Offers important advantages and unique possibilities in the development of nanomaterials and nanostructures
Sol/gel	Material (substrate) is at first treated with a coating solution, that after drying forms a polymeric or cross-linked network on the surface A: Low cost, applicable to all surfaces and coating types
Molten or semi-molten (frequently combined with plasma treatments) coating	
Laser	Laser overlay deposition process, coating material as powder or wire is applied on the surface of the base material through a melting process A: Produces coatings with high density, crack-free and non-porous microstructures (Vuoristo *et al.* 2005)
Thermal spraying	Coating adhesion is usually based on mechanical interlocking A: It can be applied on all surfaces
Welding	Process used for metals, done with coated electrodes

Table 10.3. *Surface coating methods*

Food preservation requires highly efficient gas, water vapor and aroma barriers. Up-to-date packaging coatings are based on petrochemical-derived waxes and polymers. At an industrial scale, the significant coatings that are mostly applied are polyvinylidene chloride (PVDC), aqueous or extrusion-coated ethylene vinyl alcohol (EVOH), acrylic polymers, inorganic and organic sol–gels, and aqueous poly(vinyl alcohol) (PVOH). However, with increasing petroleum costs and when searching for alternative reliable and sustainable products, renewable functional barriers are gaining market interest. This would also promote the recycling characteristics of materials used for food packaging (currently, this is difficult, as petroleum-based coatings commonly used for this purpose reduce the compostability of paper and recyclability of multilayers).

Already applicable coatings like metallized (aluminum) coatings and silicon oxides (SO_x) are applied as thin layers on some films. Diamond-like coatings (DLC) are carbon thin layer coatings used on PET bottles in order to prolong the shelf life of drinks. The most well-known coating process is tin coating (tinning) for metal packaging. Cans are usually further protected with one or more layers of organic coatings (lacquers), inside and outside the body of the container (Piergiovanni and Limbo 2016).

Another issue is that biodegradable materials have no sufficient barrier properties, so they cannot guarantee the required minimum shelf life of many food products. In particular, some of the most interesting are nanotechnological strategies, yielding high-barrier food packaging (e.g. cellulose nanocrystals, barrier lacquers on the basis of hybrid reactive polymers, etc.). Some examples of novel deposition techniques to obtain conformal coatings are given in Table 10.4 (Vasile and Sivertsvik 2019). Unfortunately, detailed descriptions are outside the scope of this chapter.

Method	Result
In situ polymerization: dispersion of fine nanofillers in monomers	Monomers interact with the nanofiller surface forming uniform suspension
Atomic layer deposition	Excellent permeation resistance, allows the integration of different functions into packaging material, improves humidity tolerance and the performance of bio-based polymers
Layer-by-layer	Used for active packaging materials, covalently bonded layers with opposite charges, cross-linking
Dispersion and chemical reaction	UV curing in the presence of a photoinitiator; cross-linked composites
Electrospinning	Various morphologies and versatility in processing, possibility of producing fiber/particle diameters within the submicron and nanorange

Table 10.4. *Novel methods in the production of coated materials*

10.3. Adhesives

Adhesives are an unavoidable part of packaging structure. According to the definition of DIN EN 923:2016-03 (DIN 2016), adhesives are non-metallic substances capable of joining materials by surface bonding (adhesion), with a bond possessing adequate internal strength (cohesion) (Figure 10.1). It follows that adhesion is the process of bonding two surfaces together if the process does not include lamination by coextrusion as an alternative. To be effective, the adhesive must establish an intimate contact with the surface to join (adhere). This is known as wetting. If the linkage is not well established, then the connection will break down and surfaces will not be tightly joined together, resulting in packaging failure.

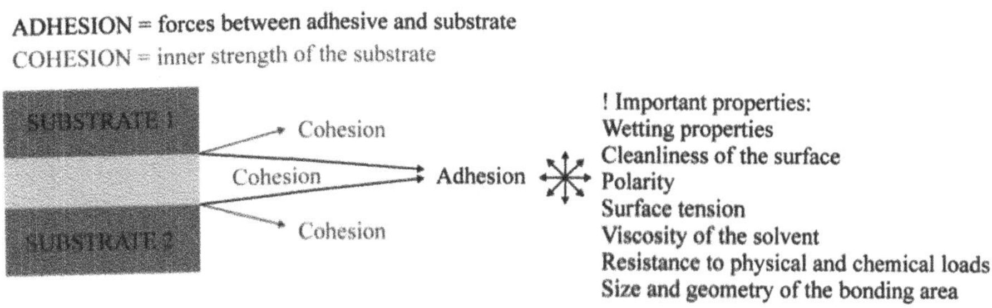

Figure 10.1. *Multilayer structure bonded with an adhesive*

According to the International Life Sciences Institute (ILSI) (Toenniessen 2018), more than 80% of food packaging materials contain at least a very small quantity of adhesives. Selecting the appropriate adhesive is an important task in order to meet the packaging performance criteria. This will depend on the food product and its packing conditions, storage, retail and use. Many types of adhesives are available for any specific application. They are used to:

→ manufacture rigid packs from cardboard (box closures);

→ seal flexible packaging – including wrappers, pouches and lidding;

→ attach labels;

→ laminate (bind together) layers of food contact materials;

→ prevent or detect tamper-evident seal removal and replacement (Farmer 2013).

Adhesives play an integral role in the production of flexible multilayer packaging materials. They are generally applied in the last steps of the converting process. They should meet rigorous requirements: bonds that exceed the strength of the film, high application speeds, excellent heat, water, chemical and UV resistance, excellent clarity, resistance to delamination, low odor and they must not contain any potential migrants or toxic additives (Dunn 2003).

Depending on the chemical composition, they are classified as natural or synthetic (Figure 10.2). Synthetic adhesives can be further grouped depending on the curing method. The most relevant examples of natural and synthetic adhesives (water-based, solvent and 100% solid-based) are given in Tables 10.5 and 10.6.

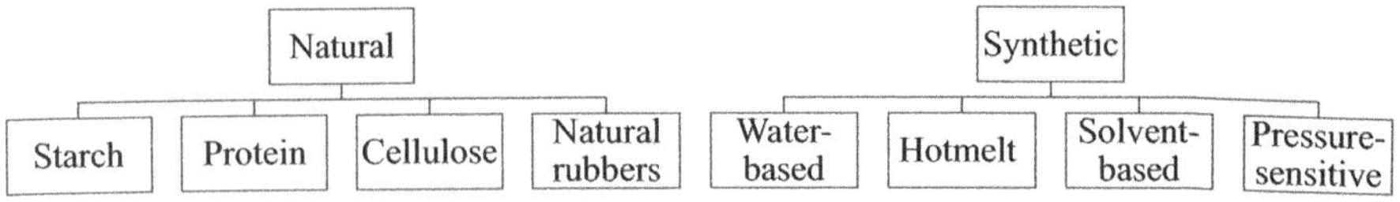

Figure 10.2. *Classification of adhesives depending on their source*

Water-based adhesives are widely used and are low cost. They can withstand a wide temperature range, but dry slowly, requiring a lot of heat energy; they do not provide shear or peel strength and usually do not have good moisture resistance. They are used to form, seal or label cartons, tubes, bags and bottles, and can be formulated to adhere to paper, glass and most plastics and metals.

Hotmelt adhesives are 100% non-volatile and non-reactive thermoplastic materials, where the setting takes place by temperature heating and without any chemical change. The bond is formed when the adhesive re-solidifies to room temperature. Their main advantage is an extremely rapid rate of bond formation (used on a high-speed packing line). A typical hotmelt sealant is composed of three primary components: polymers (30–40%); tackifying resin (30–40%); and petroleum wax (20–30%) plus antioxidants, fillers, plasticizers and blowing agents to enhance other properties. Hotmelts can be used for carton and case closing (e.g. cereals), tray erection (e.g. fruits and vegetables), wrap around boxes (e.g. chocolate) and labeling.

Solvent-based adhesives use organic solvents as the main carrier; thus, due to environmental and health concerns, they are used less and less.

Pressure-sensitive adhesives are applied with pressure upon contact with the substrate surface. Generally, no chemical reaction takes place between the substrate and the adhesive and no water, solvent or curing is necessary.

<table>
<tr><th>Source</th><th>Type</th><th>Application</th><th>Advantages (A)/ disadvantages (DA)</th></tr>
<tr><td rowspan="3">Starch-based</td><td>Starch</td><td>Corrugated board</td><td rowspan="3">A: Easy to handle, inexpensive
DA: Slow rate of bond formation, poor water resistance, limited adhesion to coatings and plastics</td></tr>
<tr><td>Modified starch and dextrin</td><td>Sealing of cartons
Winding of spiral tubes
Seaming and forming bags
Labeling metal cans</td></tr>
<tr><td>Alkaline-treated starch</td><td>Labeling glass bottles</td></tr>
<tr><td>Cellulose</td><td>Cellulose ether (methylcellulose)</td><td>Bonding paper in tissue and towel applications
Cigarillo leaf wrapping</td><td>A: Easy to prepare, water soluble at room temperature</td></tr>
<tr><td rowspan="2">Proteins</td><td>Casein</td><td>Labeling glass beer bottles
Lamination of aluminum foil to paper</td><td>A: High bond strengths, good water resistance</td></tr>
<tr><td>Animal glue</td><td>Rigid setup boxes</td><td>A: High level of hot tack, long gummy tack range</td></tr>
<tr><td>Natural rubber latex</td><td>Extract from rubber tree Hevea brasiliensis</td><td>Self-seal candy wraps
Packing of heat-sensitive food (e.g.) chocolate</td><td>A: Limited stability on high shear, high speed production lines
DA: Allergy, odor and quality variations</td></tr>
</table>

Table 10.5. *Types, applications and characteristics of naturally sourced adhesives*

Adhesives may join two surfaces either through *mechanical* or *chemical bonding*. Mechanical adhesion is the joining of two surfaces together where the adhesive enters into a porous mechanical structure of the substrate (e.g. in paper). Its porosity will affect the degree of penetration – too little or too much results in a weak bond. Chemical adhesion is established due to the formation of chemical bonds between one surface and the adhesive.

Cold seals are adhesives that require no heating to result in good adhesion, while *heat seals* combine heating and pressure. Natural rubbers, acrylic polymers and additives (wetting agents, stabilizers, anti-blocking agents, etc.) belong to cold seals.

They can be applied on a wide range of different substrates (paper, polyolefins, etc.) via printing processes or pressure reactivated systems that result in uniform and elastic film. For heat seals, at least one of the substrates has to be heat-sealable (either by an applied heat-sealable layer or by being heat-sealable itself).

Curing method	Adhesive	Application	Advantages (A)/ disadvantages (DA)
Chemical reaction	Polyurethane (PU)	Flexible packaging (laminates) Retortable pouches Hot filling	*A*: High flexibility and bonding strength, resistance to elevated temperatures. Can be waterborne or solvent-based. *DA:* Produced by a reaction between isocyanates and organic compounds with reactive hydroxyl group → possibility of migration of toxic 1°C aromatic amines from residual aromatic isocyanates
Cooling from melt (hotmelt)	Synthetic rubbers (styrene–butadiene–styrene (SBS), styrene–isoprene–styrene (SIS))	Tapes Self-adhesive labels Bottle labeling Pallet stabilization	*A*: Solid at room temperature, stickiness *DA*: Rapid fall of strength at elevated temperatures
	Waxes (plant or animal origin – beeswax; synthetic)	Adjustment of setting time, structure stabilization, reduce melting viscosity	*A*: Improve the adhesion of a hotmelt to a substrate, narrow melting range (from 40°C), very low degradation tendency
	Oils (purified naphthenic or paraffinic oils)	Tape/labeling (metal cans or glass bottles)	*A*: Infinite open time, stickiness at room temperature, paraffinic purified white mineral oil for food contact application
	Copolymers of ethylene vinyl acetate (EVA)	Carton sealing, labeling, frozen food, heat sealing	*A*: Tack-free, flexible at room temperature, UV resistance, light stability
	Polyolefins	Adhering paper, (olefin) plastic films and metal foils to a variety of substrates	*A*: Good thermal stability, range of melt viscosities, tackiness, long open times, low water vapor permeability, good chemical resistance, bonds well to nonpolar substrates

Loss of solvent	Acrylic	Flexible packaging laminates, boil-in-bags, retort packs	*DA*: Environmental and health concerns about VOC
	Vinylic		
Loss of water	Vinyl acetate polymers (PVA), ethylene/vinyl acetate (EVA)	Carton side seam and closing Bag-making	*A:* Solvent-free *DA:* At least one substrate has to be porous
	Acrylic (polyester acrylates, epoxy acrylates)	Film lamination, Pressure-sensitive coatings for labels	*A:* Low cost adhesives, low odor, good heat and chemical resistance, excellent strength, improved wettability
Energy curable: UV and electron beam	Acrylic	Wraps and shrinks Flexible packs/ Multilayer materials	*A:* No solvent (low VOCs), instantaneous cure, excellent heat, moisture and chemical resistance, applicable at room temperature *DA:* Irradiation through substrates, potential migration of remaining photo-initiator fragments, high cost equipment

Table 10.6. *Examples of curing method, adhesive type and its application with main characteristics*

Even though most of the adhesives are in indirect contact with food (except for labels for fruits and vegetables), special testing for their appliance with food contact regulation is required. It is very important to identify compounds present in the adhesives due to their possible risk as potential migrants to the food when the laminates are used as food packaging materials.

10.4. Printing inks

Various printing inks and varnishes are used for food packaging. Printing inks are mixtures of a number of substances: colorants (5–30%), binders (15–60%), solvents (20–70%) and additives, including plasticizers (1–10%) (Brede *et al.* 2003). Together with other colored or uncolored overprint varnishes and coatings or primers, they are applied on materials to form a graphic or decorative design. More than 5,000 different chemicals are used in the production of printing inks (EuPIA 2009). These components generally have strong odors, so careful selection ensures that odor or taint residues are kept at very low and acceptable levels.

Inks are prepared in four steps: 1) pre-wetting/mixing of pigments with grind varnish/medium; 2) milling/dispersion; 3) concentrating pigment dispersion; 4) addition of varnish or specific additives.

Coloring materials (colorant) are a mixture of pigments (insoluble in the medium (the vehicle or the binder)) or dyes (soluble in the medium). The coloring effects result from the presence of chromophore groups that absorb specific wavelength areas of the visible light spectrum (EuPIA 2018).

Primers or size coats are continuous coatings applied to the base substrate to provide good adhesion and printability of inks and coatings (EuPIA 2018).

Binders are polymeric resins and the film-forming component of inks, in which the coloring material is finely dispersed or dissolved (EuPIA 2018).

The aim of the solvents is to keep the binders in a soluble state until the printing ink formulation is applied onto the substrate. Inks can be water-based or solvent-based (EuPIA 2018).

Varnishes are mixtures of binders and solvents. They are added in order to enable printed designs to achieve specific functions such as ink adhesion, rub resistance, gloss and slip/friction properties (EuPIA 2018).

After applying the ink formulation to the substrate, the drying of the solvent and binder takes place. Depending on the field of application, printing inks have different mechanisms of drying, such as evaporation, absorption into substrate, oxidative drying or curing by the supply of irradiated energy, for example, in the UV systems.

The choice of ink may vary depending on the printing method used. For food packaging materials, four printing technologies and modifications are used: offset, flexography, gravure printing and ink jet printing (Table 10.7).

	Offset	Flexography	Gravure	Inkjet
Schema	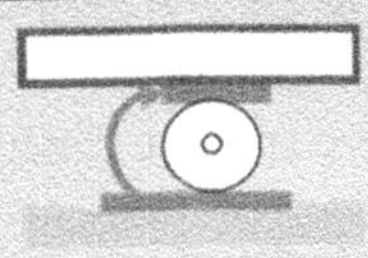	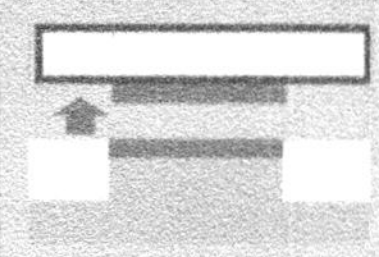		
Material	Paper, cardboard	Plastic, paper, cardboard, metal	Paper or plastic	Paper, cardboard, plastic, metal
Solvents	Non-volatile solvents Mineral/vegetable oils (boiling points 210–300°C)	Volatile solvents Water-borne	Volatile solvents Water-borne	Water-based Solvent-based Energy-curing type
Drying/ curing	Absorption of solvents into substrate UV-curing	Solvent evaporation UV-curing	Solvent evaporation	Energy-curing
Characteristics	Inked image is transferred (or "offset") from a plate to a rubber blanket, then to the printing surface	Uses flexible relief plate; high definition, low cost, flexible printed packaging, fast ink-washout	Uses printing press, high quality but more expensive than flexible and offset printing	High quality, high speed, easy to integrate

Table 10.7. *Comparison of different printing techniques used in food packaging*

Printing inks are usually applied on the non-food contact side of a food contact material (FCM), or between one of the outer layers of a multilayer material. However, a migration of ink ingredients may happen though the food contact layer.

10.5. Interaction between packaging and contents

10.5.1. *Permeation*

Knowledge on the barrier and permeation properties of food packaging materials is a vital feature for packaging performance and its efficiency in food preservation and extended shelf life. All polymeric materials are permeable to a certain extent and their barrier properties may impact suitability for a specific application. Commonly used coatings serve as barriers to gases (oxygen, carbon dioxide, nitrogen, etc.), water vapor and aroma compounds, in order to prevent the penetration or loss of given gases, while others have specific functionalities against light, VOC, irradiation (UV) and others. Permeation requirements may be divided into two categories: those requiring a gas barrier and those requiring a specific selective transport of gases (permselectivity) (Robertson 2013). In order to ensure that films provide sufficient barriers for a specific application, it is quite common to use thin barrier layers applied as coatings, rather than to increase the thickness of barrier monolayers in multi-layered materials. Measurement methods for the determination of permeation to gases and water vapors are given in various international standards. Permeation can be affected by processing parameters during production, or by ambient temperatures associated with its end-use.

10.5.2. *Migration*

An important function of food packaging is to protect packed products from any external contamination (microorganisms, dust or chemical contaminants). Chemical contaminants may already be contained in the packaging material itself. This risk becomes significant for long-term storage conditions or especially in the case of packed liquids.

Inks and adhesives might be applied to the packaging material on the external surface of primary or secondary packaging, in the middle layer or in direct contact with food items. Substances of low molecular weight used in the production of inks and adhesives may be transferred by different mechanisms (Figure 10.3):

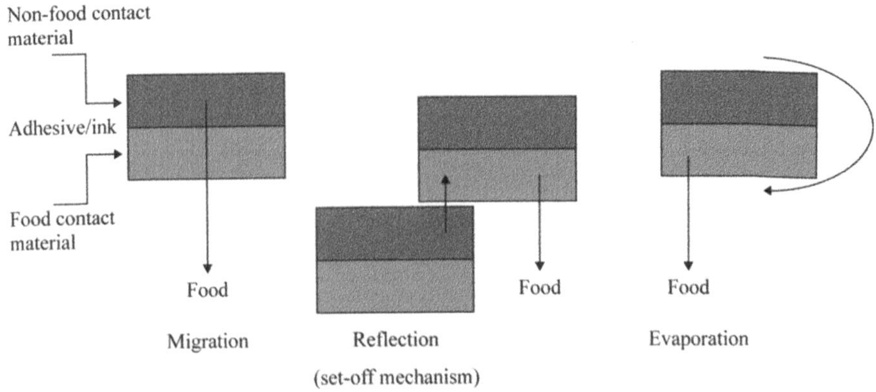

Figure 10.3. *Example of adhesives and ink migration process*

a) migration (substance passes through material and migrates to food that is in direct contact with the packaging material);

b) set-off migration (during production, the outer side of the packaging, which is initially not aimed to be in a direct food contact, is in close contact with the inner packaging layer (because it is packed in rolls or in pilled sheets) where migration occurs. Once packed, food is in contact with this “invisible” side where the transfer occurred);

c) transfer via product gas space (volatile compounds negatively influencing organoleptic food characteristics).

Migration can be defined as a global or a specific process. Global migration is defined as the transfer of all substances of a low molecular weight, from packaging to food, whereas in specific migration the amount of one specific chemical compound is measured. Global migration is determined gravimetrically, while sophisticated analytical methods (such as GC–MS) are required for specific migrants.

Aznar *et al.* (2011) described a wide variety of compounds that were found in adhesives. They concluded that their composition depends on the nature of the adhesive. About 57% of all compounds detected in the materials, initially present from material composition, migrated into the dry food simulant Tenax®. This study confirms the necessity of controlling the composition of the adhesives used for food contact materials.

Migration of inks is commonly avoided by using an overprint varnish or varying the ink position in the multilayer structure. Following good manufacturing practice (GMP), it is also favorable to select such raw materials where limits for evaluated substances are not exceeded in the finished product. The major risk lies in solvents that are used as dissolving agents in inks and adhesives and in active components (like photo-initiators in UV curing systems). This process is called tainting. Even though UV-cured inks are initially solventless, there are some residual acrylate monomers, such as benzophenone, used as photo-initiators or reaction by-products from the polymerization process, such as benzaldehyde and alkyl benzoates, that can lead to trace odors that could migrate into the food product (Ridgway *et al.* 2009). Proper drying of printing materials may completely eliminate solvent migration from adhesives and inks (Katan 2012). The legal requirements are set up by Regulatory bodies. Under the EU food-contact legislation, printing inks must legally comply with Regulation 1935/2004, Commission Regulation 2023/2006), as well as Commission Regulation 10/2011 on plastic materials and articles intended to come into contact with food, and Commission Directive 2007/42/EC on regenerated cellulose films (European Commission 2004, 2006, 2011, 2007). There are also guidelines on printing inks issued by EuPIA for industrial stakeholders (EuPIA 2011).

10.6. Future trends and challenges

The main goal for successful packaging of food is recognizing the need for packaging products and their main protection mechanisms. It relates to the nature of the product when used and the possible disposal of packaging, providing the shelf life, distribution, storage and sale to the final consumer. Selection should consider the ecological issues of environmental protection and waste management, such as the low energy consumption required to manufacture these materials. As is described in this chapter, new environmentally friendly biodegradable raw materials, coatings and inks have been developed in the last decade; however, there are great concerns about their high cost and production price, as well as barrier/protection performance and the safety of products.

10.7. References

Aznar, M., Vera, P., Canellas, E., Nerin, C., Mercea, P., and Stormer, A. (2011). Composition of the adhesives used in food packaging multilayer materials and migration studies from packaging to food. *J. Mater. Chem.*, 21, 4358.

Brede, C., Skjevrak, I., and Fjeldal, P. (2003). *Colour Substances in Food Packaging Materials*. SNT, Oslo.

Coles, R., McDowell, D., and Kirwan, M.J. (2003). *Food Packaging Technology*. Blackwell Publishing Ltd, Oxford.

DIN (Deutsches Institut für Normung) (2016). Adhesives – Terms and definitions. DIN EN 923:2016-03 [German version EN 923:2015].

Dorey, R. (2012). *Ceramic Thick Films for MEMS and Microdevices*, 1st edition. William Andrew, Oxford.

Dunn, D.J. (2003). *Adhesives and Sealants: Technology, Applications and Markets*. iSmithers Rapra Publishing, Shrewsbury.

Elshof, J.T. (2015). Chemical solution deposition techniques for epitaxial growth of complex oxides. In *Epitaxial Growth of Complex Metal Oxides*, Koster, G., Huijben, M., and Rijnders, G. (eds). Woodhead Publishing, Oxford.

EuPIA (2009). Incomplete public inventory list – Version 3. Comprising of packaging ink raw materials applied to the non-food contact surface of food packaging [Online]. Available: www.eupia.com.

EuPIA (2011). Printing inks applied to the non-food contact surface of food packaging materials and articles [Online]. Available: www.eupia.com.

EuPIA (2018). Standard glossary of food contact material inks and coatings terms [Online]. Available: www.eupia.com.

European Commission (2004). Materials and articles intended to come into contact with food and repealing Directives 80/590/EEC and 89/109/EEC. Regulation 1935/2004 [Online]. Available: https://eur-lex.europa.eu/legal-content/EN/ALL/?uri=CELEX%3A32004R1935.

European Commission (2006). Good manufacturing practice for materials and articles intended to come into contact with food. Commission Regulation 2023/2006 [Online]. Available: https://eur-lex.europa.eu/legal-content/EN/TXT/?uri=celex%3A32006R2023.

European Commission (2007). Materials and articles made of regenerated cellulose film intended to come into contact with food. Commission Directive 2007/42/EC [Online]. Available: https://eur-lex.europa.eu/legal-content/EN/ALL/?uri=CELEX%3A32007L0042.

European Commission (2011). Plastic materials and articles intended to come into contact with food. Commission Regulation 10/2011 [Online]. Available: https://eur-lex.europa.eu/legal-content/EN/ALL/?uri=CELEX%3A32011R0010.

Farmer, N. (2013). *Trends in Packaging of Food, Beverages and Other Fast-Moving Consumer Goods*. Woodhead Publishing, Oxford.

Farris, S. and Piergiovanni, L. (2012). Emerging coating technologies for food and beverage packaging materials. In *Emerging Food Packaging Technologies: Principles and Practice*, Yam, K.L. and Sun Lee, D. (eds). Woodhead Publishing, Oxford.

Holmberg, K. and Matthews, A. (2009). *Coatings Tribology: Properties, Mechanisms, Techniques and Applications in Surface Engineering*. Elsevier, Amsterdam.

HyperBioCoat Project (2020). BBI Project: Research & innovation action, value chain: VC1 – Lignocellulose, 2016–2019 [Online]. Available: http://www.hyperbiocoat.eu.

Katan, L.L. (2012). *Migration from Food Contact Materials*. Springer Science & Business Media, Berlin.

Market Research Report (2020). Can coatings market size, share & industry analysis, by type (epoxy, acrylic and others), by application (food cans, beverage cans, aerosol cans, and others) and regional forecast 2019–2026. Report FBI102537. Fortune Business Insights.

O'Brien, P. (2001). Chemical vapor deposition. In *Encyclopedia of Materials: Science and Technology*, Buschow, J., Cahnw, K.H., Flemings, R.W., Ilschner, M.C., Kramer, B., Mahajan, E.J., and Veyssiere, S. (eds). Elsevier, New York.

Oldring, P.K. and Nehring, U. (2007). Packaging materials 7. Metal packaging for foodstuffs. ILSI Report. ILSI Europe Packaging Materials Task Force.

Piergiovanni, L. and Limbo, S. (2016). *Food Packaging Materials*. Springer, New York, NY.

Rane, A.V., Krishnan, K., Abitha, V.K., and Sabu, T. (2018). Methods for synthesis of nanoparticles and fabrication of nanocomposites. In *Synthesis of Inorganic Nanomaterials*. Bhagyaraj, S.M., Oluwafemi, O.S., Kalarikkal, N., and Sabu, T. (eds). Woodhead Publishing, Oxford.

Ridgway, K., Lalljie, S.P.D., and Smith, R.M. (2009). Analysis of food taints and off-flavours – A review. *Food Add. Contam.*, 27(02), 146–168.

Robertson, G.L. (2013). *Food Packaging Principles and Practice*. CRC Press, Boca Raton, FL.

Sinha, B. (2017). Packaging coatings market by type (epoxy thermoset, urethane, UV-curable, BPA free, and soft touch UV-curable & urethane), substrate (metal, rigid plastic, glass, liquid cartons, paper-based containers, flexible packaging, and others), application (food cans, beverage cans, caps & closures, aerosols & tubes, industrial packaging, promotional packaging, and specialty packaging), and end user (food & beverages, cosmetics, pharmaceuticals, consumer electronics, and automotive components) – Global opportunity analysis and industry forecast, 2017–2023. Report. Allied Market Research.

Toenniessen, M. (2018). Packaging materials 10: Adhesives for food packaging applications. ILSI Report. ILSI Europe.

Vasile, C. and Sivertsvik, M. (2019). *Food Packaging Materials and Technologies*, printed edition. MDPI Books, Basel.

Vuoristo, P., Tuominen, J., and Nurminen, J. (2005). Laser coating and thermal spraying – Process basics and coating properties. In *Thermal Spray 2005: Thermal Spray Connects: Explore Its Surfacing Potential!* ASM International/ DVS, Dusseldorf.

11

Food Packaging Methods

Kata Galić

Faculty of Food Technology and Biotechnology, University of Zagreb, Croatia

11.1. Introduction

One of the key packaging functions is to protect the food. Depending on the food product, this task can be very simple, like wrapping, or very demanding, like in-package food processing. One way to select and define the packaging roles depends on the temperature applied during food processing (Galić *et al.* 2011). Thus, the packaging selection will depend on the food product (fresh or processed) and the process applied, either as thermal (retorting, aseptic packaging, etc.) or non-thermal (high-pressure, pulsed electric fields, etc.). Furthermore, the packaging method used will dictate the material selection, such as those based on atmospheric characteristics inside the package (vacuum, modified atmosphere, inert atmosphere, active packaging). Other packaging methods are used to communicate with consumers (such as intelligent packaging) and offer services such as gathering primary packaging.

According to ALL4PACK (2016) estimations, the global packaging market amounted to \$812 billion in 2014, and was projected to grow at a compound annual growth rate (CAGR) of 4.2% from 2010. Thus, the estimated value of \$839 billion in 2015 was predicted to reach \$998 billion in 2020. The global packaging market includes the following materials: paper and board, plastics, metal, glass and wood (Table 11.1).

For a color version of all figures in this book, see www.iste.co.uk/debeaufort/packaging.zip.

Packaging Materials and Processing for Food, Pharmaceuticals and Cosmetics, coordinated by Frédéric DEBEAUFORT, Kata GALIĆ, Mia KUREK, Nasreddine BENBETTAIEB and Mario ŠČETAR.

Packaging material	Share (%) of materials in value	
	2015 Total: $839 billion	2020 Total: $998 billion
Other	2.78	2.35
Metal	12.26	12.64
Glass	7.20	6.81
Paper and board	31.11	31.06
Rigid plastic	21.61	22.28
Flexible	25.03	24.85

Table 11.1. *Estimated values of packaging materials on a global level (ALL4PACK 2016)*

Among the processing technologies, the aseptic milk product market was calculated to grow at 12.8% a year to reach $138 billion by 2019 (Transparency Marketing Research 2014), while the high-pressure processed food market was about $9.8 billion in 2015 and is expected to reach $54.77 billion by 2025 (Marangoni and Anjos 2019). The amount of irradiated foods in Asia and the USA in 2010 was estimated at 285,200 and 103,000 tons, respectively. In 2015, the total amount of US irradiated foods was 125,000 tons which included mainly spices, fruits and vegetables, meat and poultry (Kume and Todoriki 2013; Eustice 2018). In the EU, a total of 10,211 tons of products were treated with ionizing irradiation during 2016 and 2017, and the two main commodities irradiated are frog legs (about 57%) and dried aromatic herbs, spices and vegetable seasonings (about 21%) (Koutchma *et al.* 2018; European Commission 2019).

The modified atmosphere packaging (MAP) application, which started commercially at the beginning of the 1970s, today covers a range of food products on the market (meat, poultry, fish, bacon, bread, cakes, salad vegetables, crisps and cheese) (Mullan and McDowell 2003; Galić *et al.* 2019). The MAP food sector continues to grow and is anticipated to reach $6.4 billion by 2020. Active and intelligent food packaging technologies, coupled with controlled and modified atmosphere packaging, is expected to grow at a rate of 6.9% per year (Arvanitoyannis and Kotsanopoulos 2012; Boz *et al.* 2018).

11.2. Packaging operations

The choice of packaging method depends on a number of factors such as: available machines and their requirements, characteristics of both food and packaging material, and general requirements. In terms of machine requirements, the important factors are production rate, package size, product type and work requirement of the machine. In food and packaging material characteristics, the main variations are in product sensitivity and size, packaging material requirements (such as barriers to oxygen and water vapor), in-package processing and post-packaging handling of the products (Mahalik 2014; Regattieri and Santarelli 2015).

Different packaging operations such as wrapping, filling and forming, sealing, bagging, canning, bottling, weighing, labeling, etc., are performed before the final product is ready to be placed on the market. These operations can be performed in a manual, semi-automatic or fully automatic way (Srivastava *et al.* 2018).

11.2.1. *Food wrapping*

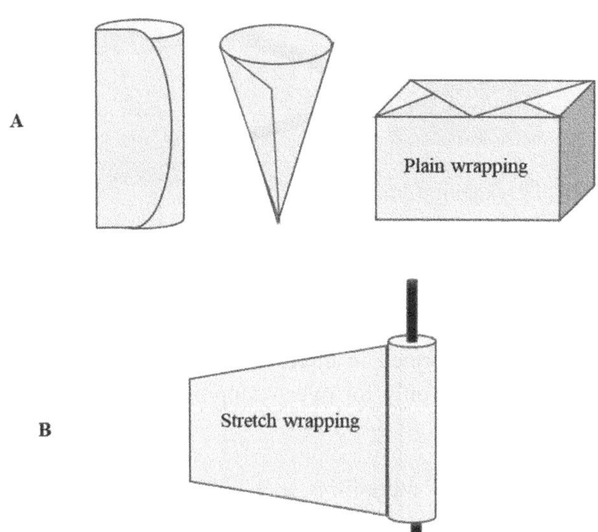

Figure 11.1. *Examples of plain (A) and stretch (B) wrappings*

Simple wrapping is used to protect individual food products using various types of paper (such as greaseproof paper) and flexible plastic film. Apart from being flexible, plastic film is a light, cheap, transparent and tough material used for food wrapping. These materials can have different types: plain (Figure 11.1(A)), stretch (Figure 11.1(B)) or shrink (Figure 11.2). Apart from paper as a traditional type of wrapping material, plastic materials are also used where the product is packed at the selling location (e.g. butcher shops). The most common polymers used for food wrapping are low-density polyethylene (LDPE), linear low-density polyethylene (LLDPE) and polyvinyl chloride (PVC) films (Kirwan *et al.* 2011).

In the stretch operation, the material is wrapped around the products (usually two or more units) while shrink wrapping is enabled by setting the polymeric cover over the product, which shrinks when heat is applied (Figure 11.2).

Stretch wrapping is a combination of two operations, wrapping and sealing. Here, the same material is used to "seal" itself; thus, no special sealer is used. The best example of stretch wrapping is in the case of pallet wrapping.

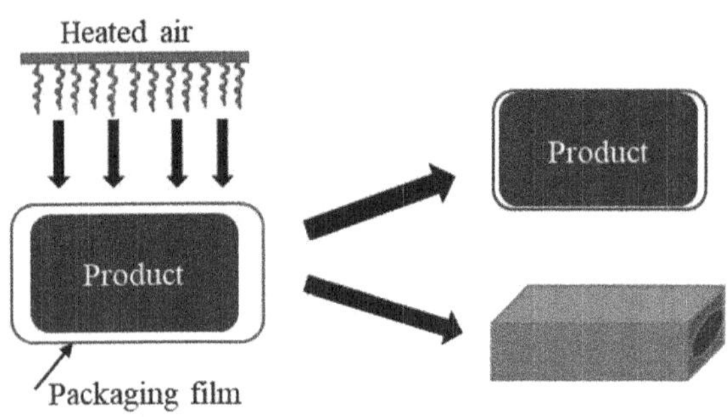

Figure 11.2. *Principle of shrink packaging*

Shrink wrapping is performed in three or four stages: (1) wrapping (sleeve or over-wrapping), (2) sealing (only for over-wrapping), (3) shrinking (application of hot air) and (4) cooling.

Shrink films are made of monofilms such as polyvinylidene chloride (PVDC), LDPE, LLDPE, polypropylene (PP), polyethylene terephthalate (PET) and PVC material (thickness between 0.040 and 0.120 mm), and are used for individual food products (e.g. cucumber packaging) or as a secondary package which contains more units of the same product (e.g. water bottles). Laminates are composed of polyolefins (PE, PP) and gas barrier layers made of PVDC or ethylene vinyl alcohol (EVOH). Thus, combinations as PP/EVOH/PP, PE/EFOH/PE and ethylene vinyl acetate (EVAC)/PVDC can be found on the market. In the case of shrink sleeves, PVC, OPP, PET and oriented polystyrene are used.

11.2.2. *Food bagging*

Plastic bagging includes product filling and sealing. There are many types of bags or pouches (open-mouth, valve-bags which have closed ends and are filled through a valve) used these days. Pouch closing can be performed by folding, twisting, heat-sealing, gluing, clipping or sewing (Figure 11.3(A)). Sealed bags and pouches are also distinguished based on the seam position (Figure 11.3(B)).

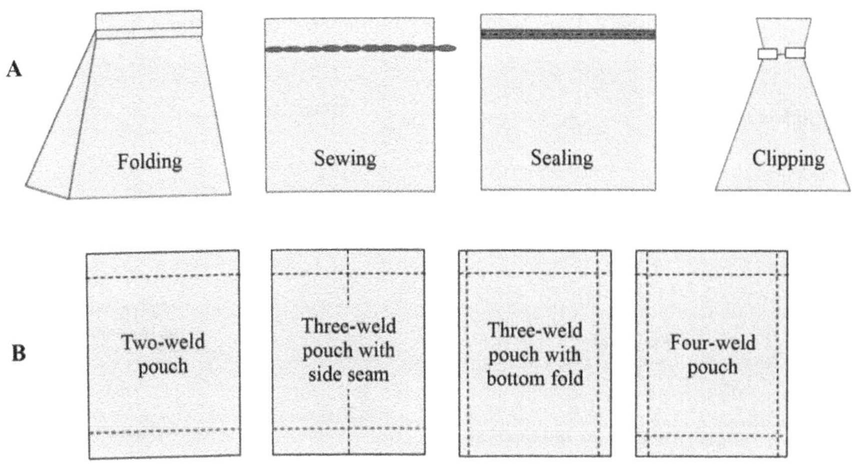

Figure 11.3. *Examples of pouch closing (A) and some types of sealed (dotted line) pouches (B)*

11.2.3. *Form-fill-seal packaging*

Form-fill-seal (FFS) represents an operation which integrates three operations: forming packaging material, filling the food into the formed package and sealing it on a single machine. Depending on the packaging material flow, horizontal (Figure 11.4) and vertical (Figure 11.5) pouch forming machines are used (Schmidt and Leiking 2005). These systems enable different kinds of food (solid, powder, liquid, viscous) to be packaged.

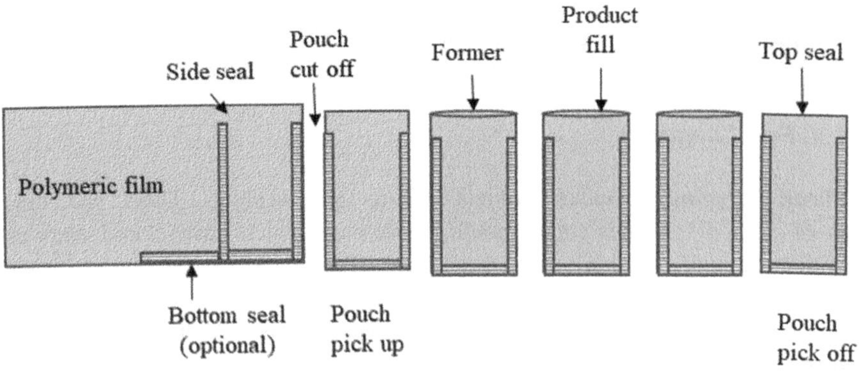

Figure 11.4. *Horizontal form-fill-seal operations*

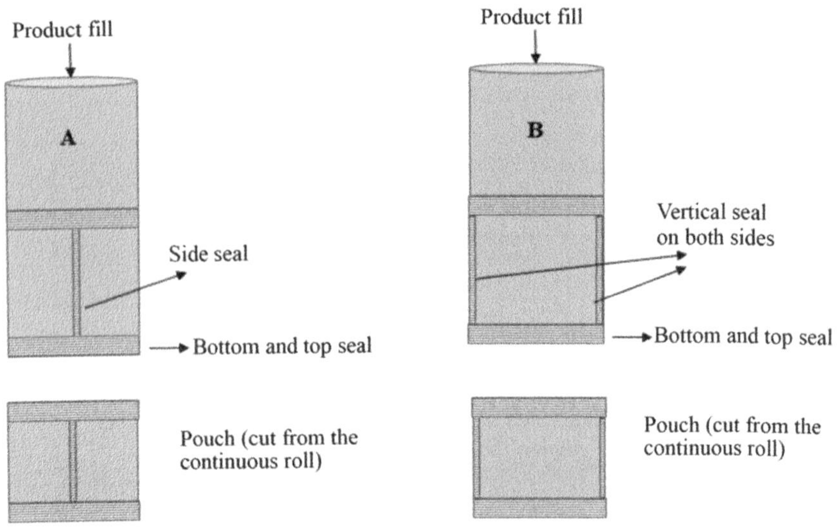

Figure 11.5. *Variations (A and B) of the vertical form-fill-seal operation*

11.2.4. *Blister packaging*

Blister packaging is a combination of preformed plastic packaging and covers, used for small consumer commodities, foods and medicaments. The preformed blister containing the product is covered with a heat-seal coated paperboard (Figure 11.6), offering good visibility of the packaged product.

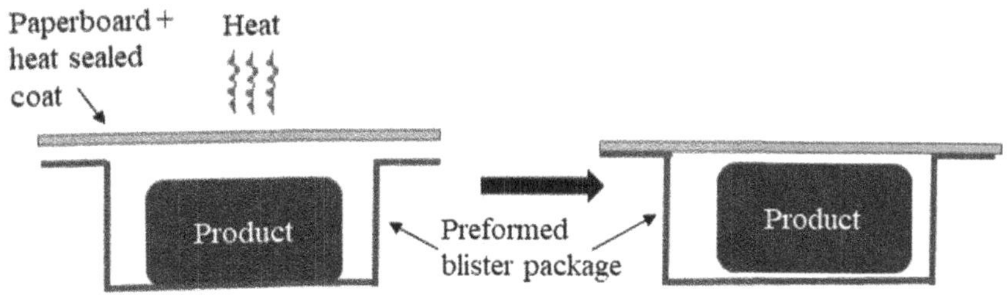

Figure 11.6. *Principle of blister packaging*

The form-fill-seal process is also used for blister packaging. The thermoformed blister with the product is sealed on the top layer of the coated aluminum film, instead of paperboard. Different materials are used for blister forms such as polypropylene (PP), PET and PVC. Blister forms are used for food (chocolate, sliced ham, sausages and sandwiches), pharmaceutical products (tablets, capsules) and other product packaging (tools, toys). Press-through-pack versions are also used for tablets and chewing gum, where the blister is sealed with the bottom layer of the sealant-coated aluminum foil, enabling easy opening (Kirwan *et al.* 2011).

11.3. Packaging for thermally processed food

Thermal processing involves retorts and heat to produce hermetically sealed food containers (e.g. cans, glass bottles and pouches). A "retort" is a pressure vessel designed for the thermal processing of food packed in hermetically sealed containers. This treatment requires heat conduction and convection (such as hot water, steam, etc.) to transfer heat into the food in order to destroy pathogens and increase food shelf life. The term "hermeticity" or "hermetically-sealed" indicates a closed can (body, both ends and seams) or glass jar that is absolutely impervious to air, microorganisms or content leakage during processing, distribution and storage.

11.3.1. *Canning*

Canning (or retorting) is the process which applies heat (>100°C) to food that is hermetically sealed in a can or jar, in order to destroy all microorganisms that can cause food spoilage. The negative effect of such a treatment is the degradation of nutritional and sensorial characteristics of the product, due to prolonged thermal exposure during heating and cooling (Holdsworth and Simpson 2007; Park *et al.* 2014; Vergara-Balderas 2016; Dantas and Dantas 2017).

The simplest retorts are those that are discontinuous (batch) that use steam as the heating medium. They are known as still retorts. The temperature in these retorts is automatically controlled. The loading and unloading of the containers from these retorts is performed by preloading containers into crates, cars, baskets or trays. "Crateless" systems drop containers into the retort vessel, which is filled with water, to ensure a cushion effect and to prevent container drainage. The water is removed before processing. The orientation of the retort depends on the type of container system used. Those that use crates or baskets and the crateless systems use vertical vessels, whereas the car system requires a horizontal orientation (Hotchkiss 2009).

Metal cans are one of the examples of how to provide package hermeticity. The quality of the double seam and side seam formed is of huge importance to ensure a hermetically sealed can (AFDO 2011; Dantas and Dantas 2017). Metal cans, which are able to prevent food recontamination, are made of tin-plated steel (tinplate), tin-free steels (TFS, chrome-plated steel) or aluminum (see Chapter 4). Food cans use both types, i.e., three-piece cans, consisting of one-bottom lid, two-body and three-top lid, and two-piece cans, consisting of one-body connected to the bottom lid and two-top lid (Figure 11.7).

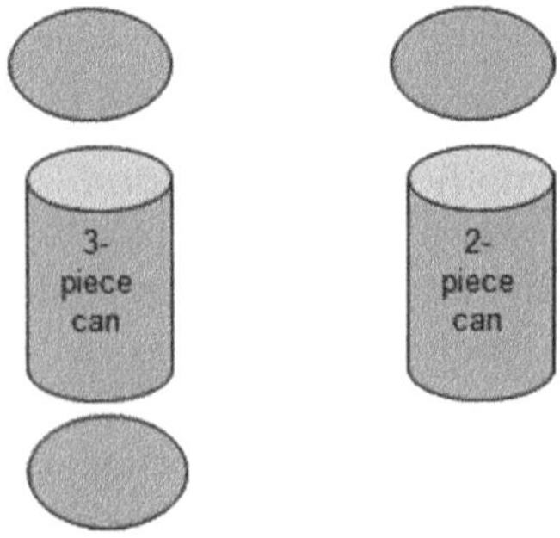

Figure 11.7. *Types of metal cans*

With regard to the presence of a protective lacquer, both plain (with no lacquer) and lacquered tinplate cans are used for food and beverage packaging. Plain cans are used for specific (bright color) food types (e.g. pineapple, pear, mushrooms and asparagus). The type of can lacquer is selected based on food characteristics (aggressiveness).

Glass jars of different shapes and sizes (Figure 11.8) are also widely used for food packaging of pickled vegetables, fruits in syrup, jams, etc. They have the advantages of being inert, with regard to the contents, and transparent which enable consumers to have a clear view of the product (see Chapter 3).

Figure 11.8. *Glass jars with metal closures*

However, glass is very fragile and requires careful processing and handling. An important part of the jar is the closure. Various types of closures are available (both venting and non-venting types) and are made from either tinplate or TFS (see Chapter 9). In order to prevent closure distortion, the correct overpressure must be maintained during retorting and the jars must be preheated prior to processing, in order to prevent shock breakage (Holdsworth and Simpson 2007).

Rigid plastic containers. The main requirement for a plastic material is that it must withstand processing conditions (heating and cooling). Here again, it is important to control the overpressure in order to maintain a pressure balance (pressure developed during processing and the pressure of the heating system). The main plastic materials used for thermal-processed foods are both mono- and multilayer (laminate) PP and PET (Figure 11.9). Biaxially oriented polypropylene (BOPP) bottles and jars are also used to replace traditional glass containers. PP and PET are usually manufactured with an oxygen barrier layer such as EVOH, PVDC and polyamide (PA) to create multilayer flexible pouches, semi-rigid and rigid containers (Holdsworth and Simpson 2007).

Figure 11.9. *Rigid plastic containers*

Flexible materials include wraps, pouches or bags, with a thickness equal to or less than 0.127 mm, while semi-rigid materials are thicker than 0.127 mm. They are usually formed as sheets. Rigid materials include trays, bottles, boxes or crates and are composed of thick or rigid polymers. They are used for beverages, pasteurized products, pickles, instant coffee, sauces, powdered milk and creams (Ebnesajjad 2013).

Glass, semi-rigid and flexible containers must be processed in still retorts designed to ensure the stability of these containers at applied temperatures (121°C) and pressures (above 103 kPa) (Hotchkiss 2009).

11.3.2. *Retortable pouches*

Retortable flexible pouches are usually laminates that are thermally processed, similar to cans. These materials are characterized by good seal integrity, toughness, puncture resistance, and the ability to withstand thermal processing conditions and provide superior barrier properties, for long shelf life products. Laminated flexible pouches can easily replace traditional materials such as glass or metal cans, due to their excellent properties.

An example of the retortable pouch is a three-ply laminate (Figure 11.10) made from: (1) an outer layer of 0.012 mm polyester film for strength, high temperature resistance, toughness and printability; (2) an adhesive laminated to a middle layer of 0.009–0.018 mm aluminum foil (which is an excellent barrier to moisture, light and gas); which is laminated to (3) the inner layer of 0.076 mm PP film as the heat seal and food-contact material (Abdul Ghani Al-Baali and Farid 2006; Holdsworth and Simpson 2007). Other pouches that are mostly used are a four-ply laminate consisting of polyester (outside layer), a polyamide (second layer), aluminum foil (third layer) and a polypropylene as an inner layer (Majumdar *et al.* 2017).

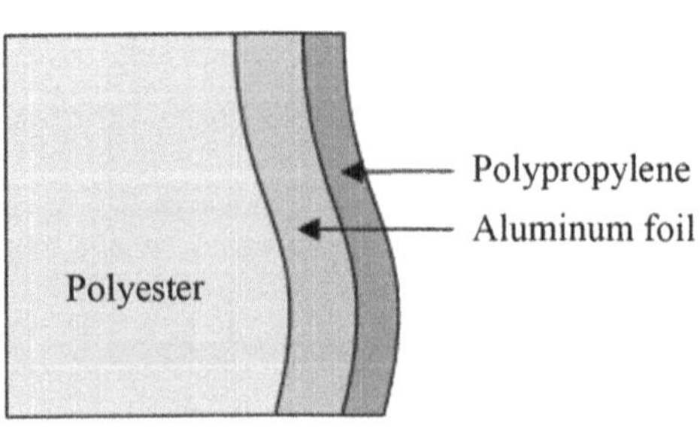

Figure 11.10. *Example of a three-ply retort pouch*

An advantage of using pouches is that they take less time to reach sterilization temperature than cans or jars, due to the thinner structure. Such a product is sterile, does not require refrigeration and is shelf-stable at room temperature (Holdsworth and Simpson 2007). Retort pouches are used for different products, such as juices, soups, sauces and gravies.

Polymer (number after each polymer is the layer thickness in μm)	**O_2 permeability (cm^3/m^2 day atm)**	**Water vapor permeability (g mm/m^2 day)**
	at 23°C, 50% RH	**at 23°C, 85% RH**
PET 12/ Alu 9/PE 50	~0	~0
PET 12/ vacuum deposited aluminum /PE 50	1-2	0.1-0.5
PET 12/ EVOH 5/PE 50	1	2-4
PET 12/ PVDC 4/PE 50	5	2
PET 12/ PVOH 3/PE 50	2	4-6
PET 12/ PE 50	15-20	4-6
	at 23°C, 50% RH	**at 23°C, 50% RH**
PET 12/SiOx	0.5-5	0.2-5
PP 20/SiOx	1-20	0.1-1
PA 20/SiOx	1	0.5
PET 12 untreated	110	15
PP 20 untreated	1500-1800	1-1.5

Table 11.2. *Barrier properties of laminates (Lange and Wyser 2003)*

In order to replace aluminum foil and construct clear pouches, new laminates with an inorganic layer (such as silicon, SiOx or aluminum oxide, AlOx) are produced (Strullera *et al.* 2019). A transparent SiOx coating is applied on polymers such as PET, PP, PA, polyethylene naphthalate (PEN) and polyvinyl alcohol (PVOH) films, having excellent oxygen and water vapor properties (Table 11.2) (Lange and Wyser 2003).

Sous-vide **cooking** (Figure 11.11) involves packaging ("*sous-vide*" means "under vacuum" in French) food before the application of low temperatures (65–95°C), heating and storing under refrigerated conditions (0–3°C). This process is known as a low temperature, long cooking time process.

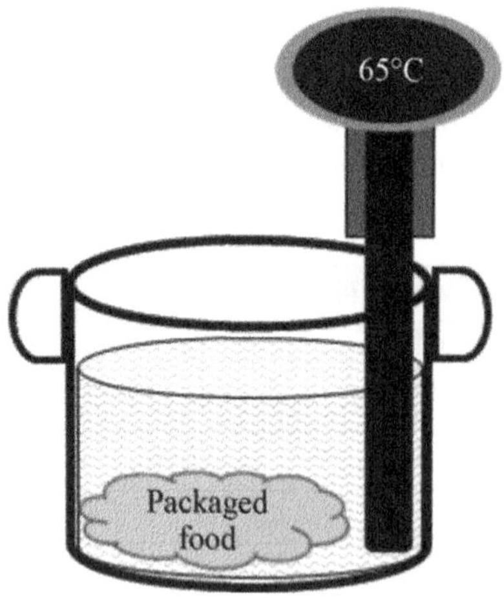

Figure 11.11. Sous-vide *immersion cooker*

This process is sometimes confused with regard to its name (vacuum), as a vacuum is not actually applied. The packaging used in this case are Ziploc-bags or bags intended for vacuum sealers (without sealing them), they can withstand temperatures higher than 70°C. Thus, laminates with different material (5, 7 or 11 layers) compositions, such as PA/EVOH/PE and PA/PE, are used.

The application of *sous-vide* has been used widely in the processing of packaged raw meat, poultry, fish and vegetable-based products to enhance sensorial and organoleptic characteristics. This technology is especially attractive to the food service industry. Some negative points are that, due to the use of mild temperatures, *sous-vide* cooking does not inactivate bacterial spores, and vacuum packaging conditions could support the potential survival of *Clostridium botulinum* spores (Baldwin 2012; Park *et al.* 2014; Neetoo 2015), i.e., only pasteurization is allowed.

11.3.3. *Aseptic packaging*

In the aseptic filling process, ultra-high temperature (UHT) treatment of food is followed by filling into sterilized packages (with a 35% solution of H_2O_2, in the form of a fine spray or as peroxide vapor or a peracetic acid solution) within a sterile environment (Figure 11.12(A)).

An example of a process used in aseptic packaging is the form-fill-seal process. It is used for sachets, trays and pots, in which the packaging material is formed into the package as part of the filling operation. The typical laminate used consists of

LLDPE, with a center layer of EVOH and black carbon for additional light protection. Other solutions include a 40% calcium carbonate (chalk) filled PP layer (for stiffness and integrity), EVOH (O_2 protection), carbon black and LLDPE as a sealing and food contact layer.

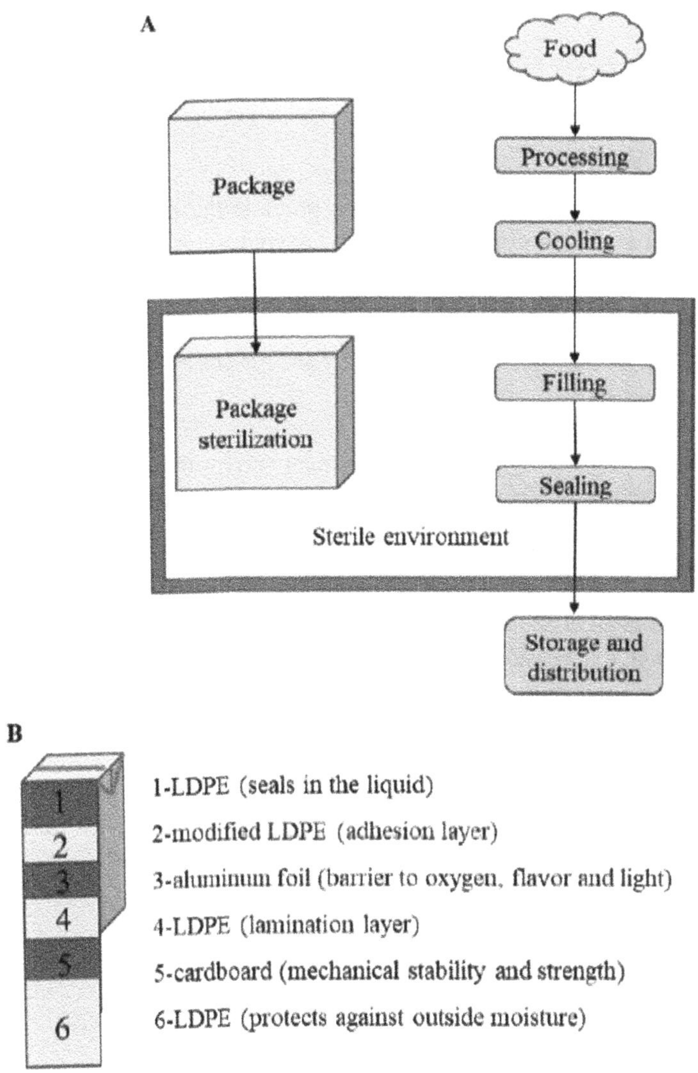

Figure 11.12. *Schematic view of the aseptic process (A) and multilayer packaging form for aseptic packaging (B)*

Aseptic packaging is also available in the form of metal cans, plastic pots, plastic bottles (HDPE, PP with or without pigments added for light protection), flexible packaging and laminated (multilayer) materials (Figure 11.12(B)). Multilayer materials are also used for bag-in-box systems containing metallized polyethylene terephthalate (PETmet) as barrier layers (e.g. OPP/PETmet/EVOH/PET/LLDPE).

The food (soups, fruit juices, milk and other liquid dairy products) is sterilized in a continuous process by traveling through a heat exchanger before being filled into the package cold.

Since air overpressure is not required in this process, the packaging does not need to be as strong as for retort-processed foods. Aseptic filling is also applied in the beverage industry using PET bottles and flexible pouches with caps (Tucker 2011; Götz *et al.* 2014; Robertson 2016).

11.3.4. *Ohmic heating*

Ohmic heating, performed by passing an electrical current through food (Figure 11.13) which is heated by molecular agitation from the motion of ions within the material, enables uniformity of heating and improvements in nutritional or organoleptic characteristics of food (Butz and Tauscher 2002; Park *et al.* 2014). This technique is especially useful in leafy spices (Mudgal 2019).

Using this method, it is possible to sterilize foods within the package (in-package sterilization). For this purpose, flexible packaging (laminates) can be used as an alternative to a rigid container.

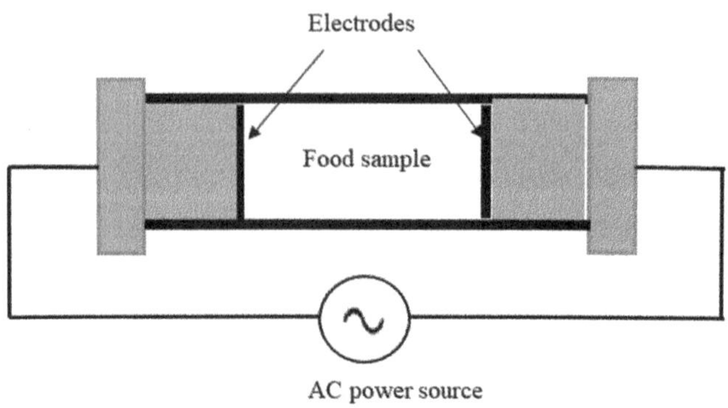

Figure 11.13. *Schematic view of ohmic heating (Lee* et al. *2016)*

In order to yield uniform and rapid heating, an electrode configuration in a pouch (Figure 11.14) was created by Jun and Sastry (2005, 2007).

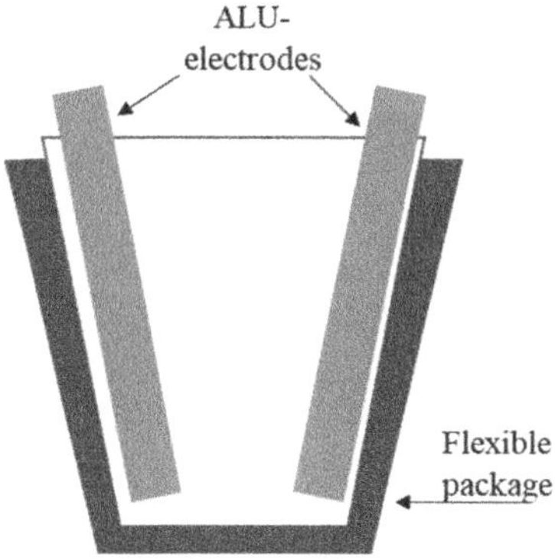

Figure 11.14. *A schematic view of a flexible package with aluminum (ALU) foil electrodes (Jun and Sastry 2005)*

A laminate is made of a thin metal (aluminum) film (0.018 cm) with a protective polymer (polyester and polyamide) film (0.010 cm) on one side for external resistance and a heat-sealable interior polyethylene (0.010 cm) film. The electrode construction, made of aluminum foil (0.005 cm in thickness), is placed between a folded laminate, with the electrodes extending out and heat-sealing the edges (Jun and Sastry 2005, 2007).

Apart from ohmic pouches with electrodes, intensive research is being carried out into conductive (Ag, Au, Cu, Zn, Ni, graphene and carbon nanotubes/ nanoparticles) packaging for the pasteurization and/or sterilization of foods, such as milk and chunky mixed fruit juice (Kanogchaipramot *et al.* 2016; Liao *et al.* 2019).

11.3.5. *Infrared treated pre-packaged food*

Infrared (IR) radiation is part of the electromagnetic spectrum that is located between the visible region and microwaves (from 0.5 to 100 mm). They are classified as: near-IR (0.75–1.4 mm), mid-IR (1.4–3.0 mm) and far-IR (3.0–1000 mm) (Krishnamurthy *et al.* 2008; Pan *et al.* 2016).

Infrared penetration causes a vibrating movement of water molecules at a frequency of 60,000–150,000 MHz, which causes heating. IR technology is characterized by high energy efficiency, low energy consumption, less water consumption, homogeneity of heating with high heat transfer rate, low heating time, improved product quality and food safety and environmental friendliness (Nowak and Lewicki 2004; Aboud *et al.* 2019). The energy is directly concentrated on the heating material and does not produce volatile organic compounds, carbon monoxide or nitrogen oxides (Dagerskog and Österström 1979; Afzal and Abe 1999; Neetoo 2015).

IR heating is used in various thermal food processing operations such as dehydration (fruits and vegetables, fish, pasta and rice); heating (flour); frying (meat); roasting (cereals, coffee and cocoa); baking (cookies and bread); blanching and peeling (fruits and vegetables); and disinfection (corn and rice). Successful IR applications exist for thawing and surface pasteurization of pre-packaged hotdogs and ready-to-eat (RTE) meats, such as turkey bologna, roast beef, corned beef and ham (Gande and Murina 2003; Huang and Sites 2008; Krishnamurthy *et al.* 2008; Rastogi 2012; Pan *et al.* 2016).

Results show that the IR heat of pre-package and post-package surface pasteurization, or a combination of the two processes, can reduce potential *Listeria* contamination on RTE deli meat surfaces, with minimal effects on product quality (Gande and Murina 2003).

An infrared heating system was developed (Aghajanzadeh *et al.* 2016) to study the effect of IR radiation on the quality of juice products (Figure 11.15).

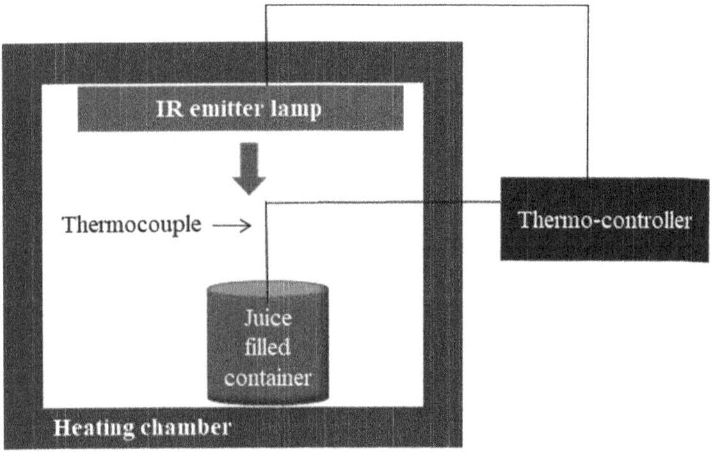

Figure 11.15. *Schematic view of an infrared heater for fruit juice processing (Aghajanzadeh* et al. *2016)*

It has also been shown that IR heating can cause less ascorbic acid degradation, lower brown pigment formation and color variation in analyzed juice, compared to conventional thermal treatment in a water bath (Aghajanzadeh *et al.* 2016).

11.3.6. *Radiofrequency treated pre-packaged food*

Radiofrequency (RF) heating is the heating of dielectric materials with electromagnetic energy, at frequencies between 1 and 300 MHz. It is used to inactivate heat-resistant spores to produce shelf-stable pre-packed foods such as pasta, meats, processed eggs, nuts, broccoli powder, etc. (Zhu *et al.* 2012; Wang *et al.* 2012; Zhao *et al.* 2017; Altemimi *et al.* 2019).

For the RF heating application (pasteurization), the food is packaged in plastic or other materials and is surrounded by a small air layer or a deionized water layer. A schematic diagram of a generic RF dielectric heating system is shown in Figure 11.16 (Zhao *et al.* 2000, 2017).

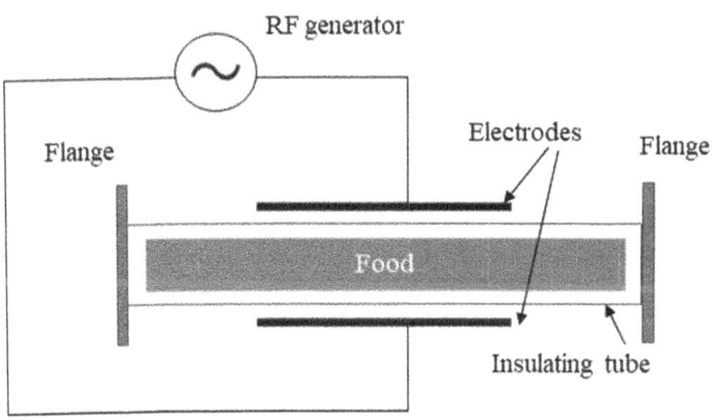

Figure 11.16. *Radiofrequency (RF) dielectric heating systems (Zhao* et al. *2000)*

One of the early applications of RF heating was applied on packaged bread (Cathcart *et al.* 1947). Loaves of wrapped, sliced white bread and wrapped brown bread were treated in a 3 kW unit at a frequency of 14–17 MHz, to provide sterile products with regard to mold. No mold developed on samples heated to 60°C or above after 10 days of storage. It was also found that bread needed to be wrapped in moisture-proof packaging material to maintain moisture during heating and storage. Wax and glassine paper were not suitable for this purpose as wax paper softened at temperatures above 38°C, while glassine papers became sticky at temperatures ranging from 52 to 60°C. Cellophane was found to be a satisfactory wrapping material as it did not show any failure during processing (Cathcart *et al.* 1947). Some other materials also showed satisfactory performance, such as polytetrafluoroethylene (PTFE), polycarbonate (PC) and borosilicate glass when using RF heating to pasteurize sausage emulsion (Houben *et al.* 1991).

RF drying applications are used for the sterilization of packaged flours as well as for drying granular foods such as coffee and cocoa beans, corn, grains and nuts. Although not all applications of RF dielectric heating in the food industry are carefully examined, there is potential for its use in the blanching of vegetables, thawing of frozen foods, post-baking snack foods, meat processing, pasteurizing and the sterilization of liquid foods and pre-packed foods (Piyasena *et al.* 2003).

11.3.7. *Microwavable packaging*

Microwaves are a type of electromagnetic radiation characterized by the wavelength and frequency of the waves used in food processing. A microwave oven (Figure 11.17) consists of a cavity surrounded by metal walls with a door at the front. The microwaves, generated by a magnetron, enter the cavity through the wave guide and the stirrer distributes the microwaves while the turntable enables energy distribution into the food (Hill 1998; Schiffmann 2017).

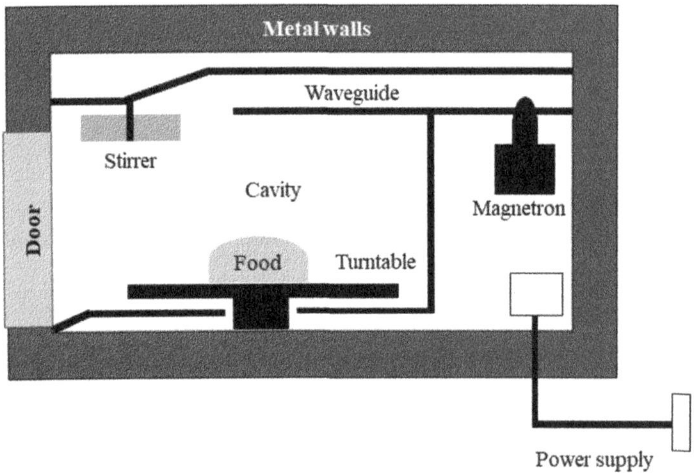

Figure 11.17. *Basic structure of a microwave oven (Hill 1998)*

Generally, microwave ovens for food applications operate at two frequencies. The household microwave oven operates at 2,450 MHz, and industrial ones at 915 MHz. As microwave energy is not heat energy, in order to heat the food, the energy must be converted to heat. Energy conversion can occur by (1) dipole rotation of polar molecules (such as water) and (2) ionic polarization (of ions present in solution).

When a microwave oven is used to heat packaged food, the package plays an active part as it can transmit (paper, plastics and glass), reflect (aluminum foil) or absorb (susceptors) microwaves, thus affecting the heating process (Neetoo 2015).

Microwave transparent materials are materials which must be wholly or partially transparent to microwaves. Such materials are paper and plastics. Plastic-coated paperboard trays are also used for microwavable frozen meals. For microwave applications, the following plastics are used: LDPE, high-density polyethylene (HDPE), PP and PET. LDPE trays are suitable for light microwave heating of frozen food because the trays tend to deform at temperatures as low as 75°C, while PP trays have a deformation temperature of about 110°C. Crystallized PET trays, the most widely used for microwaveable frozen meals, are functional at temperatures ranging from 40–220°C (Kirwan *et al.* 2011; Schiffmann 2017).

Microwave reflective (shielding) materials include plain aluminum foil (Alu) and laminates with an aluminum layer (such as Alu/plastic and Alu/plastic/paperboard). Aluminum can intensify the microwave energy locally in the package which otherwise would likely have little direct exposure to microwaves. Some precautions are needed when using aluminum foil in microwave ovens in order to prevent arcing between Alu-foil and the oven walls. Thus, the foil should be moved far from the edge to avoid arcing with the oven walls (Kirwan *et al.* 2011).

Microwave absorbent materials used for food packaging are known as susceptors, which generate surface heating and enable browning and crisping of food. A commercially available susceptor is the metallized PET film laminated to a thin paperboard (Figure 11.18). The susceptors can reach surface temperatures of over 200°C within seconds and they are used for products such as pizza, frozen hot pies and popcorn (Yam 2009; Schiffmann 2017).

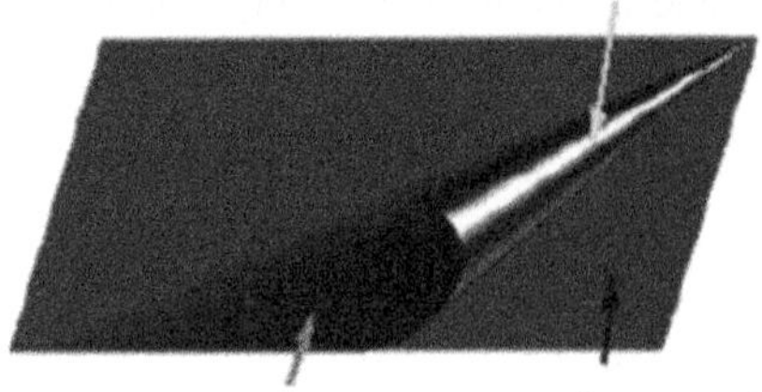

Figure 11.18. *Susceptors consist of a metallized PET film laminated to a thin paperboard (Hill 1998)*

11.4. Packaging for non-thermally processed food

Non-thermal processes have many advantages, but they also have disadvantages (Knorr and Watzke 2019). They have different mechanisms to preserve food compared to thermal processes. Thus, packaging material for non-thermal processing requires knowledge of the packaging material properties and interactions between the packaging materials and food components during the non-thermal processes. Packaging materials for non-thermal processes should have proper resistance to the particular non-thermal processing conditions (Galić *et al.* 2011).

11.4.1. *High hydrostatic pressure*

During high hydrostatic pressure (HHP), pre-packaged foods are briefly (for a few minutes) exposed to pressures of 200–600 MPa. Process temperatures during treatment can vary from sub-zero temperatures to above 100°C (Caner *et al.* 2004).

The packaging material must be able to withstand the HPP conditions (Figure 11.19), have good sealing and barrier properties, and have the ability to prevent quality deterioration during the application of pressure. In cases where pack integrity (delamination, seal perforation, etc.) is lost, food spoilage can occur easily.

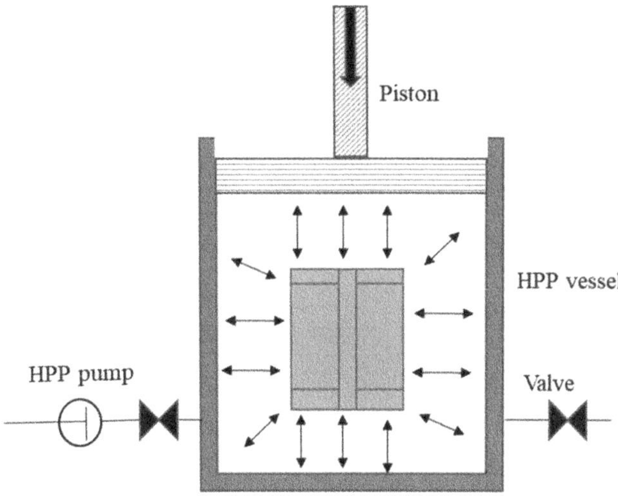

Figure 11.19. *Schematic view of the high-pressure processing (Goyal* et al. *2013)*

In order to transmit pressure, at least one interface of the package must be flexible enough. Thus, rigid metal, glass or plastic containers cannot be used. Because gases are very compressible under high pressure, the headspace area must be as small as possible to avoid package deformation (Rastogi *et al.* 2007; Guillard *et al.* 2010; Fraldi *et al.* 2014; Marangoni and Anjos 2019).

The effect of HPP treatment on the packaging material can cause a significant loss of heat sealability (as in the case of the following laminates: PA/PE, PE/PA/EVOH/PE, PET/PE/EVOH/PE) or an increase of migration levels, as observed in LDPE/EVAC/VDC laminate (Dobiaš *et al.* 2004; Rivas-Cañedo *et al.* 2009; Bello *et al.* 2014). In some cases, delamination between PP and Alu, in PE/PA/Alu/PP laminate, was observed in ready-to-eat meal pouches treated at 200 MPa and 90°C for 10 minutes (Schauwecker *et al.* 2002), and PETmet/PE after processing at 400 MPa and 60°C for 30 minutes (Galotto *et al.* 2008).

11.4.2. *Pulsed electric fields*

Pulsed electric fields (PEF) use very short pulses (in milliseconds, or in microseconds) of a high voltage (from 1 to 80 kV/cm) to inactivate microorganisms on foods placed between the two electrodes (Figure 11.20).

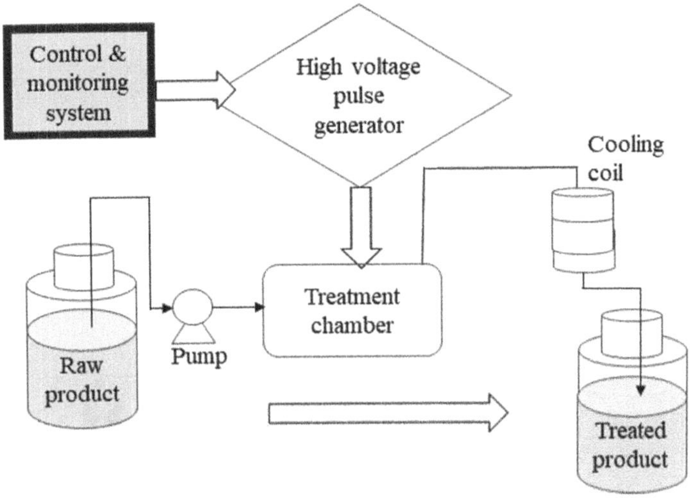

Figure 11.20. *Schematic view of the pulsed electric field processing system with basic components (Maged and Ayman 2012)*

The selection of adequate packaging materials is important to keep the quality of PEF-treated foods during storage. Packaging materials with very low O_2 permeability values should be selected for oxygen-sensitive food products. For these products, oxygen in the food and in the headspace of the package should also be minimized.

The effects of different packaging materials (glass, PET, HDPE and LDPE) on the stability of PEF processed (35 kV/cm, 59 μs) orange juice was investigated (Ayman *et al.* 2001). The retention of flavor compounds tested, vitamin C and color was significantly higher in glass and PET than in HDPE and LDPE packages. The reasons for such behavior were explained by: (1) the absorption of flavor compounds into the polymeric packaging materials tested; (2) the acceleration of degradation reactions (ascorbic acid, color and flavor) due to initial oxygen concentration and oxygen permeation through the package; and (3) increases in browning, absorption and degradation of flavor compounds with increasing storage temperatures (from 4 to 22°C). PEF has been applied on different beverages, mainly on fruit juices, tea and milk (Gabrić *et al.* 2017). The PEF-treated orange juice had a shelf life of over 16 weeks at 4°C in glass and PET bottles (Kumar and Han 2012).

11.4.3. *Irradiation*

Food irradiation is a process that involves the exposure of food to ionizing radiation (Figure 11.21), such as gamma rays emitted from the radioisotopes cobalt-60 (^{60}Co) and cesium-137 (^{137}Cs), accelerated electrons (at or below 10 MeV) and X-rays (at or below 5 MeV).

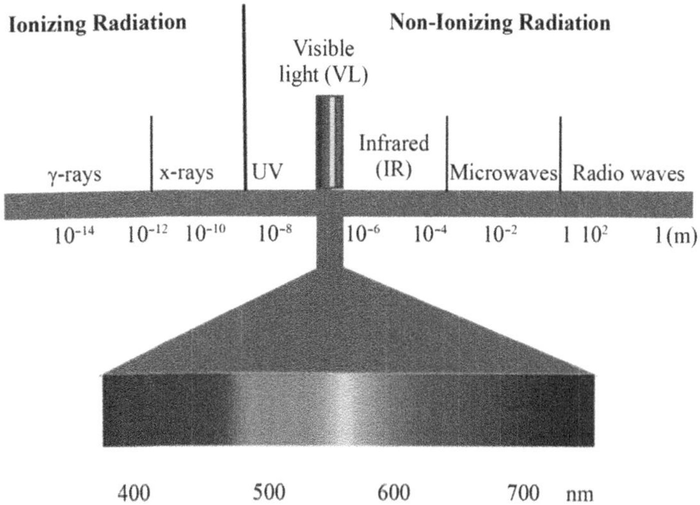

Figure 11.21. *The spectrum of electromagnetic waves*

Ionizing radiation can be applied to decontaminate pre-packaged food, either by pasteurization (1–5 kGy) or sterilization (10–74 kGy) using radiation-stable packaging material (Kumar and Han 2012; Neetoo 2015).

Among the food products, spices and herbs are the most widely irradiated items, while fruits, vegetables, fish and grains are also irradiated (Koutchma *et al.* 2018).

Packaging material may be irradiated before (dairy products, processed food, beverages, pharmaceutical and medical device) or after filling.

Packaging materials for irradiation should have chemical and physical stability under radiation without degradation (which can result in the migration of compounds from the packaging into the food) or significant changes in mechanical and barrier properties. Ionizing radiation can affect polymers in two ways: through crosslinking (mainly under vacuum or an inert atmosphere) and chain scission (degradation) in the presence of oxygen or air. Therefore, it is important to determine the stability/integrity of irradiated packaging. Some monofilms (HDPE, LDPE, PS, BOPP) were not significantly affected, while in the case of laminates, barrier properties either decrease, as in the case for BOPP/CPP and PET/PVDC/PE, or are not significantly affected, as in the case for PA/PVDC/EVAC, PET/PE/EVAL/PE and PET/PET/PE-LLD, with applied radiation doses (Ozen and Floros 2001; Haji-Saeid *et al.* 2007; Mizani *et al.* 2009; Kumar and Han 2012).

Irradiated products placed on the market must be marked with the "radura" (radurization is a process of packaged and non-packaged food irradiation) symbol (a plant in circle).

11.4.4. *Pulsed light technology*

Pulsed light (PL) technology is a method used for the sterilization of foods using very high power and very short duration pulses of light emitted (Figure 11.22) by inert gas-flash lamps (Gómez-López *et al.* 2007). The high-power pulses of radiation can be in the spectra of ultraviolet (UV), visible (VL) and infrared (IR) light.

PL has also been used for the surface decontamination of foods in plastic packaging. The sterilization of foods is possible as long as the packaging material and packaged contents are UV transparent. However, because of the opacity and irregular shape of most foods, PL is mostly effective for surface decontamination.

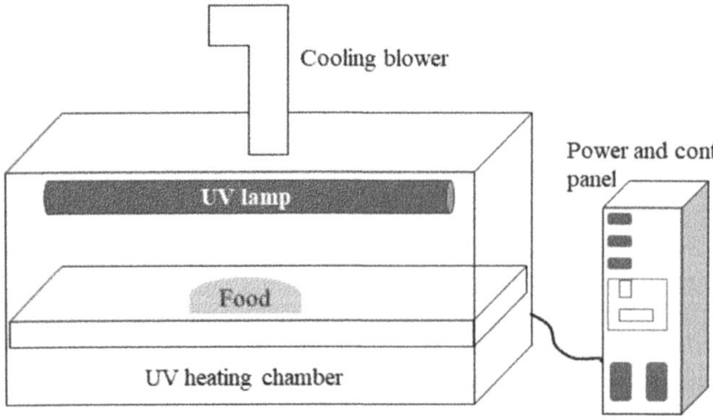

Figure 11.22. *Schematic view of pulsed UV-light food treatment (Murugesan* et al. *2012)*

Microorganisms penetrating into deep surface irregularities of food are thus protected from the PL action due to the "shadow effect" (Elmnasser *et al.* 2007;

Bhavya and Umesh Hebbar 2017). The application of PL is especially convenient for cap, tray and cup decontamination, where conventional techniques are difficult to use. UV light can reduce the contents of a range of microorganisms on packaging materials (such as black PP, white PP trays and PVC) and on food contact surfaces (such as stainless steel) and was most effective for reducing *C. jejuni* on the packaging materials mentioned (Haughton *et al.* 2010).

Generally, UV light was least effective when applied to microorganisms inoculated onto PET/PP films and onto blue PS. It was also demonstrated that UV light can interact with the packaging materials, such as PET, yielding undesirable products, or can cause the package constituents to undergo certain chemical reactions (Ozen and Floros 2001). Among PE films, the surface of HDPE oxidizes slightly faster than that of LDPE (Peeling and Clark 1983).

It is evident that the effects of novel food processing methods on packaged food can cause different failures in the packaging material, thus affecting shelf life of the packaged food. Hence, food–packaging interactions under specific processing conditions should be carefully investigated in order to select the optimal packaging material for specific novel treatment technologies (Guillard *et al.* 2010; Neetoo 2015).

11.5. Packaging with atmosphere modification

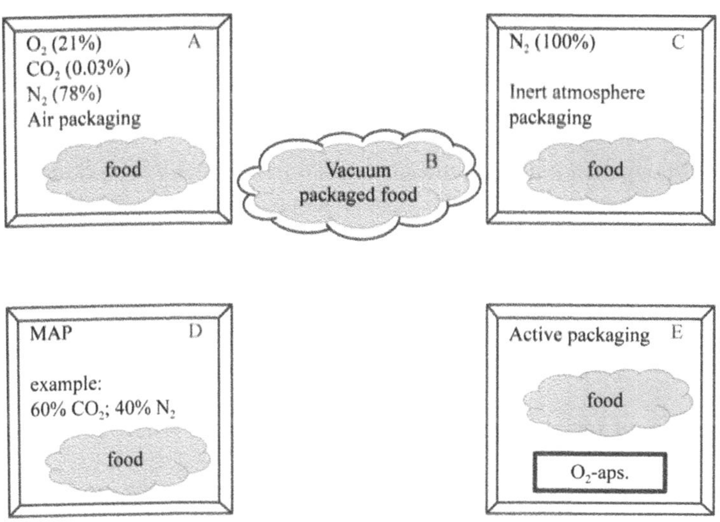

Figure 11.23. *Food packaging methods with atmosphere modification*

The main role of this method is to use reduced oxygen content inside the packaging (Figure 11.23). Reduced oxygen packaging (ROP) is a process of modifying the air (78% nitrogen, 21% oxygen, 0.93% argon, 0.03% carbon dioxide and other trace gases) atmosphere (Figure 11.23(A)) inside the package with a desired gas composition. By applying an appropriate gas composition inside the package, increased food shelf life can be achieved due to reduced microbiological growth and/or enzymatic spoilage.

ROP methods include: (1) vacuum packaging (Figure 11.23(B)); (2) gas packaging, (a) inert packaging (Figure 11.23(C)), (b) modified atmosphere packaging (Figure 11.23(D)) and (c) active packaging (Figure 11.23(E)).

11.5.1. *Vacuum and vacuum skin packaging*

Vacuum packaging (VP) refers to packaging containers (rigid or flexible) with all air having been removed prior to final sealing of the container (Figure 11.23(B)). Since it is not possible to evacuate all of the air (about 3% remains after sealing), the gaseous atmosphere of the vacuum package is changed during storage (due to microbial activity and gas permeation of the package) and therefore the atmosphere becomes modified.

Vacuum skin packaging (VSP) methods involve wrapping/packing a product with a polymeric film and applying a vacuum so that a skin is formed around the food. This allows the vacuum packaging of irregularly-shaped products and also of products that could puncture vacuum films, such as poultry and shrimp (Lagerstedt *et al.* 2011; Li *et al.* 2012).

The most common polymeric materials used for VP and VSP include monofilms (LDPE, PVC, ionomer (IO) and PP) and laminates such as PA/PE, PE/PA/PE, PE/EVOH/PE/PA, PE/EVAC/IO/EVOH/IO/EVAC, PET/SiOx and SiOx/PE (Gertzou *et al.* 2017; Van Rooyen *et al.* 2018; Mazzola *et al.* 2019).

VP products include fresh red meat, processed meat (ham, bacon, salami, frankfurters, etc.) and cheeses.

Advantages of vacuum packaging are:

1. extends shelf life: controls oxidative reactions; prevents the growth of spoilage bacteria; aerobic organisms (e.g. *Pseudomonas*) are prevented and lactic-acid bacteria (LAB) are favored;

2. reduces moisture loss: prevents movement of water out of the product into the headspace; prevents loss of moisture at the product surface and eliminates freezer burns;

3. requires minimal storage space, as product is tightly wrapped and occupies minimal space;

4. enables easy detection of leaks, as small punctures in a vacuum pack are easy to detect as the product is deformed and loses its specific "vacuum shape".

11.5.2. *Inert atmosphere packaging*

Inert atmosphere packaging (the gas-flushing process, Figure 11.23(C)) is usually performed by an FFS machine. An inert gas (mostly nitrogen) is injected into the package to replace the air. When most of the air has been replaced, the package is sealed. The typical residual oxygen levels in gas-flushed packs are 2–5%. Thus, this technique is not suitable for packaging very oxygen-sensitive foods. Flushing with nitrogen is a common industry practice to extend the shelf life of beverages, juices, some bakery products and snack foods (Wagner 2010; Jensen *et al.* 2011). Snack foods (such as potato chips) are usually packaged in metallized films (BOPP/BOPPmet, BOPP/LDPE/BOPPmet, EVAC/HDPE/BOPPmet), which provide protection from oxygen, moisture and light. Salty snacks, with a high fat content, are packaged in barrier films containing aluminum foil, a metallized polymer film or a barrier polymer such as EVOH or PVDC, and are also gas-flushed with nitrogen. Such barrier laminates are BOPP/LDPE/BOPP and BOPP/LDPE/Alu/LDPE.

The advantages of gas packaging are: (a) extended shelf life – controls oxidation reactions by replacing the oxygen with nitrogen; (b) prevents crushing of soft products (such as potato chips); and (c) retains moisture – prevents the product from drying out by containing the moisture in the package. However, water can move from the center of the product to the surface, resulting in staling (as is the case for bread).

Disadvantages of gas packaging include a) capital cost of gas packaging equipment, b) cost of gases and packaging materials and c) increased pack volume increases transport costs and retail display space.

11.5.3. *Controlled atmosphere packaging*

Controlled atmosphere packaging (CAP) and controlled atmosphere storage (CAS) are characterized by the proportion of each gas used and are controlled at the original level throughout the distribution chain. The conditions (temperature and relative humidity) adopted for storage of fresh products (fruits and vegetables) have a critical influence on the choice of gas composition and play an important role in the preservation period. CAP and CAS techniques are primarily used for long-term bulk storage and transportation of food (pears, sweet cherries, apples, cabbage, etc.) of irregular shapes, or as master packs for retail-ready products (Yang and Wang 2014). This technique requires constant monitoring and control of the gas composition (and/or other parameters such as relative humidity and temperature) within the system (Figure 11.24).

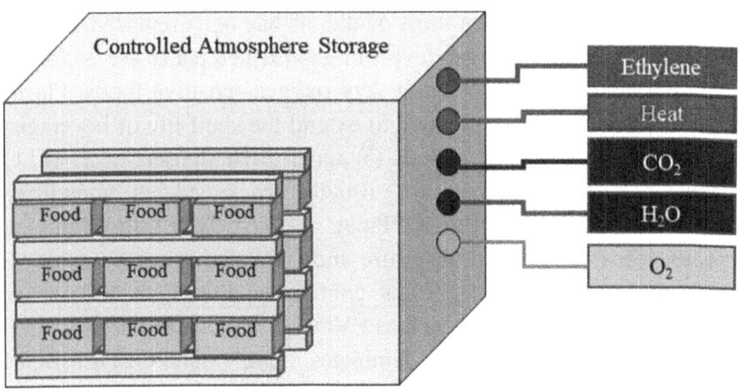

Figure 11.24. *Controlled atmosphere storage elements*

11.5.4. *Modified atmosphere packaging*

The modified atmosphere packaging (MAP) method is the replacement of air (N_2 – 78%, O_2 – 21%, CO_2 – 0.05%, etc.) in a pack, with a single gas or mixture of gases (Figure 11.25).

The proportion of each component is fixed when the mixture is introduced, and no further control is applied. Thus, the gas composition changes with time due to the diffusion of gases into and out of the product, the permeation of gases into and out of the pack, and the effects of food and microbial metabolism.

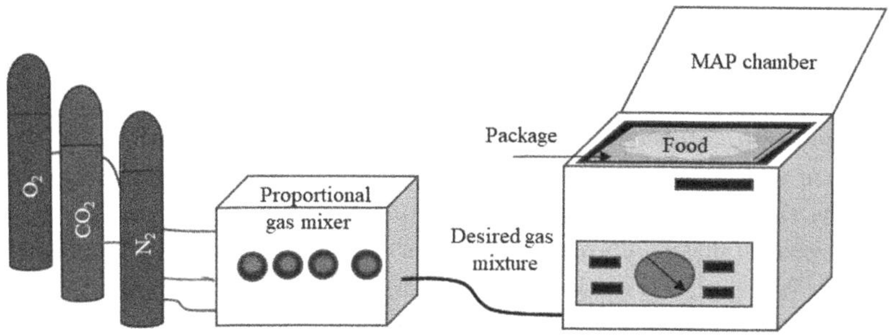

Figure 11.25. *Modified atmosphere packaging elements*

Three gases are generally used: oxygen (O_2), nitrogen (N_2) and carbon dioxide (CO_2), where a proper combination needs to be used for a particular food product's packaging (Table 11.3). Each of these gases possesses specific functions which should be taken into account.

Food	Gas content (% v/v)		
	CO_2	N_2	O_2
Red meat	15–40	-	60–85
Poultry	75	25	-
Game	30	50	20
Cooked meats	30	70	-
Processed meats	-	100	-
Apples	1–3	95–98	1–2
Lemons	0–10	80–95	5–10
Strawberries	15–20	70–80	5–10
Tomatoes	0	95–97	3–5
Cauliflower	2–5	90–96	2–5
Carrots	3–4	91–95	5
Sliced bread	100	-	-
Rye bread	100	-	-
Fresh pasta	50	50	-

Table 11.3. *MAP gas mixture for some food items (Smith and Ramaswamy 1996; Smiddy* et al. *2002, 2004)*

Oxygen exclusion is generally desirable due to the spoilage of oxygen-sensitive food. However, oxygen presence is essential for packaging fresh fruit and vegetables, as these continue to respire (i.e. consume O_2 and produce CO_2) post-harvest, and in the absence of O_2 anaerobic respiration occurs, the effect of which is to accelerate spoilage reactions. Thus, fresh produce packages are designed with perforations that enhance the removal rate of respiration heat from the packaged products, either in plastic, corrugated board or package stacking on pallets (Hussein *et al.* 2015; Wu *et al.* 2019; Mukama *et al.* 2020). O_2 is also desirable in red meat packaging as it maintains the coloring pigment myoglobin in the bright red, oxygenated form (oxymyoglobin) that consumers prefer. It is thus important to control O_2 levels within the packages. By reducing the concentration of O_2, the growth of aerobic bacteria and molds is retarded. An oxygen level of less than 1% is needed to inhibit the growth of molds in products with low water activity, such as bread. Oxidation and enzymatic browning are also restricted with the exclusion of O_2.

Carbon dioxide is responsible for the main bacteriostatic effects of the modified atmosphere (MA), by extending the lag phase of bacterial growth and decreasing the growth rate (DeWitt and Oliveira 2016). The effectiveness of CO_2 depends on the organisms present. A CO_2 concentration of 10% is necessary to inhibit molds. Gram-negative organisms are strongly influenced by higher CO_2 concentrations, while LAB growth is minimally influenced. The inhibitory effect of CO_2 increases linearly with concentrations up to 50–60%, above which there is little or no further effect on the majority of microorganisms. An amount greater than that, which will dissolve into the product, is required for optimum effect. Therefore, the package volume and the permeability and surface area of the packaging material are important parameters to be considered. The effect of CO_2 increases as temperature decreases, due to higher CO_2 solubility at lower temperatures. When dissolved, some of the gas is converted into carbonic acid (H_2CO_3), which can cause an unpleasant acidic taste when high concentrations of CO_2 are used. The absorption of the gas by the product (especially high moisture foods such as meat and seafood) causes some reduction in gas volume and consequent pack collapse. However, packaging in high CO_2 concentrations can cause fluid release in ham, a taint in fatty fish, product separation in cream, increased drip in fresh meat and physiological damage to fruit and vegetables.

Nitrogen is an inert, tasteless gas, which displays no antimicrobial activity. It is used as a cushion, balance or filter gas in order to replace O_2, either as an alternative to vacuum packaging when the product is fragile, or to limit pack collapse caused by the absorption of CO_2. By displacing O_2 in the pack, N_2 can delay oxidative rancidity and reduce the growth of aerobic microorganisms.

Carbon monoxide in concentrations as low as 1% can inhibit the growth of many bacteria, yeasts and molds. It can also delay the oxidative browning of fruits and vegetables when combined with low O_2 concentrations (2–5%). By adding small amounts of CO to red meat packaging, products such as minced beef can maintain their natural appearance throughout their shelf life, since CO reacts with the meat pigment myoglobin to create carboxymyoglobin, a bright red pigment that masks the natural aging of meat. CO has limited commercial usage, due to its toxicity and explosive nature at 12.5–74.2% in air; thus, it must be handled with special precautions (Brandenburg 2009).

Other gases that have been considered for their potential use in MAP applications include SO_2, nitrous oxide (N_2O), nitric oxide (NO), ozone (O_3), He, H_2, Ne, Ar, propylene oxide, ethylene and Cl_2. However, the use of these gases is limited by safety concerns, legislation, cost and negative effects on the organoleptic properties of packaged products.

For gases to be used as a food additive, for example as a packaging gas, they must be approved under EU law and be known by their E Numbers (e.g. E941 for nitrogen; E290 for CO_2; E948 for O_2, etc.). The EU also sets a minimum purity criterion for gases, when they are used as a food additive (FSAI 2011).

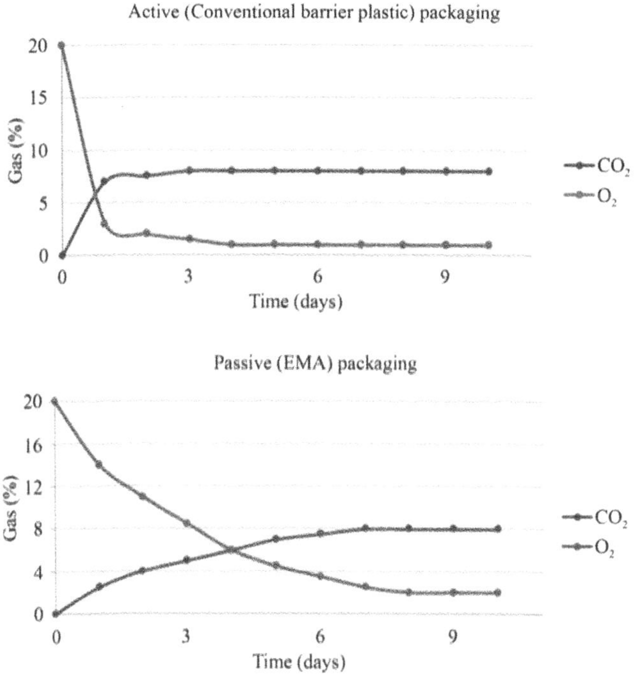

Figure 11.26. *Gas change in active and passive packaging with time*

Equilibrium-modified atmosphere (EMA) packaging is mainly used in fresh fruit and vegetable packaging. In this case, the pack is either flushed with the required gas mix (active packaging), or the produce is packaged with no atmosphere modification, which is known as passive packaging (Figure 11.26). Successive product respiration and gas permeability of the package allow an EMA (passive packaging) to be reached.

From all of the gasses mentioned above, it is obvious that the packaging material selection needs to take into account different factors, in order to obtain the desired shelf life of the product. These include general (Table 11.4) and different barrier characteristics (Table 11.5).

Properties	Materials	Abbreviation
Opacity	Aluminum foil	Alu
	Paper	PAP
	Pigmented plastic films	
Transparency	Regenerated cellulose	RC
	Poly(ethylene terephthalate)	PET
	Most plastic films	
Heat stability	Aluminum foil	Alu
	Polyester	
	High-density polyethylene	HDPE; PE-HD
	Polypropylene	PP
Low temperature stability	Aluminum foil	Alu
	High-density polyethylene	HDPE; PE-HD
	Oriented polypropylene	OPP
	Low-density polyethylene	LDPE; PE-LD
	Ethylene/vinyl acetate	EVAC (EVA)
	Polystyrene	PS
	Polyamide	PA
Heat-sealability	Low-density polyethylene	LDPE; PE-LD
	Ionomer	IO
	Poly(vinylidene chloride)	PVDC
	Polyamide	PA
	High-density polyethylene	HDPE; PE-HD

	Polypropylene	PP
	PVDC-coated oriented polypropylene	OPP/PVDC
Printability	Paper	PAP
	Regenerated cellulose	RC
	Polycarbonate	PC
	Ionomer	IO

Table 11.4. *General properties of packaging materials*

Barrier against	Packaging material	Abbreviation
O_2	Ethylene/vinyl alcohol Poly(vinylidene chloride) Poly(acrylonitrile) Metallized aluminum Laminates with silicon oxide coating Some types of cellophane Polyamide Crystalline polyethylene terephthalate	EVAL PVDC PAN Alu-met SiOx PA PET-C
CO_2	Ethylene/vinyl alcohol Crystalline polyethylene terephthalate Polyamide Laminates with silicon oxide coating	EVAL PET-C PA SiOx
Water vapor	High-density polyethylene Oriented polypropylene Poly(vinylidene chloride) Polyamide Metallized aluminum Laminates with silicon oxide coating	PE-HD; HDPE OPP PVDC PA Alu-met SiOx
Ultraviolet light	Poly(ethylene terephthalate with nanoparticles (silicon and titanium oxide), etc.	PET+SiOx PET+TiO_2
Smoke	Laminates with polyamide	

Table 11.5. *Barrier characteristics of food packaging materials*

Despite many advantages, the high cost of MAP products (Table 11.6) limits their application; thus, the incorporation of active packaging (see Chapter 8) seems to be a promising solution to overcome these problems (Boz *et al.* 2018).

Modified atmosphere packaging	
Advantages	**Disadvantages**
– Shelf life can be increased by 50–400% – Reduced economic losses due to longer shelf life – Maintains high quality product – Centralized packaging and portion control – Improved presentation – clear view of product – Little or no need for chemical preservatives – Sealed packages are barriers against product recontamination – Odorless and convenient packages	– Added costs for gases, packaging materials and machinery – Temperature control is necessary – Different gas formulations for each product type are needed – Special equipment and training are required – Increased pack volume adversely affects transport costs and retail display space – Benefit loss once the pack is opened or leaks – CO_2 dissolving into the food could lead to pack collapse, increased drip and could increase food acidity

Table 11.6. *Advantages and disadvantages of MAP*

11.6. List of abbreviations

Abbreviations	Description
Alu	Aluminum Foil
BOPP	Biaxially Oriented Polypropylene
BOPPmet	Metallized Biaxially Oriented Polypropylene
CAP	Controlled Atmosphere Packaging
CAS	Controlled Atmosphere Storage
EMA	Equilibrium-Modified Atmosphere
EU	European Union
EVAC (EVA)	Ethylene Vinyl Acetate
EVOH (EVAL)	Ethylene Vinyl Alcohol
FFS	Form-Fill-Seal
HDPE (PE-HD)	High-Density Polyethylene

HPP	High Hydrostatic Pressure
LAB	Lactic Acid Bacteria
LDPE (PE-LD)	Low-Density Polyethylene
LLDPE	Linear Low-Density Polyethylene
MAP	Modified Atmosphere Packaging
OPP	Oriented Polypropylene
PC	Polycarbonate
PEF	Pulsed Electric Fields
PEN	Polyethylene Naphthalate
PET	Polyethylene Terephthalate
PETmet	Metallized Polyethylene Terephthalate
PP	Polypropylene
PS	Polystyrene
PTFE	Polytetrafluoroethylene
PVC	Polyvinyl Chloride
PVDC	Polyvinylidene Chloride
PVOH (PVAL)	Polyvinyl Alcohol
RF	Radiofrequency
ROP	Reduced Oxygen Packaging
TFS	Tin-Free Steel
UHT	Ultra-Heat Treatment
VDC	Vinylidene Chloride
VP	Vacuum Packaging
VSP	Vacuum Skin Packaging

11.7. References

Abdul Ghani Al-Baali, A.G. and Farid, M.M. (2006). *Sterilization of Food in Retort Pouches*. Springer Science+Business Media, Berlin.

Aboud, S.A., Altemimi, A.B., Al-Hilphy, A.R.S., Lee Y.-C., and Cacciola, F. (2019). A comprehensive review on infrared heating applications in food processing. *Molecules*, 24, 4125.

AFDO (2011). A guide to can defects and basic components of double seam containers. Association of Food and Drug Officials [Online]. Available: http://www.afdo.org/resources/Documents/Committee%20Reports%202016-2017/Guide%20to%20Can%20Defects%20and%20Double%20Seam%20Containers_Nov2011%20(FINAL-compressed).pdf [Accessed 28 April 2020].

Afzal, T.M. and Abe, T. (1999). Energy and quality aspects of combined FIR convection drying of barley. *J. Food Eng.*, 42, 177–182.

Aghajanzadeh, S., Kashaninejad, M., and Ziaiifar, A.M. (2016). Effect of infrared heating on degradation kinetics of key lime juice physicochemical properties. *Innov. Food Sci. Emerg. Technol.*, 38, 139–148.

ALL4PACK (2016). Packaging: Market and challenges in 2016. The global marketplace for packaging-processing-printing-handling [Online]. Available: https://docplayer.net/44626796-Packaging-market-and-challenges-in-2016.html [Accessed 6 April 2020].

Altemimi, A., Aziz, S.N., Al-Hilphy, A.R.S., Lakhssassi, N., Watson, D.G., and Ibrahim, S.A. (2019). Critical review of radio-frequency (RF) heating applications in food processing. *Food Qual. Safe.*, 3, 81–91.

Arvanitoyannis, I.S. and Kotsanopoulos, K. (2012). Safety and quality control of modified atmosphere packaging products. In *Modified Atmosphere and Active Packaging Technologies*, Arvanitoyannis, I.S. (ed.). CRC Press, Boca Raton.

Ayhan, Z., Yeom, H.W., Zhang, H., and Min, D.B. (2001). Flavor, color, vitamin C retention of pulsed electric field processed orange juice in different packaging materials. *J. Agr. Food Chem.*, 49, 669–674.

Baldwin, D.E. (2012). Sous vide cooking: A review. *Int. J. Gastron. Food Sci.*, 1, 15–30.

Bello, E.F.T., González Martínez, G., Klotz Ceberio, B.F., Rodrigo, D., and Martínez López, A. (2014). High pressure treatment in foods. *Foods*, 3, 476–490.

Bhavya, M.L. and Umesh Hebbar, H. (2017). Pulsed light processing of foods for microbial safety. *Food Qual. Safe.*, 1(3), 187–201.

Boz, Z., Welt, B.A., Brecht, J.K., Pelletier, W., McLamore, E., Kiker, G.A., and Butler, J.E. (2018). Review of challenges and advances in modification of food package headspace gases. *J. Appl. Packag. Res.*, 10(1), 5.

Brandenburg, J. (2009). Modified atmosphere packaging. In *The Wiley Encyclopedia of Packaging Technology*, Yam K.L. (ed.). John Wiley & Sons, New York.

Butz, P. and Tauscher, B. (2002). Emerging technologies: Chemical aspects. *Food Res. Int.*, 35, 279–284.

Caner, C., Hernandez, R.J., and Harte, B.R. (2004). High-pressure processing effects on the mechanical, barrier and mass transfer properties of food packaging flexible structures: A critical review. *Packag. Technol. Sci.*, 17, 23–29.

Cathcart, W.H., Parker, J.J., and Beattie, H.G. (1947). The treatment of packaged bread with high frequency heat. *Food Technol.*, 1, 174–177.

Dagerskog, M. and Österström, L. (1979). Infrared radiation for food processing I: A study of the fundamental properties of infrared radiation. *LWT Food Sci. Technol.*, 12, 237–242.

Dantas, F.B.H. and Dantas, S.T. (2017). *Canned Food and Packaging Parameters. Reference Module in Food Sciences* [Online]. Available: https://doi.org/10.1016/B978-0-08-100596-5.03205-4.

DeWitt, C.A.M. and Oliveira, A.C.M. (2016). Modified atmosphere systems and shelf life extension of fish and fishery products. *Foods*, 5, 48.

Dobiaš, J., Voldřich, M., Marek, M., and Chudáčková, K. (2004). Changes of properties of polymer packaging films during high pressure treatment. *J. Food Eng.*, 61(4), 545–549.

Ebnesajjad, S. (2013). *Plastic Films in Food Packaging: Materials, Technology, and Applications*. Elsevier, New York.

Elmnasser, N., Guillou, S., Leroi, F., Orange, N., Bakhrouf, A., and Federighi, M. (2007). Pulsed–light system as a novel food decontamination technology: A review. *Can. J. Microbiol.*, 53, 813–821.

European Commission (2019). Report from the Commission to the European Parliament and Council on food and food ingredients treated with ionizing radiation for the years 2016–2017. COM/2019/454 final.

Eustice, R.F. (2018). Global status and commercial applications of food irradiation. In *Food Irradiation Technologies: Concepts, Applications and Outcomes*, Ferreira, I.C.F.R., Antonio, A.L., and Verde, S.C. (eds). The Royal Society of Chemistry, London.

Fraldi, M., Cutolo, A., Esposito, L., Perrella, G., Pastore Carbone, M.G., Sansone, L., Scherillo, G., and Mensitieri, G. (2014). Delamination onset and design criteria of multilayer flexible packaging under high pressure treatments. *Innov. Food Sci. Emerg. Technol.*, 23, 39–53.

FSAI (Food Safety Authority of Ireland) (2011). Guidance on food additives, Revision 1 [Online]. Available: http://www.qsd.ie/wp-content/uploads/2013/06/food-additives-2011-final1.pdf [Accessed 7 February 2020].

Gabrić, D., Barba, F., Roohinejad, S., Gharibzahedi, S.M.T., Radojčin, M., Putnik, P., and Bursać Kovačević, D. (2017). Pulsed electric fields as an alternative to thermal processing for preservation of nutritive and physicochemical properties of beverages: A review. *J. Food Process. Eng.*, e12638, 1–14.

Galić, K., Ščetar, M., and Kurek, M. (2011). The benefits of processing and packaging – Review. *Trends Food Sci Tech.*, 22(2–3), 127–137.

Galić, K., Gabrić, D., and Ćurić, D. (2019). Packaging and the Shelf Life of Bread. *Reference Module in Food Science*, 1–8 [Online]. Available: doi:http:// dx.doi.org/ 10.1016/B978-0-08-100596-5.22655-3.

Galotto, M.J., Ulloa, P.A., Hernandez, D., Fernandez-Martin, F., Gavara, R., and Guarda, A. (2008). Mechanical and thermal behaviour of flexible food packaging polymeric films materials under high pressure/temperature treatments. *Packag. Technol. Sci.*, 21(5), 297–308.

Gande, N. and Murina, P. (2003). Prepackage surface pasteurization of ready-to-eat meats with a radiant heat oven for reduction of *Listeria monocytogenes*. *J. Food Prot.*, 66, 1623–1630.

Gertzou, I.N., Karabagias, I.K., Drosos, P.E., and Riganakos, K.A. (2017). Effect of combination of ozonation and vacuum packaging on shelf life extension of fresh chicken legs during storage under refrigeration. *J. Food Eng.*, 213, 18–26.

Gómez-López, V.M., Ragaert, P., Debevere, J., and Devlieghere, F. (2007). Pulsed light for food decontamination: A review. *Trends Food Sci. Technol.*, 18, 464–473.

Götz, A., Wani, A.A., Langowski, H.-C., and Wunderlich, J. (2014). Aseptic packaging. *Encyclopedia of Food Safety*, 3, 124–134.

Goyal, A., Sharma, V., Upadhyay, N., Sihag, M., and Kaushik, R. (2013). High pressure processing and its impact on milk proteins: A review. *RRJoDST.*, 2(1), 12–20.

Guillard, V., Mauricio-Iglesias, M., and Gontard, N. (2010). Effect of novel food processing methods on packaging: Structure, composition, and migration properties. *Crit. Rev. Food Sci. Nutr.*, 50(10), 969–988.

Haji-Saeid, M., Sampa, M.H.O., and Chmielewski, A.G. (2007). Radiation treatment for sterilization of packaging materials. *Radiat. Phys. Chem.*, 76(8–9), 1535–1541.

Haughton, P.N., Lyng, J.G., Cronin, D.A., Morgan, D.J., Fanning, S., and Whyte, P. (2010). Efficacy of UV light treatment for the microbiological decontamination of chicken associated packaging, and contact surfaces. *J. Food Prot.*, 74, 565–572.

Hill, A. (1998). *Microwave Ovens*. ILSI Europe, Belgium.

Holdsworth, D. and Simpson, R. (2007). *Thermal Processing of Packaged Foods*, 2nd edition. Springer Science+Business Media, Berlin.

Hotchkiss, J.H. (2009). Canning, food. In *The Wiley Encyclopedia of Packaging Technology*, 3rd edition, Yam, K.L. (ed.). John Wiley & Sons, New York.

Houben, J., Schoenmakers, L., van Putten, E., van Roon, P., and Krol, B. (1991). Radio-frequency pasteurization of sausage emulsions as a continuous process. *J. Microw. Power Electromagn. Energy*, 26(4), 202–205.

Huang, L. and Sites, J. (2008). Elimination of *Listeria monocytogenes* on hotdogs by infrared surface treatment. *J Food Sci.*, 73(1), M27–M31.

Hussein, Z., Caleb, O.J., and Opara, U.L. (2015). Perforation-mediated modified atmosphere packaging of fresh and minimally processed produce – A review. *Food Packag. Shelf Life*, 6, 7–20.

Jensen, S., Oestdal, H., Clausen, M.R., Andersen, M.L., and Skibsted L.H. (2011). Oxidative stability of whole wheat bread during storage. *LWT Food Sci. Technol.*, 44, 637–642.

Jun, S. and Sastry, S. (2005). Modeling and optimization of ohmic heating of foods inside a flexible package. *J. Food Process Eng.*, 28, 417–436.

Jun, S. and Sastry, S. (2007). Reusable pouch development for long term space missions: A 3D ohmic model for verification of sterilization efficacy. *J. Food Eng.*, 80, 1199–1205.

Kanogchaipramot, K., Tongkhao, K., Sajjaanantakul, T., and Kamonpatana, P. (2016). Ohmic heating of an electrically conductive food package. *J. Food Sci.*, 81(12), E2966–E2976.

Kirwan, M.J., Plant, S., and Strawbridge, J.W. (2011). Plastics in food packaging. In *Food and Beverage Packaging Technology*, Coles, R. and Kirwan, M. (eds). Wiley-Blackwell, Hoboken.

Knorr, D. and Watzke, H. (2019). Food processing at a crossroad – Review. *Front. Nutr.*, 6, 85.

Koutchma, T., Keener, L., and Kotilainen, H. (2018). Global Harmonization Initiative (GHI) consensus document on food irradiation, pp. 1–18 [Online]. Available: https://www.globalharmonization.net/sites/default/files/pdf/GHI-Food-Irradiation_October-2018.pdf [Accessed 15 January 2020].

Krishnamurthy, K., Khurana, H.K., Jun, S., Irudayaraj, J., and Demirci, A. (2008). Infrared heating in food processing: An overview. *Compr. Rev. Food Sci. Food Saf.*, 7, 2–13.

Kumar, P. and Han, J.H. (2012). Packaging materials for non-thermal processing of food and beverages. In *Emerging Food Packaging Technologies, Principles and Practice*, Yam, K.L. and Dong Sun, L. (eds). Woodhead Publishing, Cambridge.

Kume, T. and Todoriki, S. (2013). Food irradiation in Asia, the European Union and the United States: A status update. *Radioisotopes*, 62, 291–299.

Lagerstedt, Å., Ahnström, M.L., and Lundström K. (2011). Vacuum skin pack of beef – A consumer friendly alternative. *Meat Sci.*, 88, 391–396.

Lange, J. and Wyser, Y. (2003). Recent innovations in barrier technologies for plastic packaging – A review. *Packag. Technol. Sci.*, 16, 149–158.

Lee, S.H., Choi, W., and Jun, S. (2016). Conventional and emerging combination technologies for food processing – Review. *Food. Eng. Rev.*, 8, 414–434.

Li, X., Lindahl, G., Zamaratskaia, G., and Lundström, K. (2012). Influence of vacuum skin packaging on color stability of beef longissimus lumborum compared with vacuum and high-oxygen modified atmosphere packaging. *Meat Sci.*, 92, 604–609.

Liao, Y., Zhang, R., and Qian, J. (2019). Printed electronics based on inorganic conductive nanomaterials and their applications in intelligent food packaging. *RSC Adv.*, 9, 29154–29172.

Maged, E.A.M. and Ayman, H.A.E. (2012). Pulsed electric fields for food processing technology [Online]. Available: http://www.intechopen.com/download/get/type/pdfs/id/38363 [Accessed 16 January 2020].

Mahalik, N.P. (2014). Advances in packaging methods, processes and systems. *Challenges*, 5, 374–389.

Majumdar, R.K., Deepayan, R., and Apurba, S. (2017). Textural and sensory characteristics of retort-processed freshwater prawn (*Macrobrachium rosenbergii*) in curry medium. *Int. J. Food Prop.*, 20(11), 2487–2498.

Marangoni Júnior, L. and Anjos C.A.R. (2019). Effect of high-pressure processing on characteristics of flexible packaging for foods and beverages – Review. *Food Res. Int.*, 119, 920–930.

Mazzola, N. and Sarantopoulos, C.I.G.L. (2019). Packaging design alternatives for meat products [Online]. Available: https://www.intechopen.com/online-first/packaging-design-alternatives-for-meat-products [Accessed 3 March 2020].

Mizani, M., Sheikh, N., Ebrahimi, S.N., Gerami, A., and Tavakoli, F.A. (2009). Effect of gamma irradiation on physico-mechanical properties of spice packaging films. *Radiat. Phys. Chem.*, 78(9), 806–809.

Mudgal, V.D. (2019). Ohmic heating: An alternative technology for spices processing. *Int. J. Seed Spices*, 9(1), 44–47.

Mukama, M., Ambaw, A., and Opara, U.L. (2020). Advances in design and performance evaluation of fresh fruit ventilated distribution packaging: A review. *Food Packag. Shelf Life*, 24, 100472.

Mullan, M. and McDowell, D. (2003). Modified atmosphere packaging. In *Food Packaging Technology,* Coles R., McDowell, D., and Kirwan, M.J. (eds). Wiley-Blackwell, Hoboken.

Murugesan, R., Orsat, V., and Lefsrud, M. (2012). Effect of pulsed ultraviolet light on the total phenol content of elderberry (*Sambucus nigra*) fruit. *Food Nutr. Sci.*, 3, 774–783.

Neetoo, H. (2015). Innovative processing for pre-packaged foods. *Int. J. Inno. Sci. Res.*, 16(1), 179–194.

Nowak, D. and Lewicki, P.P. (2004). Infrared drying of apple slices. *Innov. Food Sci. Emerg. Technol.*, 5, 353–360.

Ozen, B.F. and Floros, J.D. (2001). Effects of emerging food processing techniques on the packaging materials. *Trends Food Sci. Technol.*, 12, 60–67.

Pan, Z., Venkitasamy, C., and Li, X. (2016). Infrared processing of foods. Reference module in food sciences [Online]. Available: http://dx.doi.org/10.1016/B978-0-08-100596-5.03105-X.

Park, S.H., Lamsal, B.P., and Balasubramaniam, V.M. (2014). Principles of food processing. In *Food Processing: Principles and Applications*, 2nd edition, Clark S., Jung S., and Lamsal, B. (eds) John Wiley & Sons, New York.

Peeling, J. and Clark, D.T. (1983). Surface ozonation and photooxidation of polyethylene film. *J. Polym. Sci. Polym. Chem. Ed.*, 21, 2047–2055.

Piyasena, P., Dussault, C., Koutchma, T., Ramaswamy, H.S., and Awuah, G.B. (2003). Radio frequency heating of foods: Principles, applications and related properties – A review. *Crit. Rev. Food Sci. Nutr.*, 43(6), 587–606.

Rastogi, N.K. (2012). Recent trends and developments in infrared heating in food processing. *Crit. Rev. Food Sci. Nutr.*, 52(9), 737–760.

Rastogi, N.K., Raghavarao, K.S.M.S., Balasubramaniam, V.M., Niranjan, K., and Knorr, D. (2007). Opportunities and challenges in high pressure processing of foods. *Crit. Rev. Food Sci. Nutr.*, 47(1), 69–112.

Regattieri, A. and Santarelli, G. (2015). The important role of packaging in operations management [Online]. Available: https://www.intechopen.com/books/operations-management/the-important-role-of-packaging-in-operations-management [Accessed 28 April 2020].

Rivas-Cańedo, A., Fernández-García, E., and Nuńez, M. (2009). Volatile compounds in fresh meats subjected to high pressure processing: Effect of the packaging material. *Meat Sci.*, 81(2), 321–328.

Robertson, G.L. (2016). Ultrahigh temperature treatment (UHT): Aseptic packaging. reference module in food sciences [Online]. Available: http://dx.doi.org/10.1016/B978-0-08-100596-5.00810-6.

Ščetar, M. and Galić, K. (2018). A perspective on packaging of fruits and vegetables. In *Novel Postharvest Treatments of Fresh Produce*, Pareek, S. (ed.). CRC Press, London.

Ščetar, M., Kurek, M., and Galić, K. (2010). Trends in meat and meat products packaging – A review. *Croat. J. Food Sci. Technol.*, 2(1), 32–48.

Schauwecker, A., Balasubramaniam, V.M., Sadler, G., Pascall, M.A., and Adhikari, C. (2002). Influence of high-pressure processing on selected polymeric materials and on the migration of a pressure-transmitting fluid. *Packag. Technol. Sci.*, 15(5), 255–262.

Schiffmann, R. (2017). Packaging for microwave foods. In *The Microwave Processing of Foods*, Regjer, M., Knoerzer, K., and Schubert, H (eds). Woodhead Publishing, Cambridge.

Schmidt, T. and Leiking, L. (2005). Reliability of evaluations for the choice of system solutions at the example of automated order-picking systems for bagged goods [Online]. Available: https://www.logistics-journal.de/archive/2005/723/beutel_e.pdf [Accessed 24 April 2020].

Smiddy, M., Fitzgerald, M., Kerry, J.P., Papkovsky, D.B., O'Sullivan, C.K., and Guilbault, G.G. (2002). Use of oxygen sensors to non-destructively measure the oxygen content in modified atmosphere and vacuum packed beef: Impact of oxygen content on lipid oxidation. *Meat Sci.*, 61, 285–290.

Smith, J.P. and Ramaswamy H.S. (1996). Packaging of fruits and vegetables. In *Processing Fruits: Science and Technology*, Somogyi, L., Ramaswamy, H.S., and Hui, Y.H. (eds). Technomic Publishing, Lancaster, PA.

Smith, J.P., Daifas, D.P., El-Khoury, W., and Koukoutsis, J. (2004). Shelf life and safety concerns of bakery products – A review. *Crit. Rev. Food Sci. Nutr.*, 44, 19–55.

Srivastava, A., Kumar, V., and Kumar Singh, A. (2018). Computerized and electronic controls in food packaging [Online]. Available: https://scholarworks.rit.edu/cgi/viewcontent.cgi?article=1106&context=japr [Accessed 24 April 2020].

Struller, C.F., Kelly, P.J., and Copeland, N.J. (2019). Conversion of aluminium oxide coated films for food packaging applications – From a single layer material to a complete pouch. *Food Packag. Shelf Life*, 20, 100309.

Transparency Marketing Research (2014). UHT milk market – Global industry analysis, size, share, growth, trends and forecast 2013–2019 [Online]. Available: http://www.transparencymarketresearch.com/uht-milk-market.html [Accessed 4 June 2019].

Tucker, G.S. (2011). Food biodeterioration and methods of preservation. In *Food and Beverage Packaging Technology*, Coles, R. and Kirwan, M. (eds). Wiley-Blackwell, Hoboken.

Van Rooyen, L.A., Allen, P., Kelly-Rees, C., and O'Connor, D.I. (2018). The effects of varying gas concentrations and exposure times on colour stability and shelf-life of vacuum packaged beef steaks subjected to carbon monoxide pretreatment. *Food Packag. Shelf Life*, 18, 230–237.

Vergara-Balderas, F.T. (2016). Canning: Process of canning. Encyclopedia of food and health [Online]. Available: http://dx.doi.org/10.1016/B978-0-12-384947-2.00110-0.

Wagner, J.R. (2010). *Multilayer Flexible Packaging. Technology and Appliccations for the Food, Personal Care and Over-the-Counter Pharmaceutical Industries*. Elsevier, New York.

Wang, J., Luechapattanaporn, K., Wang, Y., and Tang, J. (2012). Radio-frequency heating of heterogeneous food – Meat lasagna. *J. Food Eng.*, 108, 183–193.

Wu, W., Cronjé, P., Verboven, P., and Defraeye, T. (2019). Unveiling how ventilated packaging design and cold chain scenarios affect the cooling kinetics and fruit quality for each single citrus fruit in an entire pallet. *Food Packag. Shelf Life*, 21, 100369.

Yam, K.L. (2009). Microwaveable foods packaging. In *The Wiley Encyclopedia of Packaging Technology*, 3rd edition, Yam, K.L. (ed.). John Wiley & Sons, New York.

Yang, X. and Wang, H. (2014). Controlled atmosphere. In *Encyclopedia of Food Microbiology*, 2nd edition, Batt, C.A. and Tortorello, M.C. (eds). Academic Press, London.

Zhao, Y., Flugstad, B., Kolbe, E., Park, J.W., and Wells, J.H. (2000). Using capacitive (radio frequency) dielectric heating in food processing and preservation – A review. *J. Food Process Eng.*, 23(1), 25–55.

Zhao, Y., Zhao, W., Yang, R., Singh Sidhu, J., and Kong, F. (2017). Radio frequency heating to inactivate microorganisms in broccoli powder. *Food Quality and Safety*, 1, 93–100.

Zhu, X., Guo, W., and Wu, X. (2012). Frequency- and temperature-dependent dielectric properties of fruit juices associated with pasteurization by dielectric heating. *J. Food Eng.*, 109, 258–266.

12

Packaging Marking and Labeling

Mia Kurek
Faculty of Food Technology and Biotechnology, University of Zagreb, Croatia

12.1. Introduction

Packaging is strongly integrated into our daily lives. Even though the main purpose of packaging is to protect the goods inside, it is also a primary link between producers and packers, on the one hand, and consumers, buyers and retailers on the other hand. Packaging can be considered "a passive" and "an active" actor in the food product lifecycle. The packaging industry does not only focus on developing an effective and economic tool that unites processing, preservation and safety instruments throughout the distribution chain, but also emphasizes the importance of its esthetic value. Packaging design and labels are considered to be strong sales tools, even ahead of concurrency. According to the latest Smithers market report, the labeling market will be worth around $41 billion by 2024 (Smithers 2019). This is why packaging has also been recognized as a silent seller and a significant communication actor (Bacarella *et al.* 2015).

Food labeling is indispensable. In the development of new products, the use of labels that are attractive in shape, color and appearance is an important contribution to a commercial success. It must be understood that marking and labeling has a dual role: to inform consumers and to ensure fair marketing. To put it simply, it is essential that labels are easy to understand and that those responsible for food package information do not take advantage of vulnerabilities.

For a color version of all figures in this book, see www.iste.co.uk/debeaufort/packaging.zip.

Packaging Materials and Processing for Food, Pharmaceuticals and Cosmetics,
coordinated by Frédéric DEBEAUFORT, Kata GALIĆ, Mia KUREK, Nasreddine BENBETTAIEB and Mario ŠČETAR.

12.2. Labeling

Labels aim to communicate with all stakeholders in the food lifecycle in any written, electronic or graphic communication (Figure 12.1).

Figure 12.1. *Types of information: a) written; b) electronic; c) graphic*

Simply, a label is a tag, wrapper, seal or in-mold printed message linked to the product. Wherever you shop, labels are in evidence. They often relate specific products to consumer interests in health, the environment and culture. Labels are produced in many shapes and sizes, often with focused and defined characteristics. They should be truthful and accurate, clearly stated and easily understandable. Sometimes this is difficult to achieve. Indeed, information is changing dramatically in the 21st century because it is considered necessary or desirable for consumers to protect their health. Labeling is developed not only to inform consumers but also to help them distinguish between various similar products on offer, and to help people select the right one depending on diet habits, allergies, personal taste or cost. It also provides information to consumers on how to store, use or prepare packed food. Labeling is now, more than ever before, a marketing tool to sell the item.

12.3. Standards

There are various types of information and symbols for package labeling that are nationally and internationally standardized. Growing consumer and industry interest in food labels presents challenges for government authorities who must ensure that the information that appears on food packages is useful, credible and presented clearly so that it does not mislead the consumer (Albert 2010). The first labeling standard, the Codex Alimentarius ("the Codex"), was established by the World Health Organization and the Food and Agriculture Organization of the United Nations (Codex Alimentarius Commission 2010). Within the same country or organization, labeling policies can be both very restrictive and very permissive in terms of the type of information that may be placed on a package. Every food label must comply with food laws and standards. As food export and import is now a routine trade, authorities often give close attention to harmonize national laws with international standards.

In order to prevent mislabeling and misleading descriptions, scientists and control laboratories coordinate control screenings. If mislabeling is done intentionally, it is treated legally as criminal fraud. In the EU, Regulation 1169/2011 (2011) provides information about fair practices. It states that "food information shall not mislead the public, particularly as to the characteristics and to the nature of the food, its shelf life and country of origin, as well as processing method".

Nowadays, labels are intended not only to inform and promote health, but also to protect the environment and promote sustainable food production and social well-being, as well as to protect culture, in relation to new technologies.

12.3.1. *Food labeling requirements*

There are general labeling requirements applicable to all foods and a great deal of specific measures are undertaken for specific foods. Regulation 1169/2011 (2011) establishes the general principles, requirements and responsibilities governing food information and, in particular, food labeling. For the labeling of pre-packed foods, EU Regulation (2011) sets out a list of mandatory particulars that must be provided to the end consumer:

1) Name of the food.
2) List of ingredients.
3) Potential allergens.
4) Quantity of certain ingredients or categories of ingredients.
5) Net quantity of the food.
6) Product dates.
7) Any special storage conditions and/or conditions of use.
8) Name or business name and address.
9) Country of origin or place of provenance.
10) Instructions for use.
11) For alcoholic foodstuffs, actual alcoholic strength by volume.
12) Nutrition declaration.

Reading food packaging labels can help buyers make good food choices. In the following, explanations are given about the information, defined as mandatory, in most countries worldwide.

Name of the food

There are three types of names (Figure 12.2) that can appear on a label: a) the legal name; b) the brand name; and c) the descriptive or elaborate name that may be used in addition to the name of the food.

Figure 12.2. *Example of food name information on a label*

List of ingredients

In most international standards, it is mandatory to list the ingredients. It must be titled "Ingredients" which should be listed primarily in weight order at the time of manufacture (Figure 12.3). Fresh fruit, fresh vegetables and single ingredient foods, where the name of the food is identical to the ingredient name, are excluded from this obligation.

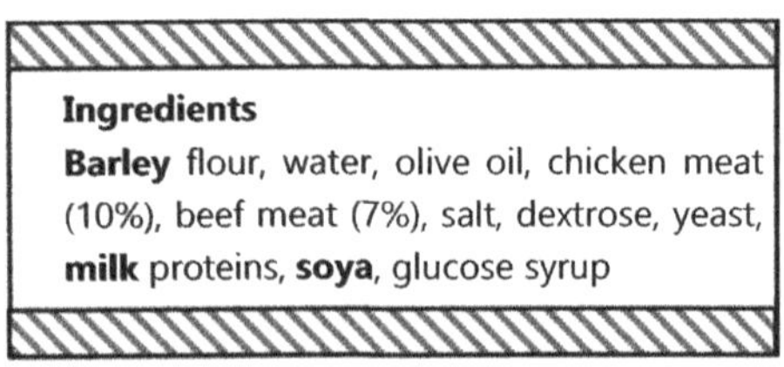

Figure 12.3. *Example of an ingredient list on a label*

Potential allergens

Adverse reactions to products due to food allergies, including potentially fatal anaphylactic shock, pose significant consumer health concerns. People need to have clear information about the ingredients and potential food allergens, or their derivatives, which are used in food recipes, in order to choose the items they can safely eat. These ingredients must be emphasized (bolded or underlined) if they are added even in small quantities (Figure 12.4). It is likely that more than 200 ingredients commonly found in food products are potential allergens, with eating patterns differing according to country. Thus, it is not surprising that the obligatory list will vary depending on the country/culture/religion. It is also possible that new allergens could be added to regulatory lists in the future, but each time this occurs, the whole process should have sufficient scientific justification.

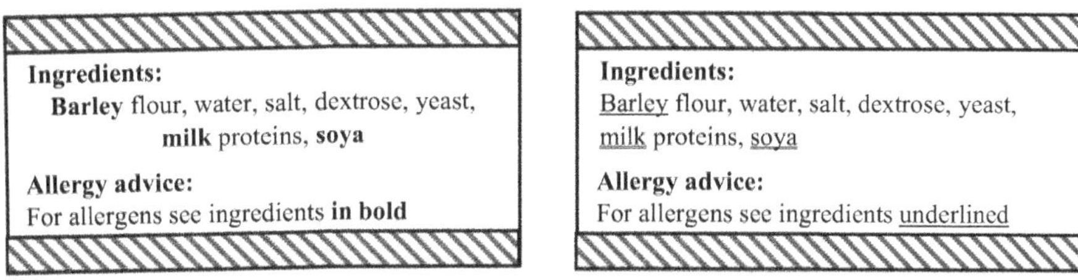

Figure 12.4. *Example of labeling of food allergens*

Quantity of certain ingredients or categories of ingredients

Marketing allows one or several ingredients on the label to be emphasized in order to categorize the food or to attract buyers' attention. The quantities of these ingredients should be given either next to the name of the food or in the ingredients list to make sure that consumers are not misled. It is often called the Quantitative Ingredient Declaration (QUID) (Figure 12.5). The QUID is mandatory if the ingredient is in the name of the food, it is emphasized on the labeling or if it is essential to describe the food in order to avoid confusing it with another product.

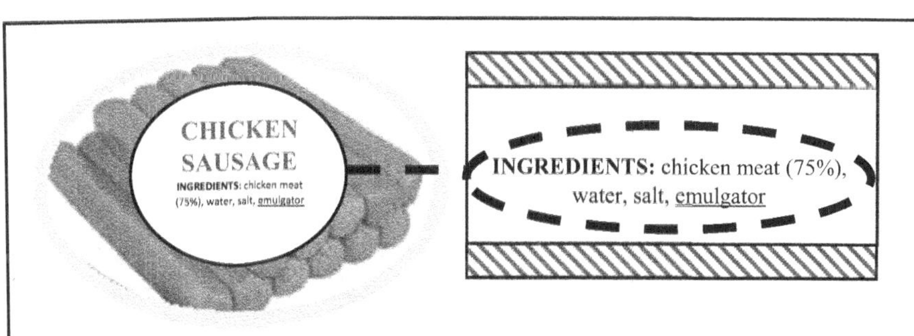

Figure 12.5. *Example of a label emphasizing the quantity of certain ingredients*

Net quantity of the food

This value gives the weight or volume of the product without packaging. Depending on the unit system in effect, it might be provided in either metric (kilograms and grams or liters, centiliters or milliliters) or imperial (pounds or ounces, gallons and pints) units.

Product dates

Dates that are actually manufacturer recommendations and that must appear on packaging labels can be classified into several categories: "use by" or "best before" markings (Figure 12.6). "Use by" informs consumers about food safety, and this date acts as a precautionary measure. Food can be eaten up until the end of this date but not after even if it looks and smells fine. It is generally used for perishable food (such as raw and cooked meat and fish, or fresh dairy products). "Best before" informs consumers about the food quality and leaves them to make their own judgments. In other words, food can be eaten after this date, but it may no longer be at its best. It is important to raise awareness among consumers about the significance of these markings as knowledge of them can diminish the risk of food poisoning, as well as prevent consumable food from being thrown away. Date marking requirements given by manufacturers vary depending on the degree of perishability, as follows:

1) for highly perishable food: day and month of expiration;

2) for food with a shelf life between 3 and 18 months: month and year of expiration;

3) for food with a shelf life longer than 18 months: year of expiration.

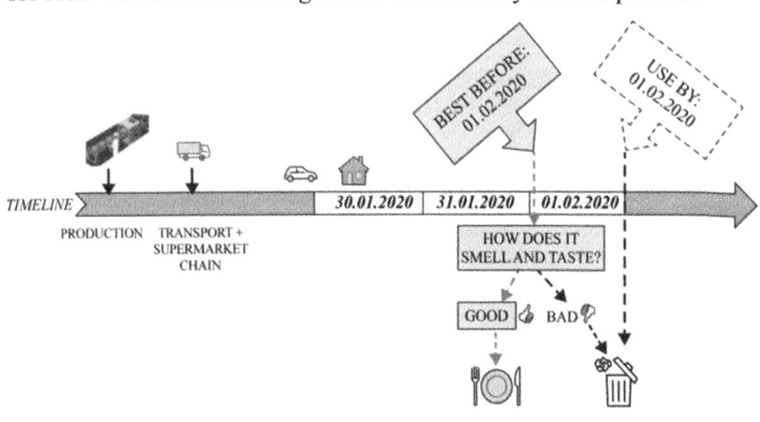

Figure 12.6. *Different date markings used for food labeling*

Any special storage conditions and/or conditions of use

In various circumstances, some products, particularly refrigerated, frozen or UV-sensitive and light perishable food, should be kept in special conditions. Thus, different instructions must be on the label (Figure 12.7). Consumers should follow written instructions.

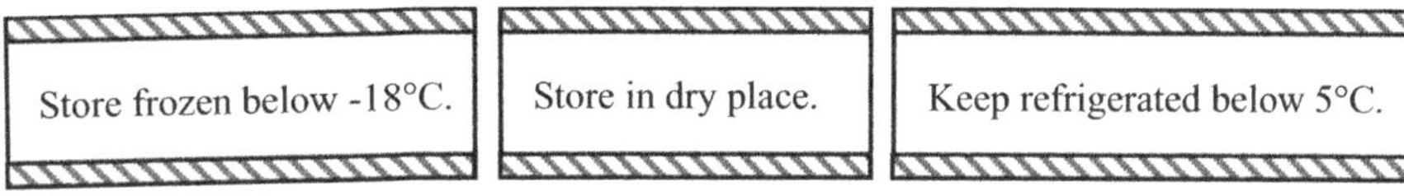

Figure 12.7. *Examples of labeling for special storage conditions*

Name or business name and address

If the consumer is not satisfied with the quality of the product or there is another problem with the item of food, then consumers should contact the food producer, whose name and address should be given on the label.

Country of origin or place of provenance

Country of origin or place of provenance on food labels states where the food is produced. It must be clearly stated.

An example: a tuna is caught in the Adriatic Sea, processed/canned in Italy and sold in Turkey. On the label there should be written: "Adriatic Sea tuna packed in Italy" or "Imported tuna packed in Italy". Labeling it as "Italian tuna" is not acceptable in this case.

Instructions for use

Here is an example of the manufacturer's instructions for preparing the food (Figure 12.8).

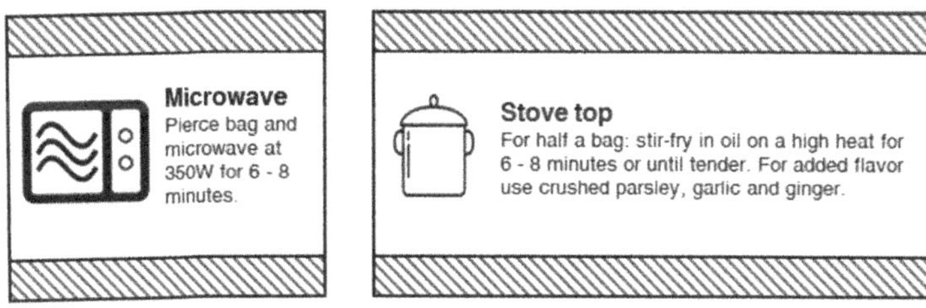

Figure 12.8. *Examples of labels with the manufacturer's instructions for preparation*

Labeling for alcoholic foodstuffs

Alcohol strength must be provided if the product contains more than 1.2% alcohol.

Nutrition declaration

Rules on nutrition and health claims are established depending on the market (EU, USA, Australia, etc.). If there is reason to make a nutrition or health claim (Figure 12.9) in relation to the packed food, then all relevant information must be provided. It is noteworthy that these claims can only be put on the label if sufficient scientific evidence exists. Health claims are defined as any statement about a relationship between food and health, while nutrition claims state, suggest or imply that food has a particular beneficial nutritional property.

Examples of health claims include those related to:

1) body function (such as growth, development or other functions) (e.g. "helps build and repair body tissues" or "vitamin C – dietary antioxidant");

2) risk reduction (e.g. "a healthy diet containing dietary fiber may reduce the risk of coronary heart disease");

3) children's development (e.g. "helps your child build immunity").

Nutrition claims might be related to:

1) energy value (e.g. "low in saturated fat" or "provides energy");

2) quantity of target nutrients (e.g. "source of iron – Nutri-Grain Iron Man Food").

Figure 12.9. *Examples of labeling with nutritional and health claims*

Voluntary labeling

There is always some information that a manufacturer or a retailer wants to put in evidence, highlighting the product on colorful store shelves. This might be the brand name, a picture of the food, recipes for the food item, coupons, the history of the company, the suitability of the food for vegetarians or vegans, etc. It is important that this does not mislead or confuse the consumer about the packed content and should be based on the relevant scientific data. As there is a large quantity of mandatory information that must be put on labels with exactly the same font size, putting additional information, that is not legally required, is quite a demanding creative challenge.

How does mandatory information need to be presented?

It is important to present mandatory food information that is: a) easily visible; b) clearly legible; and c) inedible.

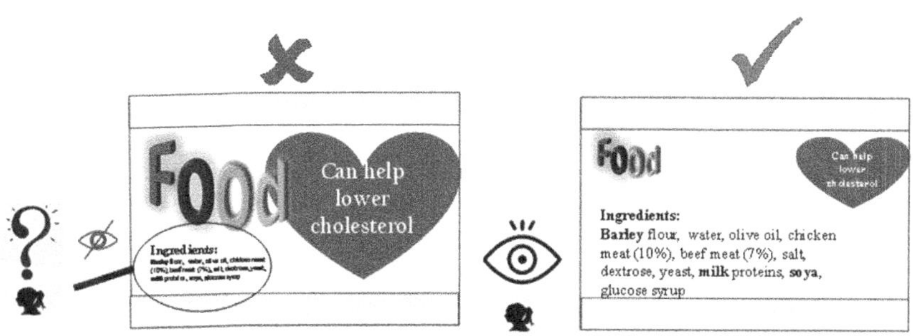

Figure 12.10. *Examples of incorrect (left) and correct (right) font size used for mandatory information on food labels*

Labels must be written using the font size that is regulated depending on the size of the package (Figure 12.10). With respect to color, good and easily visible contrasts should be used.

12.3.2. *Food packaging symbols*

Environmental and recycling symbols

Each package should have symbols explaining the type of packaging material that has been used and its disposal procedure. Packaging materials should be labeled with a triangle and number that signifies material type. On the top of the list is a food contact symbol that indicates that the material used for packing the product is considered safe for direct contact with food. There is also a long list of symbols, generally recognizable around the globe, that can appear on a food label and that describe various product characteristics such as gluten or dairy free, vegan, etc.

Due to environmental concerns related to plastic packaging disposal, great emphasis is placed on the sustainability of food/package systems. Thus, for the food to be sustainable, it is important that packaging has the lowest possible impact on the environment and can either be reused or recycled. There are numerous symbols that can be put on a package in this sense (e.g. the recycling symbol, the recycling code, the green dot).

Table 12.1. lists some of the most common symbols used to describe either food packaging material or specific properties of the food packed inside.

Symbol	Description
	Food contact material: the product is suitable for food use; may or may not have the word "food" below the wine glass and fork.
	Mobius loop – recycling: used internationally to show that the product can be recycled; the number inside the symbol indicates the percentage of recycled material used to produce the packaging.
1 PET	**Material type/recycling indicator:** the type of material used and its recycling ability.
	Tidyman: a reminder to consumers that they should dispose of their waste appropriately.
	Compostable: the registered trademark of European Bioplastics indicates that packaging is certifiably compostable in line with EN 13432:2000, 14995:2006.
FSC	**Forest Stewardship Council:** used on wood-based packaging to indicate the wood is from well-managed forests (certified according to FSC rules).
	Green dot: indicates that the producer has made contributions to a qualified national packaging recovery organization, applies to all products and does not necessarily indicate that the packaging is recyclable.
CE	**CE mark:** it indicates compliance with EN standards or EC directives.

Symbol	Description
	Biodegradable symbols: indicates that packaging can decompose under specific conditions in nature, without harmful effects.
	Vegetarian: the packed product is suitable for vegetarians.
	Ingredient free: the packed product does not contain the indicated compound.
	Suitable for freezing: the product is suitable for freezing without repacking.
	Organic (in EU): the product guarantees respect of EU rules on organic production; for processed food, at least 95% of the ingredients of agricultural origin are organic.
	Food irradiation: packed food is treated with a safe source of irradiation and is suitable for human consumption.

Table 12.1. *Some of the most common food packaging symbols used in labeling*

12.4. Material and label production

Depending on the processing, most labels can be classified as classic, in-mold or sleeve. Sometimes it is not possible to put all of the information on a single label, for example for products where a manufacturer wants to provide information in multiple languages. In these instances, multipage labels are permissible. Digital labels (smart, radiofrequency identification, digital codes, etc.) are the latest innovation within the food packaging industry that can be used by manufacturers to improve traceability, product information and food safety.

The *label material* for classic food labels is chosen depending on certain technical characteristics of the product in question. These properties dictate further conditions for application and process type. For example, if a label is intended for a glass drink bottle, which will be stored in the refrigerator, then it must meet certain criteria to prevent it from detaching from the bottle before or during use. Consideration must be given to important aspects such as whether the material is suited to damp glass surfaces, as well as its resistance to low temperatures and to submersion in water and ice. Label specifications and adhesive types will depend on the method of application. Labels can be cut, as well as being thermosensitive or pressure-sensitive. They can be supplied as sheets or reels (Paine and Paine 1992). Reel diameter, frequency of stoppages, label thickness, backing material, core dimensions and unwinding direction should be pre-defined in order to allow maximum compatibility with downstream operations.

Labels can be glue-applied or self-adhesive (pressure-sensitive). Due to faster automatic labeling machines, over the past 20 years, self-adhesive labels have overtaken market share from wet glue applications (White 2012).

12.4.1. *Self-adhesive (pressure-sensitive) labels*

Self-adhesive labels are produced on a roll-to-roll press (Figure 12.11). Normally, they are supplied on reels so that the packer or retailer can employ fast, user-friendly label application technology. Self-adhesive labels are available in permanent and peelable options, as well as water removable ones for reusable containers or recycling purposes.

Figure 12.11. *Example of a roll of self-adhesive labels*

Self-adhesive labels can be printed on paper or plastic. Even though paper is more environmentally friendly, plastics are taking over the market as they offer different benefits to paper. Printed plastics can offer better esthetic finishing (effects such as high gloss or matte, transparent or opaque or metallized) and improve the resistance of the final package.

Figure 12.12. *Example of a full-wrap label*

Full-wrap labels can be shaped around the product (Figure 12.12), offering additional space for information on the reverse side of the label. Labels can be shaped to fit perfectly, e.g. with curves and cut-outs for specific food and packages, including vacuum skin packs, modified atmosphere packs, trays and clamshells.

12.4.2. *In-mold labeling*

In-mold labels have the information put inside the core material. In other words, packaging and labeling are produced during the same process that, in the end, has the shape of the mold. Traditional plastic transformation processes can be used for their production: blow molding, injection molding and thermoforming. The main advantages of in-mold labels are their resistance to humidity, lower post-labeling costs and attractive shapes.

12.4.3. *Sleeves*

Shrink sleeve labeling involves an all-around label, applicable on any type of packaging material (glass, plastic, wood and metal), and it forms on to unique package shapes and allows for 360° design, maximizing branding opportunities. Because it is not permanently attached to the bottle or container, a sleeve label can be removed more easily for recycling purposes. Generally made from plastics, these labels are waterproof, provide scuff resistance and can serve as tamper evidence (Figure 12.13). In tamper-evident options, sleeves are extended over the closures. They are perforated in line with the closure opening and so present proof that packaging has not been opened before. Similarly, it also serves hygienic purposes. Shrink bands can also be used to combine packs together in a multi-pack as one selling unit.

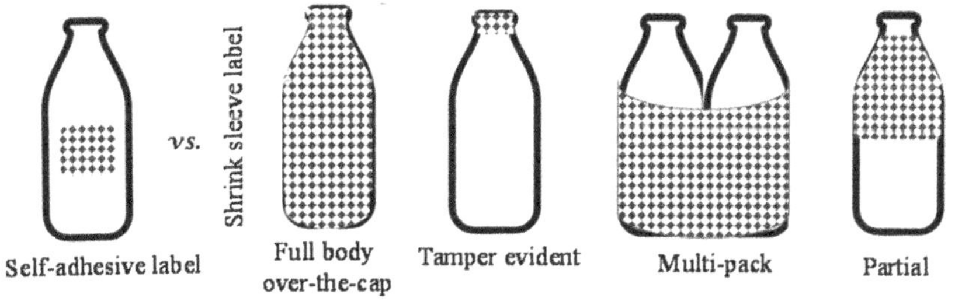

Figure 12.13. *Comparison of self-adhesive (left) and shrink sleeve (right) food labels*

Sleeves are mainly made from polyethylene terephthalate glycol (PETG), oriented polypropylene (OPP), polystyrene (PS), polylactic acid (PLA) and polyvinyl chloride (PVC). They are printed and applied in several steps (Figure 12.14). PETG is used for many products due to its widespread availability, transparency and superior shrinking capability (Durgashiam *et al.* 2019). As a result of its temperature stability, it is pasteurizable and recyclable. However, it is not good for light bottles made from PP or PE as its high shrink force can affect their shape. PS and oriented polystyrene (OPS) are more cost-effective than PETG. They offer excellent consistency and thus can be used for squeezable bottles. Its main disadvantage is low temperature stability, and therefore PS labels require more storage and transportation precautions. PLA films are novel materials used for sleeve production. They are made from renewable resources and offer high clarity and biodegradability in industrial conditions (Sin and Tuen 2019). Recently, due to the environmental and health issues related to chlorine content, there are a lot of concerns about the use of PVC as a label or food packaging material; its usage is rapidly decreasing as a result.

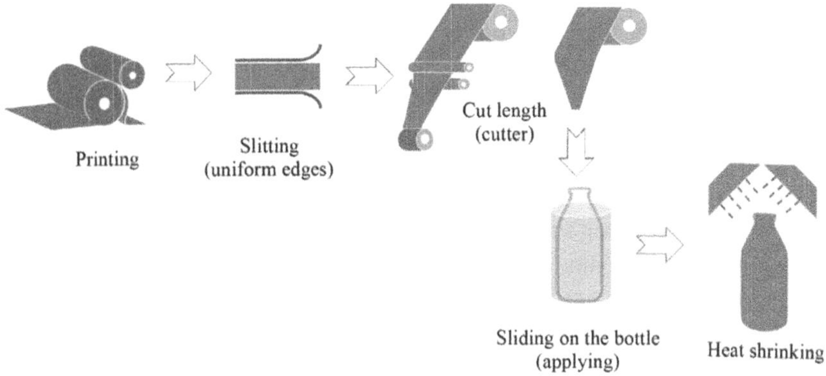

Figure 12.14. *Application steps of shrink sleeves*

12.4.4. *"Smart" and digital labeling*

Digital labeling technology (bar codes, universal product codes, smart labels and radiofrequency identification (RFID) labels) has grown considerably in the past few years (Figure 12.15). Nowadays, it is sought to be implemented for the identification of products and as part of automatic traceability in logistics and retailing in the agri-food sector (Bibi *et al.* 2017).

"Smart labels" are intended to inform buyers about the food inside. They notify consumers about product freshness or actively participate in changing the atmosphere in the package headspace (e.g. via oxygen scavenging or releasing tags, and similar). RFID tags are used in the food industry for tracking and security reasons, supply chain management and temperature monitoring. They also serve as a marketing tool and a source of additional information for the buyer. RFID technology relies on the transformation of a one-dimensional barcode into a data-rich source for suppliers, packagers, logistics workers and advertisers. RFID also enables distance identification of an object (Kumar *et al.* 2009).

Quick response (QR) codes are a type of two-dimensional barcode, composed of distinctive geometric patterns of black modules in a white square background (Ramalho *et al.* 2018). In a relatively small space, a QR code can store a huge quantity of information and can be easily and rapidly read by different electronic devices, including mobile phones. They are used for item identification, product tracking and as a simple physical tool for quick access to websites for advertising.

Figure 12.15. *Examples of digital labels: a) QR code; b) RFID code; c) bar code*

12.5. References

Albert, J. (2010). *Innovations in Food Labelling*. The Food and Agriculture Organization of the United Nations and Woodhead Publishing Limited, Oxford.

Bacarella, S., Altamore, L., Valdesi, V., Chironi, S., and Ingrassia, M. (2015). Importance of food labeling as a means of information and traceability according to consumers. *Adv. Hort. Sci.*, 29(2–3), 145–151.

Bibi, J.C.F., Guillaume, C., Gontard, N., and Sorli, B. (2017). A review: RFID technology having sensing aptitudes for food industry and their contribution to tracking and monitoring of food products. *Trend. Food Sci. Technol.*, 62, 91–103.

Codex Alimentarius Commission (2010). General standard for the labelling of prepackaged foods (CODEX STAN 1-1985). FAO, Rome.

Durgashyam, K., Indra, M., Balakrishna, R.A., and Satyanarayana, K. (2019). Experimental investigation on mechanical properties of PETG material processed by fused deposition modelling method. *Mater. Today: Proc.*, 18(6), 2052–2059.

European Normalization (2000). Packaging: Requirements for packaging recoverable through composting and biodegradation. CENELEC, EN 13432:2000.

European Normalization (2006). Plastics – Evaluation of compostability – Test scheme and specifications. CENELEC, EN 14995:2006.

European Parliament, Council of the European Union (2011). Regulation 1169/2011/EC of the European Parliament and of the Council of 25 October 2011, on the provision of food information to consumers, amending Regulations (EC) No 1924/2006 and (EC) No 1925/2006 of the European Parliament and of the Council, and repealing Commission Directive 87/250/EEC, Council Directive 90/496/EEC, Commission Directive 1999/10/EC, Directive 2000/13/EC of the European Parliament and of the Council, Commission Directives 2002/67/EC and 2008/5/EC and Commission Regulation (EC) No 608/2004. Regulation, European Union.

Kumar, P., Reinitz, H.W., Simunovic, J., Sandeep, K.P., and Franzon, P.D. (2009). Overview of RFID technology and its applications in the food industry. *J. Food Sci.*, 74(8), R101–R106.

Paine, F.A. and Paine, H.Y. (1992). *A Handbook of Food Packaging*. Springer, Dordrecht.

Ramalho, J.F.C.B., António, L.C.F., Correia, S.F.H., Fu, L.S., Pinho, A.S., Brites, C.D.S., Carlos, L.D., André, P.S., and Ferreira, R.A.S. (2018). Luminescent QR codes for smart labelling and sensing. *Optic. Laser Technol.*, 101, 304–311.

Sin, L.T. and Tueen, B.S. (2019). *Polylactic Acid: A Practical Guide for the Processing, Manufacturing, and Applications of PLA*. Plastics Design Library, William Andrew, Norwich, NY.

Smithers (2019). The Future of Label Printing to 2024. Brochure, Smithers, Akron, OH.

White, A.R. (2012). Labels for packaging. In *Packaging Technology – Fundamentals, Materials and Processes*, Emblem, A. and Emblem H. (eds). Woodhead Publishing, Cambridge.

13

Food Packaging Selection

Kata Galić

Faculty of Food Technology and Biotechnology, University of Zagreb, Croatia

13.1. Introduction

The global packaging market consumption in 2020 is split into five groups: paper and board (31.06%), plastic (flexible 24.85% and rigid 22.28%), metal (12.64%), glass (6.81%) and others (2.35%), with around 70% of all packaging used in the food industry (WPO 2008; ALL4PACK 2016). In 2015, the value of the global packaging industry was US$839 billion and was predicted to reach US$998 billion in 2020 (ALL4PACK 2016).

Glass containers are mainly used in the beverage sector (75% of the total tonnage of glass) which includes spirits, beers and ciders, flavored alcoholic beverages, still and sparkling wines, fortified wines, soft drinks, fruit juices and mineral water. The food sector (dairy products, jams and spreads, sauces and dressings, oil, vinegar, etc.) accounts for about 20% of the tonnage (mostly jars). Cosmetics, pharmaceuticals and technical product containers (mostly small bottles) account for the remaining 5% of glass container tonnage. There are many factors that can affect the market for glass containers, and the main one is alternative packaging materials, such as plastics, metals and composed materials (EC-JRC 2012).

The tendency to use lightweight and flexible packaging made plastic the most widely used material for packaging (40%) in Europe in 2016. Consequently, large amounts of plastic end up as waste (Iacovidou and Gerassimidou 2018; Plastics

---

For a color version of all figures in this book, see www.iste.co.uk/debeaufort/packaging.zip.

Packaging Materials and Processing for Food, Pharmaceuticals and Cosmetics, coordinated by Frédéric DEBEAUFORT, Kata GALIĆ, Mia KUREK, Nasreddine BENBETTAIEB and Mario ŠČETAR.

Europe 2018). Furthermore, it is estimated that inappropriate processing and packaging contributes from 25 to 50% of food waste (WPO 2008).

Taking only fresh products into account, meat processing was the largest manufacturing sector in the US in 2011. Other food industries include dairy (13%), beverages (12%), grains and oilseeds (12%), fruits and vegetables (8%) and other food products (11%) (Park *et al.* 2014).

The European market for chilled foods was estimated at 9 billion euros and is expected to grow to 15 billion euros in the coming years (Ohlsson 2002).

The consumption of frozen and packaged foods (such as meat, seafood, fruits and vegetables, and ready meals) was predicted to grow at a compound annual growth rate (CAGR) of 4.1% in the period 2015–2020 (Punathil and Basak 2016).

13.2. Food packaging material selection

All foods placed on the market are characterized with a shelf life, during which they retain their quality and safety, under defined storage conditions. It is well known that packaging materials and methods play a crucial role in obtaining the desired shelf life of a product. With a huge variety of packaging materials available, all possessing different properties, the selection of the best one for a specific product is a very challenging task.

The first step is to take into account packaging materials which will minimize reactions that affect the shelf life of the packaged food, depending on the product's sensitivity to different parameters (such as humidity, temperature, light and oxygen) during storage and distribution. The rate of transport and the effects of the above parameters are limiting factors in terms of the product's shelf life (Ahvenainen 2003; Robertson 2013). Defining the shelf life of a food is not an easy task and is often an area of intense research of a group of scientists, including experts in food technology and microbiology as well as packaging experts.

Many different food preservation treatments (both thermal and non-thermal) are applied to food (Gould 2001). Depending on the technology used, a unique characteristic of the packaging material selected should be the ability to withstand food processing conditions (Figure 13.1). Food packaging is also important in food safety, due to the possible migration of compounds from food contact materials (FCM) into packaged food.

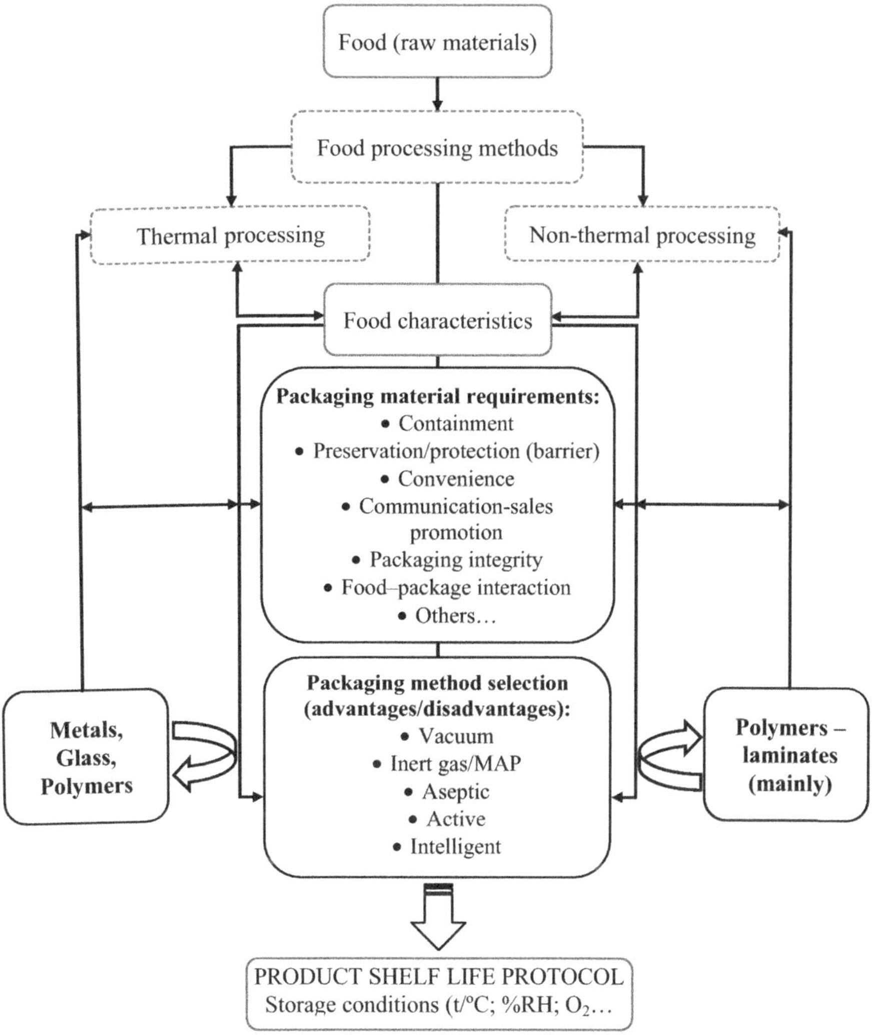

Figure 13.1. *Overview of factors for food packaging material selection*

13.2.1. *Fresh and chilled food packaging*

Optimal storage conditions can considerably increase the shelf life of fresh fruit and vegetables. As already mentioned, these conditions are the temperature, the moisture and the gas make-up of the atmosphere (oxygen, carbon dioxide and ethylene). The optimum storage conditions depend on the type of product and the processing applied to a particular product (DTIPT 2008). Ascorbic acid (vitamin C), a water-soluble vitamin, is mainly present in fresh fruits and vegetables, and it is the most sensitive under storage conditions (Tables 13.1 and 13.2). Different packaging characteristics should be taken into account (such as optical, mechanical, thermal and barrier packaging) in order to preserve food. For example, cushioning the packaging of fresh products can limit the damage caused by shocks during transport (Kitazawa 2018; Ščetar and Galić 2018). Polymeric packaging films for fruit and vegetables enable gases (O_2, CO_2, ethylene) to travel (permeate) through the film. The rate of the permeation depends on the type and the thickness of plastic, the temperature and the differences in pressure of the gases on both sides of the film.

Time (days)	Vitamin C content (mg/100 g) in capsicum (moisture content: 94 g/100 g) during storage			
	Storage temperature (°C)			
	20	5	-20	-40
0	68.0	68.0	41.5	41.5
2	39.3	44.3		
4			41.4	41.4
6	35.0	40.0		
8		38.4	34.5	41.4
10	62.0	37.2		
13	88.6	30.5		
15	58.8	23.5		
16			29.9	20.8
17	41.9	19.6		
18			11.4	25.7
21	19.5	11.8		
22			10.0	15.0
25			6.8	14.0

Table 13.1. *Vitamin C contents in capsicum during storage as a function of moisture content and temperature (Rahman* et al. *2015)*

At chilled temperatures (between 0°C and +5°C), microorganism growth occurs slowly, and food spoilage reactions are inhibited, so the safety and quality of the food are preserved for an extended period of time (days or weeks). However, chilled foods are perishable, and they deteriorate with time. Chilled food packs need to be clean but not sterile. Exceptions are found where hurdle concepts are used, for example, when using modified atmospheres or in-pack pasteurization. Chilled food packaging materials must be able to protect and preserve a range of foods, from fresh products (such as meat, fish and vegetables) to prepared meals (including those that will be reheated in-pack using a microwave or a conventional oven) (Day 2000; Tucker 2011).

Product	Vitamin C loss (%) per day (average over 7 days)		Days to frozen value	
	Ambient (20°C)	Chill (4°C)	Ambient (20°C)	Chill (4°C)
Peas	8.9	2.1	<3	14
Broccoli	8.0	0	<2	18
Green beans	7.8	10.8	<1	1
Spinach	>14.2	10.7	<1	<2
Carrots	3.8	1.4	<1	<1

Table 13.2. *Vitamin C loss in vegetables at ambient and chill storage (Favell 1998)*

One of the major problems with packaged fresh products is water condensation inside the package, due to respiration or transpiration of the fresh product, as well as the barrier characteristics (such as water vapor permeability) of the packaging material (Bovi *et al.* 2016).

As most polymeric materials used in fresh product packaging have lower water vapor permeability, taking into account the transpiration rates of fresh products, water can remain within the package. Under such conditions, even minor temperature fluctuations cause in-package condensation, leading to microbial growth (Linke and Geyer 2013). Perforation of packaging is a solution to control the atmosphere inside a package of fresh products that still respire after harvesting. Thus, O_2 can get into a package and, at the same time, CO_2 can get out. The size and number of holes (Tables 13.3 and 13.4) must be adapted to the particular product.

The shelf life of fruit and vegetables can be extended using modified atmosphere packaging (MAP). The principle of MAP is the application of reduced O_2 and/or elevated CO_2 concentrations within the package.

Fruit	Packaging film	Gas composition (%)		Temperature of storage (°C)	Storage time
		O_2	CO_2		
Pomegranate fruit (cv. Mollar de Elche)	Unperforated PP (25 μm)	8	10	2	12 weeks
		6	12	5	
Pomegranate arils (cv. Mollar)	Semi-permeable plastic bag	1	30	4	10 days
Pomegranate arils (cv. Mollar de Elche)	OPP (40 μm)	12.5	8.5	8	7 days
		13.5	7.5	4	
Pomegranate arils (cv. Primosole)	PP	6.5	11.4	5	10 days

Table 13.3. *Modified atmosphere packaging of pomegranate arils; different cultivars and types of packaging film, during storage (Caleb* et al. *2011)*

Product	Type of PM-MAP		Gas composition	Storage temp.	Shelf life
	No. of holes	Perforated area (A)/diameter (D)	(CO_2/O_2) (kPa)	(°C)	(Days)
Strawberries	7 and 9 holes	0.09 mm D	15/5	4	>14
Citrus fruit		0.002% perf. A	2–3/17–18	6	35
Bananas	4 and 10 cm	50.29 cm^2 diff. A	3.5/3	15	42
Mango	80–100 holes	~0.05–0.07 mm D	17/9	12	30
Peaches	100 holes	0.2 mm D	3.3±0.01 CO_2	20	12
Fresh sliced mushroom	2 holes	0.33 mm D	11.5/3.6	10	3
Cherry tomatoes	5 holes	0.2 mm D	4.0±0.1 CO_2	20	60

Table 13.4. *Application of perforation-mediated modified atmosphere packaging of fresh products (Hussein* et al. *2015)*

In order to extend the shelf life and maintain acceptable quality (visual and sensorial) of fresh fruits and vegetables, an active packaging (AP) system – such as oxygen (Table 13.5) and ethylene absorbers – can be used. Most of these active systems are supplied as sachets or are integrated into films (Ahvenainen 2003; Artés Calero and Gómez 2003; Ščetar and Galić 2018).

Packaging method	Storage time (weeks) at 4°C				
	0	1	2	3	4
	Color L* value				
Control: air ($21\ kPa\ O_2 + 0.03\ kPa\ CO_2$)	34.83	30.98	29.96	28.97	28.68
ATCO-100	34.83	33.63	33.23	32.75	32.09
ATCO-210	34.83	33.33	33.06	31.85	31.52
Overall	34.83	32.65	32.08	31.19	30.77
	Color a* value				
Control: air ($21\ kPa\ O_2 + 0.03\ kPa\ CO_2$)	36.07	32.31	28.83	26.02	25.68
ATCO-100	36.07	34.85	31.83	28.52	27.23
ATCO-210	36.07	34.55	31.86	28.07	27.07
Overall	36.07	33.90	30.84	27.54	26.66

Table 13.5. *Color (L*, a*) value changes of control and oxygen absorber (ATCO O_2 absorption capacity of 100 mL and 210 mL) treated strawberries during storage in a polylactic acid tray (Aday and Caner 2013)*

Fresh meat also continues to consume oxygen and emit carbon dioxide, and the result is that both the color and shelf life of the meat are affected. Because a cherry-red color of fresh meat is critical for consumer acceptance, the visual appearance affects buying decisions. Fresh meat also contains a certain number of unavoidable bacteria that grow and affect its quality and safety. Thus, application of technology that will reduce fresh meat spoilage is also important during storage and distribution. The requirements for effective packaging are very different when considering fresh meat as opposed to fruit and vegetables, for example (Fang *et al.* 2017; Dalla Rosa 2019). Generally, fresh meat is packed in an air vacuum and modified atmosphere. If meats are packed in an air atmosphere, plastic trays made of expanded polystyrene (EPS) are used with the appropriate wrapping or polymeric films (Parisi 2013).

With optimal respiratory gas composition (O_2 and CO_2) maintained within packaging, the best quality products can be obtained. MAP is used to preserve the freshness of fresh food by controlling their biochemical metabolism and can significantly extend the shelf life of packaged meat and meat products (Tables 13.6 and 13.7).

Food	Packaging method	Packaging materials	Storage conditions
Chicken meat	Nisin and EDTA + MAP (%): 65 CO_2/30 N_2/5 O_2	PE-LD/PA/PE-LD pouch (75 μm): $P(O_2)$ =52.2 cm^3/m^2 d bar at 75% RH, 25°C; WVP=2.4 g/m^2 d at 100% RH, 25°C	at 4°C for 24 days
Beef and pork cuts	N_2 atmosphere	Pack1: PS tray + shrink film: OTR=8000 mL/m^2 day at 23°C 70% RH	at 2°C for 1–3 weeks
	AGELESS FX-100 O_2-s + absorbent pad	Pack2: plastic cafeteria tray +bi-metalized, plastic laminate bag: OTR=0.55 mL/m^2 day at 23°C 70% RH	
Fresh beef	Air	PE bags: $P(O_2)$ = 1.7 cm^3/m^2 day at 23°C 75% RH	at 0, 5, 10 and 15°C for 65 days
	Vacuum		
	MAPs (%): 40 CO_2/30 N_2/30 O_2; 100 CO_2; 80 CO_2/20 air		
	Active pack: with/without oregano essential oil		
Fresh beef	N_2 atmosphere	Tray: PS, Wrap: film OTR = 8000 mL/m^2 day at 23°C 70% RH, perforated	at 1°C for 8 days
	O_2-scavengers: AGELESS FX-100, FRESHPAX R-2000		

Table 13.6. *Shelf life of combined modified atmosphere and active packaged meat (Galić 2013)*

In order to obtain an optimal gas transfer, different parameters (the type of packing material, its thickness, its surface area, etc.) should be designed in the MAP system, depending on the characteristics of the food product (Ščetar *et al.* 2010; Galić 2013; Castellanos and Herrera 2017; Dalla Rosa 2019). With regard to this, various mathematical models of modified atmosphere systems are proposed in order to ensure the desired food shelf life and quality (Finnegan *et al.* 2013; Belaya *et al.* 2016).

Freshly baked products are free of mold spores, but they soon get contaminated as a result of cross-contamination from mold spores from the air, bakery surfaces and equipment, food handlers, etc. Mold problems are more pronounced in the summer months, due to warmer and more humid storage conditions. Bread packaged in low-density polyethylene or high-density polyethylene (LDPE/HDPE) or LDPE and linear low-density polyethylene (LLDPE) bags have a shelf life of two to three days. However, if a longer shelf life is required, multilayer materials (laminates) and MAP conditions are used. Examples of laminates used include: polyamide/polyethylene (PA/PE), or a combination of these with polyvinylidene chloride (PA/PVDC/PE) or ethylene vinyl alcohol (PA/EVOH/PE). These composite structures provide strength (PA), gas and moisture vapor impermeability (EVOH or PVDC) and heat sealability through PE or ethylene vinyl acetate (EVAC) or an ionomer (Surlyn). Protection from O_2 and moisture vapor is very important as molds can tolerate and grow in quite low concentrations of O_2 and elevated levels of CO_2.

Meat	Packaging materials/methods	Shelf life
Beef and pork	Air: PE	10–11 days at 4°C
	Vacuum: PA/PE	11 days at 4°C
Beef steaks	EPS-tray + PE(80)/PA(20) MAP 1: 70% O_2/20% CO_2/10% N_2 MAP 2: 60% O_2/40% CO_2	22–28 days at 1°C
Beef loin steaks	Top: PET/PE, Bottom: PA/PE	
	Top: PET/PE; Bottom: PA/PE, Vacuum	14 days at 4°C
	MAP: 0.4% CO/60% CO_2/40% N_2	7 days at 8°C
	MAP: 70% O_2/30% CO_2	10 days at 4°C 7 days at 8°C
Pork	PA(20)/PE-E(70.5), vacuum	4 weeks at 4°C
Beef	PA(20)/PE-E(70.5), vacuum	7 weeks at 4°C

Table 13.7. *Shelf life of modified atmosphere packaged meat (Ščetar* et al. *2010)*

A combination of MAP and active packaging (AP) is also used for bakery products. The most frequently used gas compositions are CO_2/N_2: 100/0, 80/20, 70/30, 60/40 and 0/100. The main objective of MAP+AP is the inhibition of the growth of yeasts and molds. The most widely used AP in bread packaging is ageless oxygen scavengers and ethanol emitters. Using an O_2-ageless absorber for the packaging of crusty rolls, a mold free shelf life of more than 60 days was obtained, in comparison with 16–18 days for those packaged without an oxygen absorber. The shelf life of white bread packaged in polypropylene (PP) film can be prolonged from 5 to 45 days at room temperature using an O_2 absorbent sachet in the package. Pizza crust, which molds in two to three days at 30°C, can also remain mold-free for more than 14 days at this temperature when an appropriate O_2 absorber is present (Arvanitoyannis and Bosinas 2012; Del Nobile and Conte 2013).

Fresh fish and seafood are very perishable and have a shorter shelf life due to bacteria action which causes off-odors in packaged fish. The high-quality shelf life (HQL) of most seafood in chilled storage is only a few days. Packaging technologies used to extend chilled foods' shelf lives include: MAP, vacuum packaging (VP), vacuum skin packaging (VSP) and active packaging (AP). MAP, in combination with refrigeration (Tables 13.8 and 13.9), is a widely used method to extend the shelf life of these products (Lambert *et al.* 1991; Sivertsvik *et al.* 2002).

The extended shelf life will depend on the species, fat content, initial microbial content, gas mixture and the ratio between the volume of gas and volume of the food product (G/P ratio), as well as the storage temperature (Sivertsvik *et al.* 2002).

Fish and fishery products	Storage temperature (°C)	MAP CO_2:O_2:N_2	Shelf life (days)
Mediterranean swordfish	4	40:30:30	12
Pearlspot	2	60:40:0	10
Cod	-0.9	50:5:45	21
Sea bass	4	60:10:30	13
Atlantic salmon	2	90:0:10	33
Mediterranean swordfish	4	50:5:45	13
Atlantic salmon	1.2	60:0:40	15
Sea bass	4	60:0:40	18

Table 13.8. *Shelf life extension of fish and fishery products by modified atmosphere packaging (Masniyom 2011)*

Product	Storage temperature (°C)	Packaging	Shelf life (days)
Lightly preserved			
Cold-smoked salmon	5	VP	20–56
Hot-smoked cod (lightly salted)	0.5	air	24–26
	12	air	17–18
Hot-smoked mackerel (lightly salted)	0	MAP	8
	10	MAP	5–6
Brined roe (lump fish)	5	VP	75–90
Cooked and brined shrimps	0	MAP	230–>308
	25	MAP	4–6

Table 13.9. *Shelf life of seafood at different packaging conditions and temperatures (Dalgaard 2000)*

For products such as milk with a short shelf life in chilled distribution (0 to 4°C), PE-coated paperboard is used on both sides. The main plastics used in pasteurized milk packaging are high-density polyethylene (HDPE), polyethylene terephthalate (PET), polycarbonate (PC) and LDPE. Pasteurized milk has a shelf life of more than 10 days under refrigeration at 4±2°C. In order to reduce the effect of light transmission, HDPE with TiO_2 or PET is pigmented with a UV-blocker (Kontominas 2010).

MAP has also found an application in the packaging of milk and milk products with the use of CO_2, due to its antimicrobial characteristics, and because it can be added to or removed from dairy products with no harmful effects (Kirwan 2011; Singh *et al.* 2012).

For dairy-based desserts, including all yoghurt and pudding products, most of the packages are composed of plastic cups made of high impact polystyrene (HIPS) coupled with plastic or aluminum foil (Alu) closures. Other packaging materials used include HDPE bottles sealed with either Alu-laminate heat-seal closures or with LDPE snap or screw caps. Bottles made from PET are also used. The shelf life of yogurt may only be two weeks for unprotected operations, and up to six weeks or more for ultraclean operations (MacBean 2010).

Fresh cheeses (such as mascarpone, feta, cream cheese, ricotta, quark and cottage cheese), as well as whey cheeses, are susceptible to microbial spoilage and have a limited shelf life (Poças 2010). Typical packaging used for cheese includes HDPE, polypropylene (PP) and polystyrene (PS) cups. For the MAP, with the correct CO_2 level, high-barrier materials such as PA/LDPE laminates are used. Most cheeses in brine and MAP are usually packed into rigid or semi-rigid boxes with plastic or metal closures. Both portions of cheese and shredded cheese require an adequate oxygen barrier to prevent mold growth and spoilage. Ethylene vinyl alcohol (EVOH), polyvinylidene chloride (PVDC) or polyvinyl alcohol (PVOH) is used to provide the oxygen barrier. An acrylic, PVOH-coated, oriented polypropylene (OPP) film is also used in cheese packaging (Poças 2010; Ebnesajjad 2013; Parisi 2013).

Quality attributes	Storage conditions		
	3°C (air)	3°C (MA)	3–13°C (MA)
Color and appearance			
Lightness (L)	41.8	39.0	36.1
Greenness (negative a value)	-7.7	-11.1	-4.8
Infection (%)	4.5	0	6.2
Chlorophyll (mg/kg)	0.21	0.33	0.12
Texture			
Weight loss (%)	32.9	2.9	4.9
Anaerobiosis			
Ethanol (mg/kg)	52.5	58.3	157.3
Acetaldehyde (mg/kg)	47.0	45.7	121.7
Water condensation index	0	2	3

Table 13.10. *Quality attributes of broccoli stored at a constant temperature (3°C) and under temperature-fluctuating conditions (3–13°C) in different atmospheres after 30 days of storage (Tano* et al. *2007)*

Depending on the type of chilled food product, special requirements may be needed for the packaging material, such as tolerance to high temperatures during hot filling, in-pack pasteurization or reheating prior to consumption (Day 2000). For vacuum packaging, high O_2 barrier materials (less than 15 cm^3/m^2 day atm) as well as packaging materials with low water vapor transmission rates must be used. Typical VP multilayer materials are: oriented polypropylene/EVOH/PE

(OPP/EVOH/PE), PA/PE, PET/PE, OPP/PVDC/PE, OPP/PVDC/OPP and polyvinyl chloride/EVOH/polyvinyl chloride (PVC/EVOH/PVC). A further application in packaging is EPS, which provides fairly good compression strength, resistance to moisture, and is a good insulator against both temperature changes and shocks during transportation. EPS is mostly used for fast food trays, boxes for fresh fish, and for the delivery of chilled and frozen foods (Emblem 2012). In cold chains, temperature control is the most important aspect from a food safety point of view; thus, temperature fluctuation (Table 13.10) should be avoided.

13.2.2. *Shelf stable and frozen food packaging*

Shelf stable and frozen products are safe over months or years (Table 13.11), and changes in quality that occur in the product are based on the storage time (Tables 13.12 and 13.13). Modes of failure are influenced by product composition, product packaging and storage temperature. These usually include changes in color, loss of desirable aromas and flavors, formation of undesirable aromas or flavors and changes in texture.

The effects of time and temperature are evident on vitamin degradation during storage. It was shown that 50% of vitamin C degradation in frozen green vegetables occurs in 153 days at -20°C, and, for spinach, in only eight days when stored at -5°C (Giannakourou and Taoukis 2003).

Product	Practical storage life (months) at different temperatures (°C)		
	-12	-18	-24
Beef carcass (not packaged)	8	15	24
Veal carcass (not packaged)	6	12	15
Lamb carcass (not packaged)	18	24	>24
Pork carcass (not packaged)	6	10	15
Beef cuts, steaks	8	18	24
Beef minced (ground)	6	10	15
Veal cuts, steaks	6	12	15
Lamb steaks	12	18	24

Table 13.11. *Shelf life of frozen meat at different temperatures (IIR 2006)*

Parameter	Raw material before freezing	Storage (months)			
		0		12	
		-20°C	-30°C	-20°C	-30°C
Soluble solids (g/100 g)	4.88	4.93	4.94	4.98	4.97
Total sugars (g/100 g)	2.81	2.84	2.82	2.74	2.69
Pectin (g/100 g)	0.216	0.146	0.145	0.037	0.061
N-NO_3 (mg/1000 g)	32	33	34	40	38
Vitamin C (mg/100 g)	23.6	22.9	22.7	6.8	12.9
Lycopene (mg/100 g)	3.28	2.99	2.99	1.69	2.42

Table 13.12. *Changes in the levels of some physicochemical parameters in frozen tomato cubes, in a plastic box, during storage (Lisiewska and Kmiecik 2000)*

Commodity	Ascorbic acid (AA) loss (% wet weight) during storage						
	Fresh			Frozen		Canned	
	Time (days)	°C	AA loss (%)	Time (months)	AA loss (%)	Time (months)	AA loss (%)
Broccoli	21		13		50		
Carrots	84	4	5	12	0		
Green beans	16		90		45	12	
Spinach	21	4	75	1			
Green peas	7	20	60				

Table 13.13. *Losses of ascorbic acid due to storage of fresh, frozen and canned vegetables (Rickman* et al. *2007)*

Once the critical parameters of deterioration are known, product changes can be monitored and limits can be used to identify the end of the shelf life. In this case, it is of great importance to use a product's selected packaging for a shelf life study, since the packaging influences how the food changes during time.

Frozen foods are quite sensitive to dehydration (freezer burn) caused by sublimation due to temperature fluctuations during storage. Thus, moisture loss is the biggest problem in frozen food storage.

Frozen products need the barrier characteristics of the package – including light, aroma, oxygen and moisture permeability – to minimize sensory and nutritional changes of the product. Rigid packaging materials with excellent mechanical resistance to scratches in low temperature storage conditions (lower or equal to -18°C) are used. Furthermore, packaging methods that minimize initial O_2 headspace (vacuum, nitrogen flushing and oxygen-absorbing active packaging), together with a barrier material to environmental oxygen, can extend shelf life considerably for many foods. Composition changes aside, the shelf life of frozen food can be affected by moisture loss, oxidation, physical changes due to compression and vibration, which result in a loss of aroma. Furthermore, temperature fluctuations, causing thawing and refreezing, result in a loss of texture in food. Finally, packaging that provides protection from tampering can protect food from contamination (Cooksey and Krochta 2012; Parisi 2013).

Many frozen foods are packaged in either monofilms (such as PE) or multilayer materials, which are either clear or pigmented, depending on the packaged product. The most commonly used materials include LLDPE, ultra low-density polyethylene (ULDPE) and ethylene vinyl acetate (EVAC). Polyamides (PA) are mostly used for frozen food packaging, due to their strength, toughness and good heat and low temperature resistance. Due to PA sensitivity to moisture, the O_2 barrier decreases as the moisture level increases. Other high oxygen barrier materials applications include PA6 (Nylon 6) laminate with LDPE or EVAC for heat sealability. For extra high mechanical strength and heat resistance, PA6,6 (Nylon 6,6) is used (Emblem 2012).

Long shelf life products also include aseptically processed food without preservatives and refrigeration. For such products, the packaging must possess high barrier characteristics and have a tight seal in order to prevent any microorganisms from entering the package. Thus, long shelf life foods require impermeable packaging. Furthermore, aseptic packaging uses a combination of high-temperature and short-duration thermal processing to produce a shelf stable beverage at ambient temperature. Ultrapasteurized (UP) and ultra high-temperature (UHT) milk are packaged in different materials, including plastic/Alu/paperboard laminates, pigmented PET bottles, multilayer HDPE bottles, pouches composed of polyethylene/polyvinylidene chloride (PE/PVDC) or ethylene vinyl acetate/polyvinylidene chloride (EVAC/PVDC). Milk packaged in sterile multi-layered materials has an expected shelf life of 90–150 days at ambient temperature. Once opened, the food or drink must be consumed within three to five days (Kontominas 2010; Götz *et al.* 2014; Wani *et al.* 2014).

13.2.3. *Dried food packaging*

Dried foods (such as dried herbs and spices) do not support microbial growth when stored under appropriate (dry) conditions and reach an expected shelf life of months or even years. The shelf life of dry crunchy food is limited by texture changes due to moisture permeation through the packaging. Such a product will lose its crispness and become soft; hence, it will not be acceptable by the consumers. Thus, the selection of appropriate packaging materials is critical in extending the shelf life of these products. Laminated materials such as paperboard/PE are common for packaged dried foods (such as fruit and breakfast cereals). Hygroscopic foods also require high protection from moisture uptake. In this case, HDPE polymers are used to provide a moisture barrier. Sealant polymers, such as polyolefin, EVAC, ionomer or combined materials, are used for low-temperature seals, easy-opening seals and form-fill-seal (FFS) packaging. Food requiring packaging material with aroma barrier properties contains either PA or EVOH polymers. New grades of PP films possessing high stiffness, clarity and good elastomeric properties are widely used, replacing both plastic and traditional packaging materials (glass, metal). Moisture-scavenging AP systems are used for moisture-sensitive foods and pharmaceuticals. In these systems, desiccant materials (such as silica gel and activated clay) are included in the packaging in the form of a sachet.

Packaging material for milk powder must protect the product from exposure to moisture, O_2 and light, as well as environmental factors (temperature, relative humidity and physical hazards) during storage. Dried milk is an oxygen-sensitive food product. Therefore, reducing the O_2 content in the package also reduces oxidative rancidity. Milk powder is packed into either metal cans (epoxy-phenolic coated tinplate cans) or multilayer pouches (Alu/plastic) and has a shelf life of five years. For long-term storage, MAP, gas flushing and O_2 absorbers are used (Arab Tehrany and Sonneveld 2010; Tucker 2011; Singh *et al.* 2012).

Roasted coffee release volatiles and CO_2 during roasting, which expands the beans due to the internal increase of gas. However, as the beans become very brittle, they lose their ability to retain CO_2 and volatiles, which are partially released during storage. For this reason, a degassing step is performed on ground coffee before packaging. With this action, the swelling and possible bursting of the packages during storage is avoided. Instant coffee has a very long shelf life (Table 13.14) if proper packaging is applied. Traditional packaging for instant coffee consists of hermetically closed glass jars or tinplate cans of various shapes and sizes. The most commonly used flexible materials are LDPE, BOPP/BOPP, metalized PET/LDPE and PET/Alu/LDPE. These packages can be equipped with a one-way valve that opens at a preset pressure to release accumulated gases.

Package structure (thickness, μm)	WVTR (g/m^2 day)	Shelf life (days)	
		25 g product	50 g product
LDPE (30)	6.1	15	16
BOPP/BOPPP (20/40)	1.2	84	94
PETmet/LDPE (12/70)	0.9	108	112

Table 13.14. *Shelf life of instant coffee in different packaging materials (Alves and Bordin 1998)*

Once packaged coffee is opened, degradation reactions during consumption occur at higher rates than when unopened; thus, a secondary shelf life is calculated. Secondary shelf life represents the length of time after opening the package during which coffee maintains acceptable quality. The end of secondary shelf life is about 20 days, at water activity (a_w) values lower than 0.36. At higher a_w values, the secondary shelf life greatly decreases, for example, to about 13 days at a_w of 0.44 (Nicoli *et al.* 2010; Ebnesajjad 2013; Robertson 2013).

13.2.4. *Beverages packaging*

Beverage products (water, tea, coffee, juices, carbonated soft drinks, beer and wine) require special packaging to preserve their quality. Bottled water is packaged into glass and plastic bottles (PET, PP, PVC, HDPE or polycarbonate (PC)). The major concern in packaged bottled water is to avoid microbial growth and the formation of off-flavors. For carbonated soft drinks and fruit juices, packages with a strong barrier for microorganisms, oxygen and light are used. For that purpose, a wide variety of materials are available: plastics (both rigid and flexible), cardboard-based laminates, rigid metals (aluminum, Alu-laminates, tinplate, tin free steel (TFS)) and glass. These materials are produced in different packaging types and shapes (bottles, cans, pouches and others) (Wani *et al.* 2014; Ramos *et al.* 2016).

For carbonated soft drinks, which often have added flavors, the barrier against permeation of carbon dioxide and flavor substances is an important issue. Although glass and metal containers are traditionally used for premium beverages, the gas barrier properties of plastic packaging materials have improved considerably, with the introduction of new multi-layer materials, including oxygen-absorbing layers and new barrier-coating techniques. Among plastic containers, PET and polyethylene naphthalate (PEN) are used, as are multilayer materials for all types of beverages (Ramos *et al.* 2016). PET is also used for edible oils due to its transparency, excellent mechanical properties, low price, low weight, good O_2 and flavor barrier properties. Multilayer plastic bottles require a certain level of rigidity that allows the bottle to be squeezed, a good barrier to moisture and O_2 to ensure product quality, transparency to clearly show the contents and a reclosing system to allow repeated uses by consumers. Plastic combination, such as a PET/EVOH system, provides rigidity and flexibility with the required adequate barrier (PET), while an inner layer of EVOH provides an oxygen barrier to prevent spoilage of the product (for this purpose, AP systems can be used as oxygen scavengers).

Fruit juices are usually packed in glass bottles, metal cans (both tinplate and aluminum), cardboard–aluminum–plastic laminates or plastic packaging materials. Glass is mainly used for high-quality fruit and vegetable juices, beer and wine. The greatest advantage of glass is that it is an absolute barrier and an inert material. The weakest point of glass packaging is associated with the point of closure. Metal cans give protection similar to glass containers and are also a good barrier to light. Because metal cans are susceptible to corrosion, appropriate lacquer formulations for the interior coating must be selected to avoid corrosion problems. Cardboard laminates with plastic and aluminum foil are used for aseptic packaging of juices, and the shelf life of these products is four to six months. The major consideration of the use of plastic packaging materials for beverages is to achieve low oxygen permeability, in order to prevent the oxidation of sensitive ingredients, like vitamin C (Table 13.15), which avoids changes in color or the formation of off-flavors.

Types of package	Ascorbic acid content (mg/100 mL) Storage time (weeks)					Ascorbic acid retention (%)
	0	3	5	7	8	
PE/cardboard/Alu	12.68	11.48	10.70	9.77	9.38	74
PE/cardboard/EVOH	12.68	10.77	9.53	8.18	7.61	60
PE	12.68	10.43	8.95	7.26	6.59	52
PET	12.68	9.78	7.81	7.86	4.94	39

Table 13.15. *Vitamin C retention of aseptically packaged pomegranate juices in different packaging during storage at 4°C (Fathy 2014)*

Nowadays, active antioxidant packaging materials are developed for the manufacture of beverage packages by applying the recent advances in encapsulation and new packaging technologies. Scavengers are introduced into an inner layer of a multilayer system, dispersed in the polymer matrix or added in closure liners (Busolo and Lagarón 2012). Oxygen scavengers, such as Ageless Omac®, Oxyguard® and Shelfplus O_2® are some examples of packaging materials for beverage products found on the market (Yildirim *et al.* 2018).

Antimicrobial packaging is another form of AP technology, which can be incorporated into or coated onto plastic packaging in order to control microbial contamination. The principles of antimicrobial packaging action are based on (1) controlling the release of biocide species either to the liquid phase and/or to the headspace, (2) absorption of essential components for microbial growth and (3) by direct contact (Lagarón *et al.* 2012).

Beer is sensitive to O_2 ingress and loss of CO_2. Loss of CO_2 from beer in metal cans (aluminum and tinplate) is successfully prevented by the construction of the can. Beer packaged in glass or plastic bottles (PET) can experience some CO_2 loss. This can be prevented by using barrier coatings, layers and blends to prevent the ingress of O_2 and minimize the loss of CO_2. The amount of O_2 that beer is exposed to increases due to a combination of the filling operation and ingress into the beer package. An O_2 content of 0.25–1.2 ppm is identified as the concentration leading to the end of shelf life between 80 and 120 days (Robertson 2013). For an O_2 ingress of maximum 1 ppm, the resulting shelf life for the beer is 4–13 months. In order to decrease O_2 ingress, various O_2 scavengers have been developed for bottle closures. PET bottles can transmit O_2 from the environment into the beer and CO_2 from the beer into the environment at rates that produce large changes in flavor and carbonation loss, respectively, over a relatively short time. The shelf life of beer in a standard monolayer PET bottle can only be three weeks before the ingress of 500 ppb O_2, and four weeks before 10% loss of CO_2 occurred at 21°C. Polyethylene naphthalate (PEN), which has better barrier properties than PET, has also been investigated for beer bottles, usually as a copolymer with PET, because of its high cost. Other high-barrier interior layers in PET bottles investigated include ethylene-vinyl alcohol copolymer (EVOH) and MXD6 polyamide (Bamforth and Krochta 2010).

Bottled wine contains O_2 from the original bulk wine, with the headspace O_2 remaining after the closure application. The O_2, SO_2 and CO_2 permeation rates of the package, including closures, will impact the shelf life of the wine. There are many variables (closure type, flushing of empty bottles with inert gas, filler characteristics and pre-closure headspace flushing) that affect the dissolved oxygen content in bottled wine. Glass is preferred for bottling wine because of its high impermeability to gases and vapors, transparency and inertness. Glass wine bottles have been closed with traditional bark cork. Other non-cork closures include Stelvin®, Zork®, Nomacorc® and the "Vino-Lok" glass closure, designed for use with table wine. The shelf life of a table wine is directly related to the O_2 content, as oxidation can produce significant organoleptic modifications in the flavor and color of wine, the loss of aromatic freshness, the degradation of anthocyanins, etc. Besides the traditional glass bottle, wine is packaged in other containers, including bag-in-box (BIB), PET bottles, multilayer laminates and aluminum cans. PET with an extra high barrier Saran layer was shown to be as effective for wine stability as glass for up to 10 months. There were similar findings with PET bottles containing oxygen scavengers. Although not well accepted by consumers, aseptically packaged wine in cardboard/Alu/LDPE laminates has also been on the market for many years (Mentana *et al.* 2009; Reeves 2010).

13.2.5. *Snack and confectionary food packaging*

Snacks are a group of foods (such as potato and corn chips, crisps, saltines, pretzels, popcorn, extruded puffed and baked or fried products, rice-based snacks, etc.) and drinks (soft drinks, milk drinks, sports drinks, energy drinks, etc.) that are eaten between meals. Dried snack foods being crushed during distribution is the major cause that affects their quality, while texture loss due to moisture pick-up and rancidity due to the oxidation of fats and oils affect their shelf life. Thus, the packaging materials for these products have to meet requirements for oxygen and moisture barrier properties. Among packaging methods, vacuum packaging and inert gas (nitrogen) flushing are often used. Flexible packaging, such as sachets and pouches, and composite cans are forms used to package snack foods. Traditional packaging materials include multilayer compositions of cardboard/Alu/LDPE flushed with nitrogen (for fried snacks) and recycled cellulose fibers (RFC) coated with LDPE or PVDC with a glassine layer (for cookies). Nowadays, the most-used materials are metalized films or laminates with a polymer barrier, such as EVOH or PVDC, which provides an oxygen, moisture and light barrier. Other packaging films that are used are based on biaxially oriented polymers, due to increased barrier performance with material orientation. Such materials include oriented polypropylene (OPP), oriented polyethylene terephthalate (OPET) and oriented HDPE. Among the bio-based materials, polylactic acid (PLA) has recently been used for snack product packaging (Reilly and Man 2000; Ebnesajjad 2013; Chinnadurai and Sequeira 2016).

Polymer (thickness/μm)	Oxygen permeability (cm^3/m^2 day atm)	Water vapor permeability (g mm/m^2 day)
	at 23°C, 50% RH	**at 23°C, 85% RH**
PET 12/Alu 9/PE 50	0	0
PET 12/EVOH 5/PE 50	1	2–4
PET 12/PVDC 4/PE 50	5	2
PET 12/PVOH 3/PE 50	2	4–6
PET 12/PE 50	15–20	4–6
	at 23°C, 50% RH	**at 23°C, 50% RH**
PET 12/SiOx	0.5–5	0.2–5
PP 20/SiOx	1–20	0.1–1
PA 20/SiOx	1	0.5
PET 12 untreated	110	15
PP 20 untreated	1500–1800	1–1.5

Table 13.16. *Comparison of barrier properties of different flexible packaging films (Lange and Wyser 2003)*

Transparent packaging barrier coatings, such as aluminum oxide (AlOx) and silicon oxide (SiOx), have also been used for the packaging of dry foods such as snack foods, confectionary, cookies or cereal bars. When applied onto polymeric films (such as PET or BOPP), these high barrier coatings (Table 13.16) not only offer product visibility but can also be used in different conditions, such as microwave ovens (Lange and Wyser 2003; Struller *et al.* 2014, 2019).

Confectionery products (candy) include chocolate-based or sugar-based snack food. Apart from sweeteners, these products contain other compounds such as flour, colorants and flavorings. Because of the high level of sugar present in these products, they are less prone to microbiological spoilage and have a relatively long shelf life (Table 13.17). Therefore, the deterioration of flavor, odor, texture or color of these products is the main causes of spoilage (Martin 1994).

Product	Major deteriorative changes	Typical shelf life at room conditions (months)
Plain chocolate bar	Fat bloom, sugar bloom, stale flavor	24
Milk chocolate bar	Fat bloom, sugar bloom, stale flavor	16
White chocolate bar	Fat bloom, sugar bloom, stale flavor	16
Milk chocolate-coated peanuts	Fat bloom, peanut rancidity, sugar bloom, stale chocolate	12
Chocolate bars with raisins	Fat bloom, stale chocolate flavor, raisins drying out	12
Chocolate-coated wafer	Wafer becoming stale, fat bloom, sugar bloom, stale chocolate flavor	12
Chocolate-coated fondant	Fat bloom, sugar bloom, fondant drying out	18
Chocolate shells with soft caramel center	Sugar bloom, change in caramel texture, fat bloom caused by fat migration	12
Chocolate shells with praline center	Fat bloom caused by fat migration, shell softening, nut paste rancidity	12

Table 13.17. *Deteriorative changes and typical shelf lives for chocolate products (Martin 1994)*

However, some confectionery products have a shorter shelf life due to the presence of unstable ingredients (e.g. cream), making them susceptible to microbial spoilage. The packaging of confectionery depends on the equilibrium relative humidity of the product. Depending on how the product gains or loses moisture from the surrounding atmosphere, this can adversely affect the product. In order to protect them from the atmospheric moisture and from sticking, some products are wrapped individually (such as caramels, chocolate, hard candies, toffees and gum). Traditionally, wrapping materials for these products include waxed paper, waxed glassine and waterproof plasticized RCF. Since these materials offer low water vapor resistance, an additional (outer) package is used, such as metal containers, glass jars and metallized laminates. In this case, biaxially oriented films are also widely used (Subramaniam 2010; Ebnesajjad 2013; Robertson 2013). Among polymeric films, PE, PP and PVC are mostly used for confectionery products, either as transparent films, allowing visibility of the product, or with other packaging materials (laminates), providing adequate strength, flexibility and increased barrier protection. Polymer-coated papers or films are popular packaging materials due to their light weight and biodegradability. Metal foils, such as tin or aluminum, have been a traditional wrapping material for chocolate and other confectionery products, due to their high barrier properties against water vapor, gases and light, as well as fats. Biaxially oriented materials (such as OPP) are also mostly used here nowadays.

If the integrity of the package is violated, resulting in pore formation, a moisture transmission into the package is enabled. In order to prevent this, edible coatings or films (lipid, protein or cellulose-based) are often used to protect the product against moisture (Ergun *et al.* 2010; Blanco-Pascual and Gómez-Estaca 2016).

13.3. List of abbreviations

Abbreviations	Description
Alu	Aluminum Foil
AP	Active Packaging
BOPA	Biaxially Oriented Polyamide
BOPPP	Biaxially Oriented Perlized Polypropylene
BOPP	Biaxially Oriented Polypropylene
EPS	Expanded Polystyrene
EVAC (EVA)	Ethylene Vinyl Acetate
EVOH (EVAL)	Ethylene Vinyl Alcohol
HDPE	High-Density Polyethylene
HIPS	High Impact Polystyrene
HQL	High-Quality Shelf Life
LDPE	Low-Density Polyethylene
LLDPE	Linear Low-Density Polyethylene
MAP	Modified Atmosphere Packaging
MXD6	Meta-Xylylene Diamine/Adipic Acid Polyamide
OPET	Oriented Polyethylene Terephthalate
OPP	Oriented Polypropylene
PA	Polyamide (nylon)
PC	Polycarbonate
PE	Polyethylene

PEN	Polyethylene Naphthalate
PET	Polyethylene Terephthalate
PETmet	Metallized Polyethylene Terephthalate
PLA	Polylactic Acid
PP	Polypropylene
PS	Polystyrene
PVC	Polyvinyl Chloride
PVDC	Polyvinylidene Chloride
PVOH (PVAL)	Polyvinyl Alcohol
RCF	Recycled Cellulose Fiber
TFS	Tin-Free Steel
UHT	Ultra High Temperature
ULDPE	Ultra Low-Density Polyethylene
UP	Ultrapasteurized
VP	Vacuum Packaging
VSP	Vacuum Skin Packaging

13.4. References

Aday, M.S. and Caner, C. (2013). The shelf-life extension of fresh strawberries using an oxygen absorber in the biobased package. *LWT-Food Sci. Technol.*, 52, 102–109.

Ahvenainen, R. (2003). Active and intelligent packaging. In *Novel Food Packaging Techniques*, Ahvenainen R. (ed.). CRC Press, Woodhead Publishing Limited, Cambridge.

ALL4PACK (2016). Packaging: Market and challenges in 2016. The global marketplace for packaging-processing-printing-handling. Report, ALL4PACK.

Alves, R.M.V. and Bordin, M.R. (1998). Estimativa da vida útil de café solúvel por modelo matematico. *Ciencia e Tecnologia de Alimentos,* 18, 19–24.

Anonymous (2014). PM food and dairy consulting. In *World Cheese Market 2000–2020*, P. Mikkelsen (ed.). Denmark.

Arab Tehrany, A. and Sonneveld, K. (2010). Packaging and the shelf-life of milk powders. In *Food Packaging and Shelf Life – A Practical Guide*, Robertson, G.L. (ed.). Taylor and Francis Group, Boca Raton, FL.

Artés Calero, F. and Gómez, P.A. (2003). Active packaging and colour control: The case of fruit and vegetables. In *Novel Food Packaging Techniques*, Ahvenainen R. (ed.). CRC Press, Woodhead Publishing Limited, Cambridge.

Arvanitoyannis, I.S. and Bosinas, K. (2012). Bakery products. In *Modified Atmosphere and Active Packaging Technologies*. Arvanitoyannis, I.S. (ed.). CRC Press, Taylor & Francis Group, Boca Raton, FL.

Ashurst, P.R. (2010). Packaging and the shelf-life of water and carbonated drinks. In *Food Packaging and Shelf Life – A Practical Guide*, Robertson, G.L. (ed.). Taylor and Francis Group, Boca Raton, FL.

Bamforth, C.W. and Krochta, J.M. (2010). Packaging and the shelf-life of beer. In *Food Packaging and Shelf Life – A Practical Guide*, Robertson, G.L. (ed.). Taylor and Francis Group, Boca Raton, FL.

Belaya, Z.A., Caleb, O.J, and Opara, U.L. (2016). Modelling approaches for designing and evaluating the performance of modified atmosphere packaging (MAP) systems for fresh produce: A review. *Food Pack. Shelf Life*, 10, 1–15.

Blanco-Pascual, N. and Gómez-Estaca, J. (2016). Production and processing of edible packaging – Stability and applications. In *Edible Food Packaging, Materials and Processing Technologies*, Cerqueira, M.Â.P.R., Pereira, R.N.C., da Silva Ramos, Ó.L., Teixeira, J.A.C., and Vicente, A.A. (eds). CRC Press, Taylor & Francis Group, Boca Raton, FL.

Bovi, G.G., Caleb, O.J., Linke, M., Rauh, C., and Mahajan, P.V. (2016). Transpiration and moisture evolution in packaged fresh horticultural produce and the role of integrated mathematical models: A review. *Biosyst. Eng.*, 150, 24–39.

Busolo, M.A. and Lagarón, J.M. (2015). Antioxidant polyethylene films based on a resveratrol containing clay of interest in food packaging applications. *Food Packag. Shelf Life*, 6, 30–41.

Caleb O.J., Opara U.L., and Witthuhn C.R. (2011). Modified atmosphere packaging of pomegranate fruit and arils: A review. *Food Bioprocess Technol.*, DOI:10.1007/s11947-011-0525-7.

Castellanos, D.A. and Herrera, A.O. (2017). Modified atmosphere packaging: Design and optimization strategies for fresh produce. *Postharvest Handling*, 85–106 [Online]. Available at: http://dx.doi.org/10.5772/intechopen.68498.

Chinnadurai, K. and Sequeira, V. (2016). Packaging of cereals, snacks, and confectionery. *Reference Module in Food Sciences*. Elsevier Inc. [Online]. Available at: http://dx.doi.org/10.1016/B978-0-08-100596-5.03223-6.

Cooksey, K. and Krochta, J.M. (2012). Introduction to frozen food packaging. In *Handbook of Frozen Food Processing and Packaging*, Da-Wen Sun (ed.). CRC Press, Taylor & Francis Group, Boca Raton, FL.

Dalgaard, P. (2000). Fresh and lightly preserved seafood. In *Shelf-life Evaluation of Foods*, 2nd edition. Man, C.M.D. and Jones, A.A. (eds). Aspen Publishing Inc. Gaithersburg, MD.

Dalla Rosa, M. (2019). Packaging sustainability in the meat industry. In *Sustainable Meat Production and Processing*, Galanakis, C.M. (ed.). Academic Press, London.

Day, B.P.F. (2000). Chilled food packaging. In *Chilled Foods – A Comprehensive Guide*, Stringer M. and Colin D. (eds). Woodhead Publishing Limited, Cambridge.

Del Nobile, M.A. and Conte, A., (2013). *Packaging for Food Preservation*. Springer, New York.

Dodd, M. and Bouwer, J. (2014). The supply value chain of fresh produce from field to home: Refrigeration and other supporting technologies. In *Postharvest Handling: A Systems Approach*, Florkowski, W., Prussia, S., Shewfelt, R., and Brueckner, B. (eds). Academic Press, London.

DTIPT (2008). Packaging Fresh Fruit and Vegetables. Guide, Danish Technological Institute Packaging and Transport.

Ebnesajjad, S. (2013). *Plastic Films in Food Packaging – Materials, Technology, and Applications*. Elsevier Inc., Waltham.

EC-JRC (2012). Best Available Techniques (BAT) Reference Document for the Manufacture of Glass, European Integrated Pollution Prevention and Control Bureau (EIPPCB) [Online]. Available: http://eippcb.jrc.ec.europa.eu/ [Accessed 2 May 2012].

Emblem, A. (2012). Plastics properties for packaging materials. In *Packaging Technology – Fundamentals, Materials and Processes*, Emblem, A. and Emblem, H. (eds). Woodhead Publishing Limited, Cambridge.

Ergun, R., Lietha, R., and Hartel, R.W. (2010). Moisture and shelf-life in sugar confections. *Crit. Rev. Food Sci. Nutr.*, 50(2), 162–192.

Fang, Z., Zhao, Y., Warner, R.D., and Johnson, S.K. (2017). Active and intelligent packaging in meat industry. *Trends Food Sci. Technol.*, 61, 60–71.

Fathy, A.M. (2014). Quality and shelf-life of pomegranate juice aseptically packed in different packaging materials. *Middle East J. Appl. Sci.*, 4(2), 409–415.

Favell, D.J. (1998). A comparison of the vitamin C content of fresh and vegetables. *Food Chem.*, 62(1), 59–64.

Finnegan, E., Mahajan, P.V., O'Connell, M., Francis, G.A., and O'Beirne, D. (2013). Modelling respiration in fresh-cut pineapple and prediction of gas permeability needs for optimal modified atmosphere packaging. *Postharvest Biol. Technol.*, 79, 47–53.

Fuertes, G., Soto, I., Carrasco, R., Vargas, M., Sabattin, J., and Lagos, C. (2016). Intelligent packaging systems: Sensors and nanosensors to monitor food quality and safety. *J. Sensors*, 4046061.

Galić, K. (2013). Requirements for innovative food packaging. In *Innovation in Healthy and Functional Foods*, Ghosh, D., Das, S., Bagchi, D., and Smarta, R.B. (eds). CRC Press, Taylor & Francis Group, Boca Raton, FL.

Giannakourou, M.C. and Taoukis, P.S. (2003). Application of a TTI-based distribution management system for quality optimization of frozen vegetables at the consumer end. *J. Food Sci.*, 68(1), 201–209.

Götz, A., Wani, A.A., Langowski, H.-C., and Wunderlich, J. (2014). Aseptic packaging. In *Encyclopedia of Food Safety*, Motarjemi, Y., Moy, G., Todd, E. (eds). Academic Press, Amsterdam. doi:10.1016/B978-0-12-378612-8.00274-2.

Gould, G.W. (2001). New processing technologies: An overview. *Proc. Nutr. Soc.*, 60, 463–474.

Hussein, Z., Caleb, O.J., and Opara, U.L. (2015). Perforation-mediated modified atmosphere packaging of fresh and minimally processed produce – A review. *Food Packag. Shelf Life*, 6, 7–20.

Iacovidou, E. and Gerassimidou, S. (2018). Sustainable packaging and the circular economy: An EU perspective. *Reference Module in Food Sciences*. Elsevier Inc. [Online]. Available at: https://doi.org/10.1016/B978-0-08-100596-5.22488-8.

IIR (2006). *Recommendations for the Processing and Handling of Frozen Foods*, 4th edition, Bogh-Sorensen, L. (ed.). International Institute of Refrigeration (IIR), Paris.

Kirwan, M.J. (2011). Paper and paperboard packaging. In *Food and Beverage Packaging Technology*, Coles, R. and Kirwan, M. (eds). Blackwell Publishing Ltd, Chichester.

Kitazawa, H. (2018). Cushioning packaging for fresh produce. *Reference Module in Food Sciences*. Elsevier Inc. [Online]. Available at: https://doi.org/10.1016/B978-0-08-100596-5.22483-9.

Kontominas, M.G. (2010). Packaging and the shelf-life of milk. In *Food Packaging and Shelf Life – A Practical Guide*, Robertson, G.L. (ed.). Taylor and Francis Group, Boca Raton, FL.

Lagarón, J.M., Ocio, M.J., and López-Rubio, A. (2012). Antimicrobial packaging polymers. A general introduction. In *Antimicrobial Polymers*, Lagarón, J.M., Ocio, M.J., and López-Rubio, A. (eds). John Wiley & Sons, Hoboken, NJ.

Lambert, A.D., Smith, J.P., and Dodds, K.L. (1991). Shelf-life extension and microbiological safety of fresh meat – A review. *Food Microbiol.*, 8(4), 267–297.

Lange, J. and Wyser, Y. (2003). Recent innovations in barrier technologies for plastic packaging – A review. *Packag. Technol. Sci.*, 16, 149–158.

Linke, M. and Geyer, M. (2013). Condensation dynamics in plastic film packaging of fruit and vegetables. *J. Food Eng.*, 116, 144–154.

Lisiewska, Z. and Kmiecik, W. (2000). Effect of storage period and temperature on the chemical composition and organoleptic quality of frozen tomato cubes. *Food Chem.*, 70, 167–173.

MacBean, R.D. (2010). Packaging and the shelf-life of yogurt. In *Food Packaging and Shelf Life – A Practical Guide*, Robertson, G.L. (ed.). Taylor and Francis Group, Boca Raton, FL.

Martin, A.V. (1994). Chocolate confectionery. In *Shelf-Life Evaluation of Foods*, Man, C.M.D. and Jones, A.A. (eds). Springer Science+Business Media, Dordrecht.

Masniyom, P. (2011). Deterioration and shelf-life extension of fish and fishery products by modified atmosphere packaging. *Songklanakarin J. Sci. Technol.*, 33(2), 181–192.

Mentana, A., Pati, S., La Notte, E., and del Nobile, M.A. (2009). Chemical changes in Apulia table wines as affected by plastic packages. *LWT – Food Sci. Technol.*, 42, 1360–1366.

Nicoli, M.C., Manzocco, L., and Calligaris, S. (2010). Packaging and the shelf-life of coffee. In *Food Packaging and Shelf Life – A Practical Guide*, Robertson, G.L. (ed.). Taylor and Francis Group, Boca Raton, FL.

Ohlsson, T. (2002). Minimal processing of foods with thermal methods. In *Minimal Processing Technologies in the Food Industry,* Ohlsson, T. and Bengtsson, N. (eds). Woodhead Publishing Limited, Cambridge.

Parisi, S. (2013). *Food Industry and Packaging Materials: Performance-oriented Guidelines for Users*. Smithers Rapra Technology Ltd, Shrewsbury.

Park, S.H., Lamsal, B.P., and Balasubramaniam, V.M. (2014). Principles of food processing. In *Food Processing: Principles and Applications*, Clark S., Jung, S., and Lamsal, B. (eds). John Wiley & Sons, Hoboken, NJ.

Plastics Europe (2018). *The benefits of plastic packaging plastics Europe* [Online]. Available: https://www.plasticseurope.org/en/about-plastics/packaging [Accessed 4 June 2019].

Poças, M.F. and Pintado, M. (2010). Packaging and the shelf-life of cheese. In *Food Packaging and Shelf Life – A Practical Guide*, Robertson, G.L. (ed.). Taylor and Francis Group, Boca Raton, FL.

Punathil, L. and Basak, T. (2016). *Microwave Processing of Frozen and Packaged Food Materials: Experimental. Reference Module in Food Sciences*. Elsevier Inc. [Online]. Available: http://dx.doi.org/10.1016/B978-0-08-100596-5.21009-3.

Rahman, M.S., Al-Rizeiqi, M.H., Guizani, N., Al-Ruzaiqi, M.S., Al-Aamri, A.H., and Zainab, S. (2015). Stability of vitamin C in fresh and freeze-dried capsicum stored at different temperatures. *J. Food Sci. Technol.*, 52(3), 1691–1697.

Ramos, M., Valdés, A., and Garrigós, M.C. (2016). Packaging for drinks. *Reference Module in Food Sciences*. Elsevier Inc. [Online]. Available at: http://dx.doi.org/10.1016/B978-0-08-100596-5.03224-8.

Reeves, M.J. (2010). Packaging and the shelf-life of wine. In *Food Packaging and Shelf Life – A Practical Guide*, Robertson, G.L. (ed.). Taylor and Francis Group, Boca Raton, FL.

Reilly, A. and Man, C.M.D. (2000). Potato crisps and savoury snacks. In *Shelf-Life Evaluation of Foods*, Man, C.M.D. and Jones, A. (eds). Aspen Publishers, Inc., USA.

Rickman, J.C., Barrett, D.M., and Bruhn, C.M. (2007). Nutritional comparison of fresh, frozen and canned fruits and vegetables. Part 1. Vitamins C and B and phenolic compounds. *J. Sci. Food Agric.*, 87, 930–944.

Robertson, G.L. (2013). *Food Packaging: Principles and Practice*, 3rd edition. CRC Press, Boca Raton, FL.

Ščetar, M. and Galić, K. (2018). A perspective on packaging of fruits and vegetables. In *Novel Postharvest Treatments of Fresh Produce*, Pareek, S. (ed). CRC Press, Taylor & Francis Group, London.

Ščetar, M., Kurek, M., and Galić, K. (2010). Trends in meat and meat products packaging – A review. *Croat. J. Food Sci. Technol.*, 2(1), 32–48.

Schaefer, D. and Cheung, W.M. (2018). Smart packaging: Opportunities and challenges. *51st CIRP Conference on Manufacturing Systems*, 72, 1022–1027.

Singh, P., Wani, A.A., Karim, A A., and Langowski, H.-C. (2012). The use of carbon dioxide in the processing and packaging of milk and dairy products: A review. *Int. J. Dairy Technol.*, 65(2), 161–177.

Sivertsvik, M., Jeksrud, W.K., and Rosnes, J.T. (2002). A review of modified atmosphere packaging of fish and fishery products – Significance of microbial growth, activities and safety. *Int. J. Food Sci. Technol.*, 37(2), 107–127.

Struller, C.F., Kelly, P.J., and Copeland, N.J. (2014). Aluminum oxide barrier coatings on polymer films for food packaging applications. *Surf. Coat. Tech.*, 241, 130–137.

Struller, C.F., Kelly, P.J., and Copeland, N.J. (2019). Conversion of aluminium oxide coated films for food packaging applications – From a single layer material to a complete pouch. *Food Packag. Shelf Life*, 20, 100309.

Subramaniam, P. (2000). Confectionery products. In *The Stability and Shelf-Life of Food*, Kilcast, D. and Subramaniam, P. (eds). Woodhead Publishing Limited, Cambridge.

Tano, K., Oulé, M.K., Doyon, G., Lencki, R.W., and Arul, J. (2007). Comparative evaluation of the effect of storage temperature fluctuation on modified atmosphere packages of selected fruit and vegetables. *Postharvest Biol. Technol.*, 46, 212–221.

Tucker, G.S. (2011). Food biodeterioration and methods of preservation. In *Food and Beverage Packaging Technology*, Coles, R. and Kirwan, M. (eds). Blackwell Publishing Ltd, Chichester.

Wani, A.A., Singh, P., and Langowski, H-C. (2014). *Packaging. Encyclopedia of Food Safety*, 3, 211–218. Elsevier Inc. [Online]. Available: doi:10.1016/B978-0-12-378612-8.00273-0.

WPO (2008). *Market Statistics and Future Trends in Global Packaging*. World Packaging Organisation/PIRA International Ltd [Online]. Available: www.worldpackaging.org [Accessed 6 April 2020].

Yildirim, S., Röcker, B., Pettersen, M.K., Nilsen-Nygaard, J., Ayhan, Z., Rutkaite, R., and Coma, V. (2018). Active packaging applications for food. *Compr. Rev. Food Sci. Food Saf.*, 17, 165–199.

List of Authors

Nasreddine BENBETTAIEB
Institute of Technology
University of Burgundy
Dijon
France

Frédéric DEBEAUFORT
Institute of Technology
University of Burgundy
Dijon
France

Kata GALIĆ
Faculty of Food Technology and Biotechnology
University of Zagreb
Croatia

Mia KUREK
Faculty of Food Technology and Biotechnology
University of Zagreb
Croatia

Mario ŠČETAR
Faculty of Food Technology and Biotechnology
University of Zagreb
Croatia

Index

C

D

E

M

N, O

P